AF356136

Traffic Calming Measures as an Instrument for Revitalizing the Urban Environment

Traffic Calming Measures as an Instrument for Revitalizing the Urban Environment

Editors

Salvatore Leonardi
Natalia Distefano

Basel • Beijing • Wuhan • Barcelona • Belgrade • Novi Sad • Cluj • Manchester

Editors
Salvatore Leonardi
Department of Civil
Engineering and Architecture
University of Catania
Catania
Italy

Natalia Distefano
Department of Civil
Engineering and Architecture
University of Catania
Catania
Italy

Editorial Office
MDPI
St. Alban-Anlage 66
4052 Basel, Switzerland

This is a reprint of articles from the Special Issue published online in the open access journal *Sustainability* (ISSN 2071-1050) (available at: www.mdpi.com/journal/sustainability/special_issues/Traffic_Calming_Measures_Revitalizing_Urban_Environment).

For citation purposes, cite each article independently as indicated on the article page online and as indicated below:

Lastname, A.A.; Lastname, B.B. Article Title. *Journal Name* **Year**, *Volume Number*, Page Range.

ISBN 978-3-7258-0410-8 (Hbk)
ISBN 978-3-7258-0409-2 (PDF)
doi.org/10.3390/books978-3-7258-0409-2

Contents

About the Editors

Salvatore Leonardi

Prof. Salvatore Leonardi, Associate Professor of Roads, Railways, and Airports at the University of Catania, can look back on a career that began after he graduated in Civil Engineering in 1994. The author of more than 150 publications, including four monographic volumes, Prof. Leonardi has worked extensively on topics such as traffic safety, roundabout design, and the analysis of the safety and functionality of road intersections. These areas of research highlight his ongoing commitment to innovation in civil engineering and transportation, with a particular focus on road safety and efficiency. He has held key institutional positions, including Chair of the Civil Engineering for Water and Transportation master's degree, and has led numerous research projects. His teaching focuses on courses on urban road infrastructure and road intersections, with a particular interest in road infrastructure safety. He is Deputy Editor-in-Chief of an international scientific journal and a member of the editorial board of renowned scientific journals.

Natalia Distefano

Dr. Natalia Distefano is currently a researcher of Roads, Railways, and Airports at the University of Catania. Her research project (STARTGREEN project) aims to improve safety at urban intersections to promote sustainable mobility. In the field of teaching, she has extensive experience as she has been teaching courses and seminars since 2010. Her lectures primarily revolve around transportation infrastructure design, with a particular focus on road infrastructure in urban and metropolitan areas, intersection and roadway safety, and infrastructures for the mobility of vulnerable users. In terms of research interests, Natalia Distefano has focused primarily on road safety and the effectiveness of transportation infrastructure. She has been involved in numerous research projects and has published several articles on topics such as the safety of roundabouts and the effects of traffic calming measures on vehicle speeds in urban areas. Her publications demonstrate a thorough and analytical approach to the safety and efficiency of transportation infrastructure.

Preface

As urban areas continue to expand, they encounter significant hurdles such as increased traffic, escalating pollution, and the shrinking of public spaces, all of which detrimentally affect the quality of urban life. Traffic calming measures have emerged as key solutions to these issues, focusing on enhancing safety and enriching the urban environment. These initiatives, which include the creation of pedestrian-friendly zones, the encouragement of bicycle use, and the implementation of speed limits, are designed to foster a safer, more sustainable, and comfortable environment for everyone, with particular attention to the needs of vulnerable groups like children and the elderly. The influence of these initiatives extends beyond road safety, impacting public health, social cohesion, and the overall urban experience. They encourage a shift towards healthier lifestyles and reinforce community bonds. The range of these measures is diverse, featuring pedestrian pathways that promote social interaction, bike lanes that improve mobility, and intersections tailored for the safety of pedestrians and cyclists. Each of these contributes to the development of more habitable urban environments, where the focus shifts from simply controlling traffic to enhancing dynamic urban living. Implementing these measures requires a collaborative effort among local authorities, citizens, and the private sector. While there are obstacles in this process, the long-term advantages are substantial, offering innovative solutions to urban issues and promoting sustainability, health, and vitality in urban settings. These strategies encompass road safety within a larger context of improving urban lifestyles for all residents. This Special Issue, focused on "Traffic Calming Measures as an Instrument for Revitalizing the Urban Environment", delves into various approaches used to enhance safety and sustainability in urban traffic management. This exploration extends past conventional traffic control, underscoring the role of innovative measures in transforming urban spaces. Topics addressed in this Special Issue include the effectiveness of traffic circles, roundabouts, and turbo-roundabouts in residential areas, providing insights into their impact on reducing speeds and aiding sustainable development. This Special Issue advocates for the use of a comprehensive approach to urban design, focusing on safety, efficiency, and the enhancement of urban areas. This includes proposals for sustainable complete streets, shared spaces, and the integration of connected vehicles, reflecting a forward-thinking perspective in urban planning.

Salvatore Leonardi and Natalia Distefano
Editors

Editorial

Traffic-Calming Measures as an Instrument for Revitalizing the Urban Environment

Salvatore Leonardi * and Natalia Distefano

Department of Civil Engineering and Architecture, University of Catania, Viale Andrea Doria 6, 95125 Catania, Italy; natalia.distefano@unict.it
* Correspondence: salvatore.leonardi@unict.it; Tel.: +39-0957382202

Citation: Leonardi, S.; Distefano, N. Traffic-Calming Measures as an Instrument for Revitalizing the Urban Environment. *Sustainability* **2024**, *16*, 1407. https://doi.org/10.3390/su16041407

Received: 16 January 2024
Accepted: 5 February 2024
Published: 7 February 2024

With the advent of rapid urbanization, cities are confronted with increasingly complex challenges. Increases in traffic and air pollution, and gradual reductions in public space threaten the quality of urban life [1]. In this scenario, traffic-calming measures have proven to be a potentially effective tool with which to address traffic safety issues and also to improve the urban environment as a whole [2].

These measures extend beyond the mere regulation of car traffic. They represent an integrated approach that includes the creation of welcoming pedestrian zones, the promotion of bicycle mobility, and the restriction of speed limits [3,4]. The main goal of these measures is to create an urban environment characterized by safety, sustainability, and comfort for the entire resident population, with special attention being paid to the most vulnerable road users, such as children and the elderly [5].

The effectiveness of such measures cannot be limited to road safety alone. They have multidimensional impacts and influence public health, social cohesion, and the overall quality of urban life. A well-designed urban space with moderate traffic serves as a catalyst for healthier lifestyles and better-connected communities [6].

The diversity of these measures is invaluable. From the implementation of pedestrian walkways that encourage social interaction to the creation of bike lanes that improve mobility and road intersections that ensure safe and comfortable crossing for pedestrians and cyclists, every measure contributes to the creation of a more livable urban environment [7,8]. This not only concerns restricting traffic, but also producing an environment in which urban life can develop and progress.

Of course, implementing traffic-calming measures comes with challenges that require effective collaboration between local authorities, citizens, and the private sector [9]. However, these challenges are surmountable, and the long-term benefits outweigh the difficulty of initial efforts.

In summary, traffic-calming measures are not only a means of improving road safety, but also a means of revitalizing cities. They represent an innovative response to modern challenges and promote the sustainability, health, and vitality of cities. Through these measures, it is possible to create an urban future in which road safety is seamlessly integrated into the establishment of a rich and fulfilling lifestyle for the entire population [10].

In this context, this Special Issue, entitled "Traffic calming measures as an instrument for revitalizing the urban environment", takes a critical look at various strategies for improving safety and sustainability in urban traffic. Based on the concept of revitalization, the discussions in this Special Issue extend beyond conventional traffic management and emphasize the role of innovative measures in transforming urban spaces. Topics include pedestrian safety, vehicle speed control, and the dynamic interaction between infrastructure and traffic. The effectiveness of traffic circles, roundabouts, and turbo-roundabouts in residential areas is also examined, offering insights into their role in reducing speed and promoting sustainable development. Comprehensive analyses included in this Special Issue advocate for a holistic urban design that not only ensures safety and efficiency but also contributes to the overall enhancement of urban areas. Proposals for sustainable complete streets, shared spaces, and the integration of connected vehicles highlight a forward-thinking approach to urban planning. The overarching theme of the Special Issue highlights the close link between effective traffic-calming measures and the overall revitalization of urban landscapes.

The Special Issue opens with a paper by Natalia Distefano and Salvatore Leonardi on the application of an innovative method for evaluating the effectiveness of traffic-calming measures (TCMs). The SPEIR (speed profile, effectiveness indicators, and results) methodology is presented to evaluate the performance of TCMs in urban areas, especially in 30 km/h zones. The methodology aims to provide a comprehensive analysis of the effectiveness of traffic-calming measures, taking into account speed limit compliance and the consistency of speed profiles. Three case studies are presented to demonstrate the applications of SPEIR and to illustrate the fluctuating speed profiles resulting from the installation of TCMs. The authors suggest that SPEIR facilitates the comparison of different design solutions, creating a potential international database for the effectiveness of TCMs. Systematic applications of the method can help urban street managers optimize design decisions and achieve sustainable mobility goals by identifying corrective measures to improve TCM effectiveness in managed urban areas (Contribution 1).

The second article in the Special Issue is by Stanisław Majer and Alicja Sołowczuk. This article investigates pedestrian safety islands in Poland, where different traffic-calming measures in a two-way street are analyzed. The study examines the effects of diagonal and parallel parking, safety islands, horizontal deflection, and a one-way chicane. Using an heuristic method, the study identifies the main factors influencing speed on approach to the traffic island, including visibility, pedestrian visibility, and the traffic island's environment. Comparative analyses categorize TCMs as effective, moderately effective, or ineffective. The conclusions highlight the inadequacy of existing design guidelines and emphasize the importance of factors such as sight distance, traffic volume, and cityscape. The aim is to develop comprehensive design guidelines to ensure safe pedestrian safety lanes and effective speed management in urban streets (Contribution 2).

In the next paper, Mauro D'Apuzzo, Azzurra Evangelisti, Daniela Santilli, Sofia Nardoianni, Giuseppe Cappelli, and Vittorio Nicolosi address the need for effective speed control measures in urban areas, focusing on vertical traffic-calming devices. The study investigates the dynamic interaction between vehicles and road profiles, and develops a mathematical model and simulation to evaluate the acceleration associated with different vehicles and traffic-calming devices. Experimental investigations conducted on a raised crosswalk show initial results indicating different dynamic responses depending on vehicle type. The paper discusses acceleration thresholds and proposes the vibration dose value as a descriptor for the vibration experience of drivers. The theoretical approach developed is promising for setting thresholds for human vibration exposure in the development of new vertical traffic-calming devices (Contribution 3).

Following the same themes as those of the previous paper, the following is the fourth contribution of the Special Issue. Giuseppe Cantisani, Maria Vittoria Corazza, Paola Di Mascio, and Laura Moretti examine a series of traffic-calming measures to improve pedestrian safety in urban areas. The study discusses both physical (hard) and psychological (soft) measures that include vertical and horizontal devices as well as landscaping. The study highlights the need for a comprehensive multi-criterion analysis that considers factors such as pedestrian levels, access for residents and emergency vehicles, drainage, snow issues, parking, and environmental objectives. The study recognizes the challenges of implementing traffic-calming measures in consolidated areas and highlights the importance of regulatory support and standardized specifications. The study proposes supranational regulation to ensure the consistent enforcement and widespread adoption of traffic-calming measures across Europe (Contribution 4).

The fifth paper is also based on the idea of a radical transformation of urban streets into important public spaces that accommodate all users of a transportation system. Alfonso

Montella, Salvatore Chiaradonna, Alessandro Claudi de Saint Mihiel, Gord Lovegrove, Pietro Nunziante, and Maria Rella Riccardi introduce the concept of "sustainable complete streets". The proposed design criteria integrate socio-environmental considerations relating to aesthetics, environment, quality of life, and safety, and provide a framework for the creation of intuitive multimodal networks. The case study in Naples, Italy, focuses specifically on the urban redevelopment of the "Mostra d'Oltremare" area and demonstrates the practical application of the proposed criteria. The study highlights the importance of incorporating sustainability and street integrity criteria during the planning, design, and operational phases to promote healthier, safer, and more sustainable urban development. The eco-design approach, which includes ecological and architectural rehabilitation, highlights the potential benefits, such as reductions in traffic, energy consumption, noise and air pollution, as well as increases in safety and improvements in the aesthetic quality of public spaces (Contribution 5).

The exploration of the Special Issue extends to Thessaloniki, Greece, where the dynamics between pedestrians and cyclists in shared spaces are investigated through a web-based questionnaire survey. Using descriptive and inferential statistics, latent variable models and path models, Chrysanthi Mastora, Evangelos Paschalidis, Andreas Nikiforiadis, and Socrates Basbas analyze the behavior, perceptions, and preferences of users. The study concludes that crosswalks, although they do not physically separate users, contribute to a sense of order in interactions. It shows correlations between perceived safety, aggressive behavior, and preferences for interventions. The paper suggests that a degree of separation is desirable and emphasizes the importance of user beliefs and respect in shared spaces (Contribution 6).

The seventh paper is by Stanisław Majer and Alicja Sołowczuk, who examine the effectiveness of traffic circles as sustainable elements in residential areas using the example of a village in Poland. The study analyzes speed reduction within the residential zone, focusing on traffic circle design parameters, location, road function, and their impact on sustainable development factors such as fuel consumption and air pollution. Statistical analyses using non-parametric tests confirm the effectiveness of traffic circles as traffic-calming measures in residential areas. The conclusions highlight the importance of center island height, cross slope, and design considerations based on pedestrian traffic and road characteristics. Recommendations include one-way traffic, green infrastructure, and fixed barriers between traffic circles to maintain desired speed zones (Contribution 7).

The infrastructures with a rotating circulation are also the protagonist of the eighth contribution to the Special Issue. Salvatore Leonardi and Natalia Distefano examine the operational and safety performance of turbo-roundabouts compared to multi-lane roundabouts in a case study on an urban arterial road in eastern Sicily, Italy. The study uses simulation software to evaluate both operational (AIMSUN Next 20.0.1) and safety (SSAM 3.0) aspects. The results of this study indicate that multi-lane roundabouts have better operational performance at medium/low traffic volumes than do turbo-roundabouts, but the latter show significant improvement under high traffic volumes by reducing the number of stops and delays. The safety parameters highlight the benefits of turbo-roundabouts, including the presence of fewer conflict points and that of fewer rear-end and lane change conflicts compared to those observed with multi-lane roundabouts. The study suggests the introduction of turbo-roundabouts in Italy and the revision of legislation to establish them as a viable solution alongside modern roundabouts (Contribution 8).

In the ninth paper, Maria Luisa Tumminello, Elżbieta Macioszek, Anna Granà, and Tullio Giuffrè address the transformation of transportation due to connected and automated vehicles (CAVs). Focusing on roundabouts in Palermo, Italy, the study proposes a simulation-based framework to assess CAV impacts on safety and efficiency. By dedicating a lane to CAVs, the study reports reduced travel times and conflicts compared to those observed under mixed traffic scenarios. However, safety benefits decrease in scenarios with CAVs only, indicating a need for further research to address the methodological limitations of this approach and to incorporate conflict characteristics into decision support tools (Contribution 9).

The last paper in the Special Issue also deals with the topic of revitalizing urban contexts through technologically advanced tools. Rachid Belaroussi, Margherita Pazzini, Israa Issa, Corinne Dionisio, Claudio Lantieri, Elena Díaz González, Valeria Vignali, and Sonia Adelé focus on the evaluation of user perception in the urban redevelopment of the canal in the port of Rimini, Italy, using virtual reality (VR). This study engages participants through VR scenarios depicting the current state of development and proposed future redevelopments. Two questionnaires assess the correspondence between real and virtual scenarios and users' perceptions of specific elements such as green spaces and access points. The results highlight the benefits of involving users early in the planning process and emphasize the importance of attraction points and aesthetics in promoting sustainable transport. The study demonstrates the effectiveness of VR in assessing the appropriateness of projects and provides insights for the benefit of future detailed assessments (Contribution 10).

Conflicts of Interest: The authors declare no conflicts of interest.

List of Contributions:

1. Distefano, N.; Leonardi, S. Evaluation of the Effectiveness of Traffic Calming Measures by SPEIR Methodology: Framework and Case Studies. *Sustainability* **2022**, *14*, 7325.
2. Majer, S.; Sołowczuk, A. Traffic Calming Measures and Their Slowing Effect on the Pedestrian Refuge Approach Sections. *Sustainability* **2023**, *15*, 15265.
3. D'Apuzzo, M.; Evangelisti, A.; Santilli, D.; Nardoianni, S.; Cappelli, G.; Nicolosi, V. Towards a New Design Methodology for Vertical Traffic Calming Devices. *Sustainability* **2023**, *15*, 133813.
4. Cantisani, G.; Corazza, M.; Di Mascio, P.; Moretti, L. Eight Traffic Calming "Easy Pieces" to Shape the Everyday Pedestrian Realm. *Sustainability* **2023**, *15*, 7880.
5. Montella, A.; Chiaradonna, S.; Mihiel, A.; Lovegrove, G.; Nunziante, P.; Rella Riccardi, M. Sustainable Complete Streets Design Criteria and Case Study in Naples, Italy. *Sustainability* **2022**, *14*, 13142.
6. Mastora, C.; Paschalidis, E.; Nikiforiadis, A.; Basbas, S. Pedestrian Crossings as a Means of Reducing Conflicts between Cyclists and Pedestrians in Shared Spaces. *Sustainability* **2023**, *15*, 9377.
7. Majer, S.; Sołowczuk, A. Traffic Circle—An Example of Sustainable Home Zone Design. *Sustainability* **2023**, *15*, 16751.
8. Leonardi, S.; Distefano, N. Turbo-Roundabouts as an Instrument for Improving the Efficiency and Safety in Urban Area: An Italian Case Study. *Sustainability* **2023**, *15*, 3223.
9. Tumminello, M.; Macioszek, E.; Granà, A.; Giuffrè, T. A Methodological Framework to Assess Road Infrastructure Safety and Performance Efficiency in the Transition toward Cooperative Driving. *Sustainability* **2023**, *15*, 9345.
10. Belaroussi, R.; Pazzini, M.; Issa, I.; Dionisio, C.; Lantieri, C.; González, E.; Vignali, V.; Adelé, S. Assessing the Future Streetscape of Rimini Harbor Docks with Virtual Reality. *Sustainability* **2023**, *15*, 5547.

References

1. Ghafghazi, G.; Hatzopoulou, M. Simulating the air quality impacts of traffic calming schemes in a dense urban neighborhood. *Transp. Res. D Transp. Environ.* **2015**, *35*, 11–22. [CrossRef]
2. Morrison, D.S.; Thomson, H.; Petticrew, M. Evaluation of the health effects of a neighbourhood traffic calming scheme. *J. Epidemiol. Community Health* **2004**, *58*, 837–840. [CrossRef] [PubMed]
3. Huang, H.F.; Cynecki, M.J. Effects of Traffic Calming Measures on Pedestrian and Motorist Behavior. *Transp. Res. Rec.* **2000**, *1705*, 26–31. [CrossRef]
4. Gonzalo-Orden, H.; Pérez-Acebo, H.; Unamunzaga, A.L.; Arce, M.R. Effects of traffic calming measures in different urban areas. *Transp. Res. Procedia* **2018**, *33*, 83–90. [CrossRef]
5. Ess, J.; Antov, D. Unified methodology for estimating efficiency of traffic calming measures—Example of Estonia. *Balt. J. Road Bridge Eng.* **2016**, *11*, 259–265. [CrossRef]
6. Chng, S.; Chang, C.; Mosquera, K.; Leong, W.Y. Living in a Silver Zone: Residents' perceptions of area-wide traffic calming measures in Singapore. *Transp. Res. Interdiscip. Perspect.* **2022**, *16*, 100710. [CrossRef]

7.   Damsere-Derry, J.; Ebel, B.E.; Mock, C.N.; Afukaar, F.; Donkor, P.; Kalowole, T.O. Evaluation of the effectiveness of traffic calming measures on vehicle speeds and pedestrian injury severity in Ghana. *Traffic Inj. Prev.* **2019**, *20*, 336–342. [CrossRef]

8.   Canale, S.; Distefano, N.; Leonardi, S. Comparative analysis of pedestrian accidents risk at unsignalized intersections. *Balt. J. Road Bridge Eng.* **2015**, *10*, 283–292. [CrossRef]

9.   De Borger, B.; Proost, S. Road tolls, diverted traffic and local traffic calming measures: Who should be in charge? *Transp. Res. B Methodol.* **2021**, *147*, 92–115. [CrossRef]

10.   Nathanail, E.G.; Adamos, G.; Karakikes, I. Traffic Calming Measures as a Tool to Revitalise the Urban Environment: The Case of Serres, Greece. In *Advances in Mobility-as-a-Service Systems*; Springer International Publishing AG: Cham, Switzerland, 2020; Volume 1278, pp. 770–779.

sustainability

MDPI

Article

Evaluation of the Effectiveness of Traffic Calming Measures by SPEIR Methodology: Framework and Case Studies

Natalia Distefano and Salvatore Leonardi *

Department of Civil Engineering and Architectural, University of Catania, Viale Andrea Doria, 6-95125 Catania, Italy; ndistefa@dica.unict.it
* Correspondence: salvatore.leonardi@unict.it; Tel.: +39-095-7382202

Abstract: The speed value of 30 km/h should not be exceeded in urban areas, both to ensure safety requirements for all categories of users and to improve the overall quality of life in urban areas. Moreover, it is necessary not only to comply with the prescribed maximum speed, but also to ensure a uniform speed by limiting the variations in relation to the average value within an acceptable range of variation. An original analysis methodology is therefore proposed, useful for both technicians and administrators to verify the effectiveness of traffic calming measures, especially in areas where these measures are widely used, such as Zones 30. This methodology, called SPEIR (acronym for Speed Profile, Effectiveness Indicators and Results, which are the keywords of the three steps into which the proposed methodology is divided), is divided into three operational steps necessary to both verify the effectiveness of existing traffic calming measures in a given context and to plan new traffic calming measures to be implemented in specific urban sectors to be requalified and revitalized. Finally, three case studies are presented where the application of the SPEIR methodology is useful not only for understanding the operational steps in the application of the methodology itself, but also for understanding the differences in terms of the safety performance that the various traffic calming measures provide to the users of the urban streets where such measures are present.

Keywords: Zones 30; speed management; speed uniformity; speed profile; speed tables; chicane

Citation: Distefano, N.; Leonardi, S. Evaluation of the Effectiveness of Traffic Calming Measures by SPEIR Methodology: Framework and Case Studies. *Sustainability* **2022**, *14*, 7325. https://doi.org/10.3390/su14127325

Academic Editor: Wann-Ming Wey

Received: 28 May 2022
Accepted: 13 June 2022
Published: 15 June 2022

Publisher's Note: MDPI stays neutral with regard to jurisdictional claims in published maps and institutional affiliations.

1. Introduction

In recent years, in Italy, but also in Europe and the rest of the world, sustainable urban mobility planning has become increasingly established as a new approach to transport planning and mobility management in urban areas in a sustainable and comprehensive way. Stimulated by the constant increase in motorized traffic and its associated negative impacts, a new paradigm of sustainable mobility has, therefore, gradually emerged [1].

While mobility has brought positive economic and social impacts such as prosperity, international cooperation and exchange, there are also negative aspects such as the high proportion of urban land occupied by infrastructures of transport, urban sprawl, congestion, traffic noise, energy consumption, and social and environmental problems [1–4]. Moreover, the greatest negative impacts are mainly due to private cars. Its intensive use has been shown to reduce physical activity, increase the likelihood of traffic accidents, negatively impact health and the residential environment, and reduce opportunities for social interaction [1,5].

Sustainable urban mobility planning addresses these challenges. In line with explicit urban transport sustainability strategies, the world community goals call on urban planners and designers to incorporate road safety as an essential requirement to ensure sustainable mobility. Indeed, the goal of sustainable, safe road transport is to avoid accidents and reduce the likelihood of serious injuries to (almost) zero [6].

In this context, the risks to pedestrians on the road and their high vulnerability to serious injury as a result of traffic crashes have become a major concern for policy makers and health professionals [7]. The progressive aging of the population, associated with the decline in the abilities of older pedestrians, is also an important factor affecting the level of safety that urban areas provide for vulnerable users [8,9], which must also be taken into account when determining the most appropriate safety measures [10,11]. It is estimated that about 12 million road accidents involving pedestrians occur each year, killing about 270,000 people worldwide (about 23% of all road fatalities worldwide [12]). In 2019, 4628 pedestrian fatalities were reported in Europe, accounting for 20% of total traffic fatalities. Although the absolute number of pedestrian fatalities decreased from 5952 to 4628 between 2010 and 2019 (-22%), the total number of traffic fatalities decreased at the same rate (-23%), with the proportion of pedestrians in the total number of traffic fatalities remaining constant [13].

In Italy, the situation is even more serious: every year, about 20,000 traffic accidents occur with at least one pedestrian. According to Italian data published by ACI-ISTAT, 534 pedestrians were killed and more than 21,000 injured in traffic accidents in 2019 [14]. Although the data on traffic accidents for 2020 are already available, it was preferred not to use them because they are representative of a situation strongly influenced by the COVID-19 pandemic. In any case, the 409 pedestrians who died in 2020 (-23.4% compared to the previous year) are representative of the fact that the pedestrian category was one that experienced the smallest decrease in road accident mortality during the pandemic period: just consider that victims in cars decreased by 27.9% in 2020, those on bicycles by 30.4%, and those on mopeds by 33% (again, compared to 2019) [15].

Since high speeds are one of the main key factors in traffic accidents between motorized vehicles and vulnerable road users, the implementation of speed reduction measures would improve safety on urban roads [16]. It is well known that there are potential interventions in the road network that lead to a significant reduction in speed. In particular, both conventional and unconventional roundabouts can influence motorist behavior by limiting their approach speed [17]. However, in purely residential areas, it is not always feasible to install modern roundabouts, and even less feasible to install unconventional systems such as turbo-roundabouts, which would potentially be even more effective at achieving speed reductions due to their significant size [18]. In urban contexts with residential development, it is therefore necessary to resort to various strategies to achieve significant speed reduction. For this purpose, low speed zones, such as 30 km/h zones (so-called Zones 30), are a possible strategy that is increasingly used in many countries.

A low-speed zone is an important solution to reduce traffic and revitalize some urban areas, making them safer and more attractive for recreation, with a view to much desired sustainable mobility. A low-speed zone is simply defined as an area where the speed indicated on the sign (e.g., 30 km/h) cannot be exceeded [19]. A Zone 30 is a delineated area designed to improve the safety of vulnerable road users through traffic calming measures. It is well known that the placement of speed limit signs alone at the entrances to the zone is not sufficient to achieve the desired speed reduction. According to Kempa [20], engineering measures are needed to force drivers to obey the speed limit. In order to prioritize vulnerable road users, the implementation of Traffic Calming Measures (TCMs) has become very topical. TCMs are specific treatments and/or designs of the roadway whose main function is to force drivers to behave correctly. These measures (vertical deflection, horizontal deflection, physical obstacles, and signs and lane markings) work both toward reducing vehicle speed [21] and toward reducing the ability to reach certain areas [1]. Recent studies have shown that traffic calming groups must be used to ensure low speeds along an entire route or in an urban area [18–22].

Other studies have shown that the effectiveness of traffic calming groups varies depending on the geometric characteristics of the measures and the distance between them [23–25].

The establishment of traffic calming groups must provide good uniformity of speed. The lack of uniformity of speed can lead to frequent and dangerous deceleration and acceleration maneuvers [26]. These maneuvers cause higher fuel consumption and noise emissions, which have a negative impact on the environment [27].

Nowadays, speed management is a priority in urban areas. Several studies show that 30 km/h should be the maximum speed in residential areas. This is because below 30 km/h, the risk of a pedestrian fatality in a traffic accident is quite low (5–10%). At a speed of 20 km/h, the risk of serious injury is about 10%. For a pedestrian, the risk of being involved in a traffic accident (with fatal or serious injuries) increases significantly when the speed exceeds 30 km/h [28]. Injuries to cyclists show a similar pattern, with the likelihood of a fatal crash increasing with higher vehicle speeds. In high-speed environments, the risk of collisions also increases for children and the elderly as their motion perception skills are underdeveloped and diminish and they are unable to properly assess speed and the time available to cross the road [29–31]. Speed limits also have the potential to increase physical activity primarily by encouraging walking and cycling, reduce sedentary behavior, and improve the livability of an area. Speed limits to 30 km/h also reportedly have the potential to reduce fuel consumption and air pollution by reducing standing traffic, which allows for more efficient use of available road space and more effective merging and filtering at intersections, thereby reducing queues [32–34]. Therefore, speed limits of 30 km/h may have a significant impact not only on road safety but also on public health. Therefore, evaluating the effectiveness of speed limits is warranted not only from the perspective of sustainable mobility, but also from the perspective of the overall environmental sustainability of typical residential areas.

Ultimately, all technical and scientific literature agrees that 30 km/h is the speed value that should not be exceeded in urban environments, firstly to ensure safety requirements for all categories of users, and secondly to improve the overall quality of life in urban areas, which in this way would be revitalized and made more attractive, especially for pedestrians and cyclists. In this context, the authors of this work strongly believe that in order to achieve sustainable mobility, it is necessary not only to comply with the prescribed maximum speed, but also that traffic in urban areas is carried out at a uniform speed, limiting the variations in relation to the average value within an acceptable range. This has been highlighted in literature studies, as mentioned above, but not sufficiently and without the emphasis that this issue requires; to date, there is no procedure that allows a systematic evaluation of the effectiveness of traffic calming measures from the point of view of "uniformity of speed". The aim of this study is, therefore, to propose an original and simple procedure, useful for both technicians and administrators, to verify the effectiveness of traffic calming measures, mainly in areas where these measures are widely used, such as Zone 30 or any other urban context where the speed is to be kept constant at or below the 30 km/h threshold.

This study, therefore, presents the SPEIR methodology, which is divided into three operational steps required to both verify the effectiveness of existing traffic calming measures in a given context (and, if necessary, prepare corrective interventions to improve their effectiveness) and plan new traffic calming measures to be implemented in specific urban sectors to be requalified and revitalized. This work also presents three case studies where the application of the SPEIR methodology is useful not only to understand the operational steps in the application of the methodology itself, but also to understand the differences in terms of the safety performance that the various traffic calming measures offer to the users of the urban streets where such measures are present.

2. Framework of SPEIR Methodology

The proposed methodology for evaluating the effectiveness of traffic calming measures is divided into three basic steps:

(1) determination of speed profiles for the survey site;
(2) estimation of effectiveness indicators;
(3) analysis and interpretation of results.

The framework of the proposed analysis method is shown in Figure 1. The keywords of the framework steps are the following three: Speed Profile, Effectiveness Indicators, and

Results. From the initial letters of the aforementioned keywords, the acronym SPEIR was defined, which will be used to refer to the proposed methodology throughout this work.

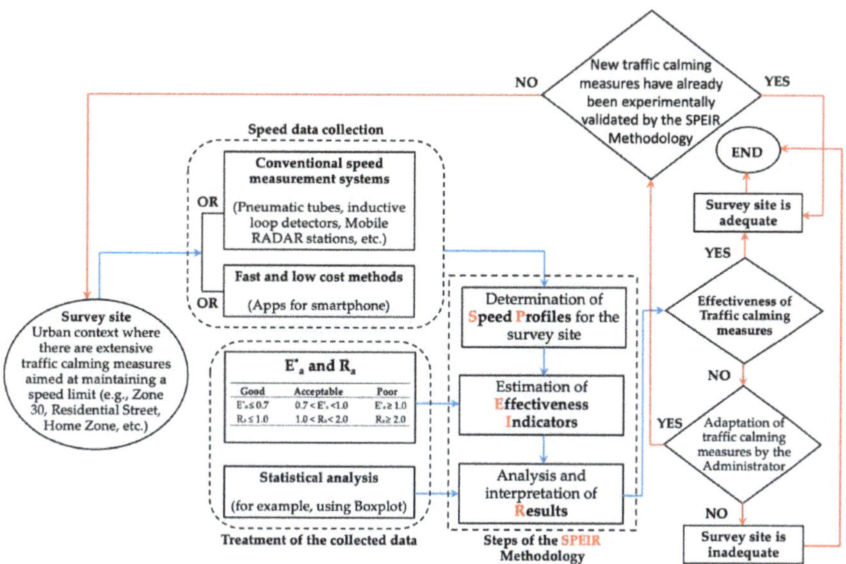

Figure 1. Framework of SPEIR methodology.

2.1. Determination of Speed Profiles for the Survey Site

Speed profile reflects the physical nature of the road. It will indicate an approximately constant speed in road sections with fairly uniform geometric characteristics. This aspect is often missing in areas where TCMs are present. This is because each TCM has a "zone of influence" in which it exerts a speed-reducing effect [35]. Therefore, it is essential to verify that the realization of TCMs does not lead to high-speed differences.

Recording speed profiles at locations where traffic calming measures are in place is, therefore, the first step in verifying the effectiveness of the measures themselves. There are various measurement tools for assessing speed. The conventional ones are listed and briefly described below [36].

Speed measurement systems for a specific road section and for aggregated traffic counts:

- Pneumatic tubes with automatic traffic counting systems: They are used for long surveys of vehicle speeds and annual average daily traffic counts. They are sufficiently accurate and inexpensive but lack flexibility in practice.
- Inductive loop detectors: Generally inexpensive and adaptable systems. However, they are not very flexible because they are permanently installed.
- Fixed RADAR stations: They can be installed on existing poles or on specially designed roadside structures. They are moderately expensive and can simultaneously measure the speeds of different vehicles in both directions of travel.

Speed measurement systems with the possibility of continuous surveys:

- Mobile RADAR stations and LiDAR guns: they have high versatility, precision and immediate correspondence between vehicles and measured speeds. However, they are very expensive.
- On-board diagnostic (OBD) black boxes: they allow knowing the exact location and speed of the vehicle through GPS systems. They are characterized by sufficient accuracy if the device is able to receive data from a sufficient number of satellites. However, a large number of devices may be required to obtain statistically significant samples.

For several years, in addition to conventional methods of measuring speed, new, inexpensive and fast methods have been developed based on the use of special apps for smartphones with the most popular operating systems such as Android and iOS.

Smartphones, which are available to drivers and have many built-in sensors (accelerometer, magnetometer, gyroscope, and GPS), have been used in the transportation, infrastructure, and automotive industries in recent years to collect sensory data from various sensors and then process it on a central server for further analysis [37–39]. Berloco et al. [36] have shown that these devices provide reliable results comparable to more expensive conventional instruments, especially when detecting low speeds such as those associated with traffic calming measures.

The application of methods based on smartphone apps requires the selection of a significant sample of test drivers, who must thus travel the road sections where the traffic calming measures are present. Each test must necessarily be conducted in the presence of a smartphone in the vehicle of each test driver, as well as an operator who must coordinate the survey operations (start the driving test, start the app, stop the app). It is advisable that this operator influences the driver as little as possible so that all driving tests can be as natural as possible. This can generally be carried out when the operator is in the back seat of the vehicle driven by the test driver.

2.2. Estimation of Effectiveness Indicators

It is now well known that a uniform speed environment that meets driver expectations avoids abrupt changes in operating speeds and creates a safe operating environment. Uniformity of speed brings some main benefits: less impact on the environment, better quality and lower driver stress, and improved safety [40].

To evaluate the effectiveness of traffic calming measures, this study proposes the use of two indicators based on the continuous speed profile. These indicators were defined by Polus et al. [40] as a measure of the consistency of a highway section and subsequently used by other researchers as surrogate measures of safety in urban areas [41,42]. The first index (E_a) evaluates the accumulated speeding along the entire road segment, while the second index (R_a) evaluates the accumulated speed uniformity.

E_a is the normalized relative area (per unit length) between the speed profile values that are above the speed limit and the speed limit line. The measure can be applied to individual speed profiles or to an operating speed profile. Given a speed limit, the areas between the speed profile and the speed limit line must be determined. Only the areas above the speed limit line (A_{si}) must be considered in the measurement (Figure 2a). The accumulated speeding must be calculated using Equation (1) as the sum of the areas divided by the length of the segment (L). Consequently, E_a is directly related to the cumulative speeding.

$$E_a = \frac{\sum As_i}{L} \tag{1}$$

where:

- E_a: accumulated speeding (m/s);
- $\sum As_i$: sum of areas bounder between the speed profile and the speed limit where speed in higher than speed limit (m^2/s);
- L: road section length (m).

Moreno et al. [41] found that the individual accumulated speeding values did not follow a normal distribution; therefore, the variable was transformed into the square root of the accumulated speeding. Therefore, the variable to be analyzed was calculated as the square root of the accumulated speeding (E_a^*).

$$E_a^* = \sqrt{E_a} \tag{2}$$

The E_a^* thresholds for good, acceptable, and poor efficiency proposed by Moreno et al. [41] are shown in Table 1.

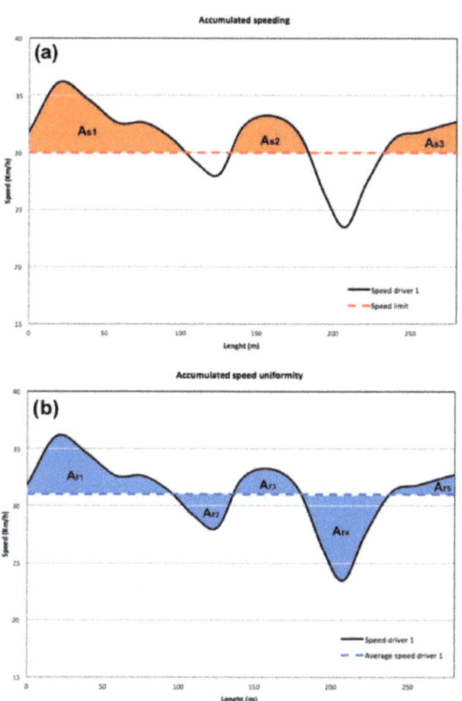

Figure 2. Example of evaluation of the areas for the definition of the effectiveness indicators: (**a**) accumulated speeding; (**b**) accumulated speed uniformity.

Table 1. Thresholds for effectiveness indicators [40,41].

Effectiveness Indicators	Good	Acceptable	Poor
Accumulated speeding E_a^* (m/s)	$E_a^* \leq 0.7$	$0.7 < E_a^* < 1.0$	$E_a^* \geq 1.0$
Accumulated speed uniformity R_a (m/s)	$R_a \leq 1.0$	$1.0 < R_a < 2.0$	$R_a \geq 2.0$

R_a is defined as the normalized relative area (per unit length) lying between the speed profile and the line of average speed. The measure can be applied to individual speed profiles or to the operating speed profile. The first step is to calculate the average speed of the speed profile along the road. In this way, the areas lying between the speed profile and the average speed line (A_{ri}) are obtained (Figure 2b). The consistency measure is the sum of the absolute values of the areas divided by the length of the segment (L). Therefore, R_a is inversely related to the accumulated speed uniformity. Equation (3) must be applied.

$$R_a = \frac{\sum Ar_i}{L} \tag{3}$$

where:

- R_a: accumulated speed uniformity (m/s);
- $\sum Ar_i$: sum of areas in absolute values bound between the speed profile and the average speed (m^2/s);
- L: road section length (m).

The R_a thresholds for good, acceptable, and poor design are shown in Table 1 [40].

2.3. Analysis and Interpretation of Results

For each speed profile acquired by one of the methods mentioned in Section 2.1, the pair of effectiveness indicators consisting of E_a^* and R_a is determined.

Depending on the number of profiles evaluated for a given study site, a more or less extensive database is created, which must be subjected to detailed statistical analyses in order to understand whether the traffic calming measures in the site in question are indeed, as a whole, suitable to ensure the desired effectiveness requirements, or whether they present anomalies that could be mitigated by appropriate corrective measures.

For this purpose, the use of the box and whisker plot (or boxplot) is considered particularly appropriate. It is a convenient method for visually representing the distribution of data by their quartiles. Boxplots have the advantage of taking up little space, which is useful when comparing distributions between many groups or datasets.

From the display of a boxplot, the following observations in particular can be made: (a) what key values there are, such as average, median, 25th percentile, etc.; (b) whether there are outliers and what their values are; (c) whether the data are symmetrical; (d) how closely the data are grouped; (e) whether the data are skewed, and if so, in what direction.

The results of the statistical analysis applied to the different values of the effectiveness indicators calculated for the traffic calming measures of a given survey site may ultimately allow the determination that the site in question is suitable or unsuitable to ensure the necessary safety requirements. If, after applying the SPEIR methodology, the site is found to be inadequate, the administrator must decide whether and when to take corrective action to improve the safety performance that the site itself must guarantee to the various users of the street. The administrator may also decide not to intervene. However, if the administrator contemplates adjusting the site, it would be advisable for him/her to have an appropriate tool to identify the corrective actions that must be implemented on the site to ensure compliance with the quantifiable requirements with the E_a^* and R_a indicators. The simplest solution, but one that could prove ineffective or only partially effective, would be to modify the traffic calming measures at the study site and verify their effectiveness by reapplying the SPEIR methodology. In this case, there could be a risk that the new traffic calming measures installed at the survey site would still be insufficient to provide the desired safety performance.

It would, therefore, be desirable for the manager to have the option of selecting design solutions that have already been experimentally "validated" in other urban contexts through the SPEIR methodology and, therefore, have a high probability of ensuring the effectiveness requirements. This would minimize the manager's risk of expending financial resources without achieving the desired results. This possibility, shown in the right part of the framework in Figure 1, is discussed in more detail in the final part of this paper when the potential of the proposed methodology and future research developments are presented.

3. Case Studies and Results

The analysis methodology proposed by the authors was applied to evaluate the effectiveness of different traffic calming measures in three Zones 30.

3.1. Selection of Zones 30

The experimental analysis was carried out in two municipalities in the hinterland of the metropolitan city of Catania, S. Agata Li Battiati and Tremestieri Etneo. These are two small municipalities (in both cases the population is less than 20,000), with predominantly residential buildings. In both municipalities, most of the inhabitants tend to move to the metropolitan capital to work, while all the needs, foodstuffs, and small trade products, as well as the children's schooling, are met in the municipalities themselves.

Thus, the two communities have many similarities, and the administrations pursue similar strategies to improve the quality of life of the residents.

Three Zones 30 in the above municipalities were selected for this work. In two of them, there are vertical misalignments (Speed Tables, ST), while one of them is formed by horizontal misalignments (chicane and center island).

Zone 1 is a 447 m road section on which there are five speed tables placed on specific sections of the road, selected to attract the attention of drivers and cause them to slow down, and therefore, spaced at different distances from each other. All speed tables have a height of 8 cm and a width of 3 m. Zone 2, on the other hand, is a 196 m long section of road in which there are 4 speed tables placed at even distances from each other (about 45 m). All speed tables have a height of 4 cm and a width of 3.20 m. Zone 3 is a 280 m road section where a 130 m chicane and a 110 m horizontal deviation of the roadway by a center island have been created. Figure 3 shows the configurations of the studied sites.

Figure 3. Schemes of traffic calming measures in the studied sites: (**a**) Zone 1; (**b**) Zone 2; (**c**) Zone 3.

3.2. Speed Data Collection

The speed measurement method used in this work is based on the use of a smartphone and a dedicated application to record the data collected. As mentioned in Section 2.1, a measurement campaign of this type requires the selection of a sample of test drivers with a sufficiently large number for the investigation carried out. Twenty drivers were recruited by the University of Catania to conduct the test drives. An advertisement was placed on the University of Catania website that included information about the study and a questionnaire to recruit participants. Participants were selected from all those who responded to the advertisement. Drivers had to be between twenty-five and sixty years old and had held a driver's license for at least three years. The twenty test drivers were equally divided between male and female. The test drivers were selected from among those who were not regular users of the three survey sites. Indeed, one of the questions in the online recruitment questionnaire asked whether the usual trips were in the city center of Catania or in the hinterland communities. Since the three survey sites are located in two municipalities in the hinterland of Catania, only those users who indicated that they traveled mainly in the center of Catania were included. The drivers were called from time to time by the authors of this study, who played the role of the coordinators of the corresponding activities. Each driver performed the test in their own car, always accompanied by one of the coordinators who sat in the back seat during the test so as not to influence the behavior of the test driver. The coordinator took care of the correct positioning of the smartphone on the car dashboard (see Section 3.3) and assessed whether each test, once completed, was valid for the purposes of the study. The study was conducted in accordance with the Declaration of Helsinki, and the protocol was approved by the DISS-Center for Road Safety of the University of Parma (Deliberation of the Steering Committee—prot. 211112/2021 of 22 February 2021).

Participants gave their informed consent to participate in the experiment. They were informed that all data collected would be kept confidential and used for research purposes only. Participants were also informed that they would not be assessed on their skills as drivers and that the sole aim of the study was to analyze the behavior of a group of drivers to draw conclusions about drivers in general. For the purposes of the study, it was deemed necessary to collect speed profiles from non-regular users of the survey sites in order to evaluate the traffic calming effect of the measures in the Zones 30 as objectively as possible. Before the actual test, each driver was invited to drive through the Zone 30 accompanied by the coordinator. This allowed drivers to familiarize themselves with the specifics of the route and avoid the uncertainties that could arise from having to navigate a completely unfamiliar route. However, the actual test was performed by each driver only once for each survey location. The start of each test drive was always given by the coordinator present in the test driver's car as soon as the conditions of the test route were judged to be optimal. For research purposes, the second coordinator, who was at the end of the path being investigated, informed the other coordinator of any anomalies in Zone 30. Speed data were collected from August 2020 to May 2021 during off-peak periods, during daylight hours, and under good weather conditions. Speed data were collected in free-flowing traffic to ensure that the measured operating speeds were influenced only by Zone 30 characteristics. Only in seven cases was it necessary for the test driver to perform more than one test, since there were episodes during the test that affected the driver's behavior and invalidated the test (e.g., the sudden appearance of a slow vehicle or the presence of a pedestrian who expressed their intention to cross the road). An integrated system using smartphone applications was used to record, collect, store, analyze, and visualize speed data. The test drives were carried out using the test drivers' cars and a high-end smartphone equipped with high-quality sensors (GPS, accelerometer, gyroscope) and with the GEO Tracker application for Android systems installed. The smartphone was placed in the car (on the front dashboard) with the Y-axis of the accelerometer pointing towards the car. The collected dataset refers to both one-dimensional indicators (e.g., X, Y, Z accelerations, speed, etc.) and high spatial resolution (at 0.3 s intervals).

3.3. Determination of Speed Profiles

At the end of the survey work, 60 speed profiles were created (20 profiles for each of the three Zones 30 selected as a case study). Figures 4–6 show the speed profiles for the three areas studied. It should be noted that these profiles vary considerably, and from a preliminary analysis of these profiles, the following considerations can be made:

- In Zone 1, there are significant deviations of the speed from the average value. The latter coincides with the speed limit of 30 km/h. In particular, it can be observed how the speed peaks reached by the users are significantly reduced when the distances between the speed tables are reduced (e.g., between ST3 and ST4).
- In Zone 2, the speed profiles of the test drivers are quite homogeneous and there are no significant variations. However, it is obvious that the average speed values are higher than those prescribed in this Zone 30.
- In Zone 3, the speed profiles show little variation compared to the average, which is just above the 30 km/h speed limit. It is particularly noticeable that the speed

profiles at the chicane fall much more below the average speed of the zone than on the central island.

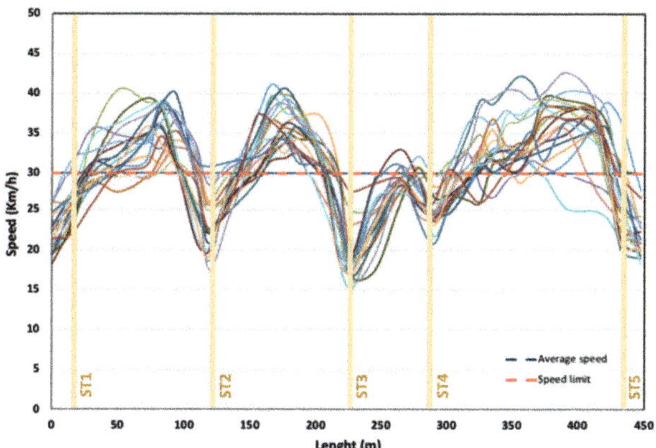

Figure 4. Speed profiles for Zone 1.

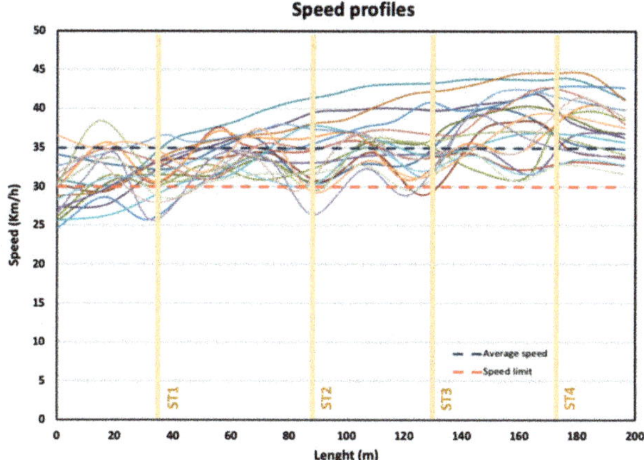

Figure 5. Speed profiles for Zone 2.

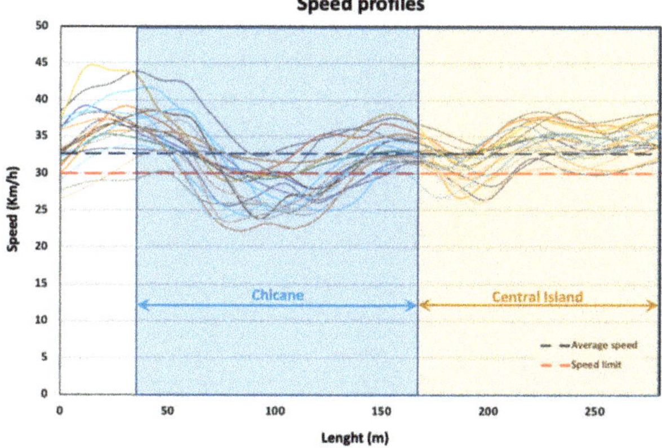

Figure 6. Speed profiles for Zone 3.

3.4. Estimation of Effectiveness Indicators

The effectiveness indicators (E_a^* e R_a) were calculated in relation to the speed profiles of the 20 test drivers and for each zone that was the subject of the experimental study. They were calculated according to the procedure outlined in Section 2.2.

Table 2 summarizes the results of the calculations performed. In particular, it contains the minimum and maximum values of the two indicators, as well as the values of the 85th percentile (i.e., the values exceeded by only 25% of the test drivers). Table 2 also contains summary information for each of the Zones 30 considered, as well as the average speeds (S_{ave}) assumed by the users who carried out the different driving tests.

Table 2. Summary of the values of the effectiveness indicators for the investigated sites.

Site	Speed Limit [km/h]	Length [m]	N. of Speed Profiles	S_{ave} [km/h]	$R_{a(min)}$ [m/s]	$R_{a(max)}$ [m/s]	$R_{a(85)}$ [m/s]	$E_{a(min)}^*$ [m/s]	$E_{a(max)}^*$ [m/s]	$E_{a(85)}^*$ [m/s]
ZONE 1	30	447	20	30.00	0.78	1.66	1.32	0.51	1.04	0.91
ZONE 2	30	196	20	34.89	0.25	1.28	0.88	0.66	1.61	1.39
ZONE 3	30	280	20	32.84	0.36	1.03	0.87	0.55	1.41	1.06

3.5. Results

Normality tests of Kolmogorov–Smirnov and Shapiro–Wilk were preliminarily performed for the distributions of E_a^* and R_a associated with the three Zones 30 analyzed. Both tests show that the three distributions of the variable E_a^* and the variable R_a are always statistically comparable to normal distributions (the tests always yield values of *p*-value > 0.05).

Subsequent statistical analyses were performed using the box and whisker plot (or boxplot). Table 3 shows all the parameters resulting from the statistical analysis of the three distributions of E_a^*. Figure 7 shows the boxplots associated with the distributions of E_a^* for the three zones studied. This plot has been divided into three areas delimited by the thresholds given in Table 1.

Table 4 shows all the parameters resulting from the statistical analysis of the three distributions of R_a, while Figure 8 shows the boxplots associated with the distributions of R_a for the three zones analyzed. This plot has been divided into three areas delimited by the thresholds given in Table 1.

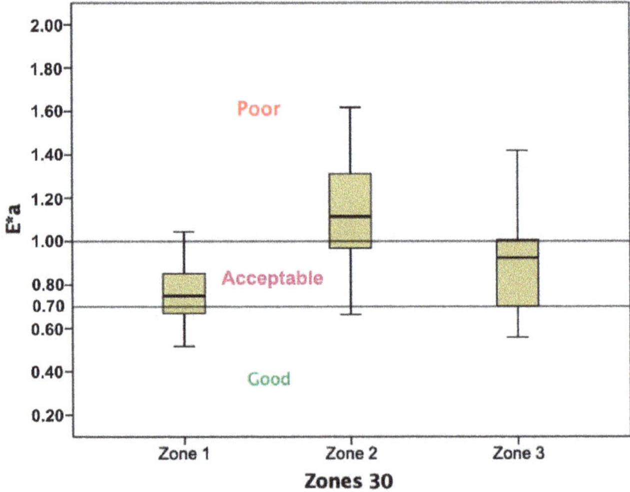

Figure 7. E_a^* boxplots for the three Zones 30 analyzed.

Table 3. Statistical parameters related to the distribution of E_a^* values.

E_a^* Distribution (ZONE 1)	Statistic	Std. Error
Mean	0.769990	0.0312493
Median	0.749350	
Variance	0.020	
Std. Deviation	0.1397513	
Minimum	0.5154	
Maximum	1.0438	
Range	0.5284	
E_a^* Distribution (ZONE 2)	Statistic	Std. Error
Mean	1.138506	0.0547420
Median	1.113268	
Variance	0.060	
Std. Deviation	0.2448137	
Minimum	0.6640	
Maximum	1.6167	
Range	0.9527	
E_a^* Distribution (ZONE 3)	Statistic	Std. Error
Mean	0.894378	0.0483706
Median	0.923961	
Variance	0.047	
Std. Deviation	0.2163197	
Minimum	0.5567	
Maximum	1.4184	
Range	0.8617	

Table 4. Statistical parameters related to the distribution of R_a values.

R_a Distribution (ZONE 1)	Statistic	Std. Error
Mean	1.129056	0.0527978
Median	1.155217	
Variance	0.056	
Std. Deviation	0.2361188	
Minimum	0.7866	
Maximum	1.6668	
Range	0.8801	
R_a Distribution (ZONE 2)	Statistic	Std. Error
Mean	0.656615	0.0607425
Median	0.620068	
Variance	0.074	
Std. Deviation	0.2716487	
Minimum	0.2557	
Maximum	1.2856	
Range	1.0300	
R_a Distribution (ZONE 3)	Statistic	Std. Error
Mean	0.692028	0.0414654
Median	0.720154	
Variance	0.034	
Std. Deviation	0.1854391	
Minimum	0.3681	
Maximum	1.0318	
Range	0.6637	

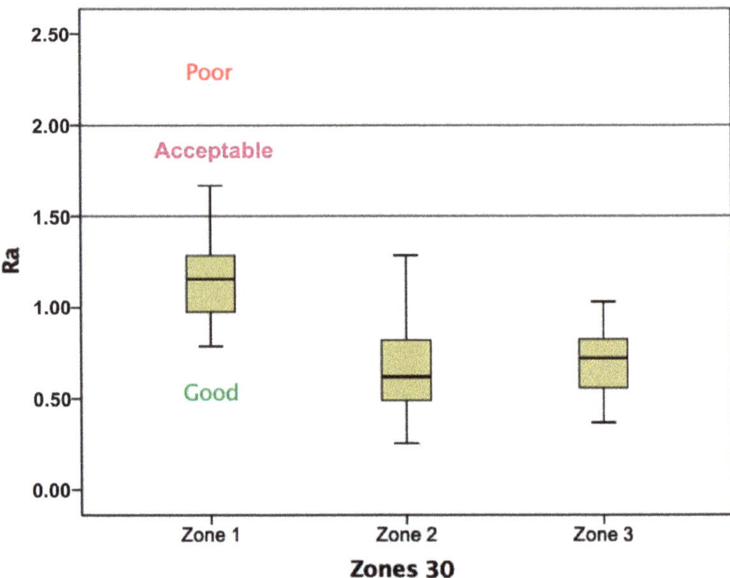

Figure 8. R_a boxplots for the three Zones 30 analyzed.

4. Discussion

The SPEIR methodology, proposed to analyze the effectiveness of traffic calming measures in a given study area, makes it possible to verify whether an urban infrastructure complex in which various measures have been implemented (e.g., low-speed zones, residential streets, home zones, etc.) is actually capable of ensuring compliance with speed limits and adherence to a uniform speed regime by road users.

The case studies presented in this work are fundamental results for the understanding of the methodology mentioned above and allow us to make considerations aimed at comparing the results obtained with those of other case studies in the technical and scientific literature.

In particular, the following observations can be derived from considering Table 3 and Figure 7:

1. No outliers were found in any of the three Zones. So, in any case, the TCMs do not lead to significant anomalies in the test drivers who participated in the experiment in terms of too high or too low speeds of the imposed limit.
2. All three Zones have levels of effectiveness that exceed the threshold labeled as "good". However, it should be noted that Zone 1, characterized by a median value of E_a^* of 0.75, performs largely acceptably and ensures that the values between the second and third quartiles, representing 50% of the total, are kept within a narrow range (dispersion of 0.53) and between the thresholds for "good" and "acceptable". Only in a few cases, as shown by the final value of the upper whisker of 1.04, does it border on what is considered a "poor" performance.
3. Zone 3 provides an overall acceptable level of performance. The box is essentially bounded by the two threshold values (lower limit equal to 0.69 and upper limit equal to 1.014). However, the value at the end of the upper whisker of 1.40 shows how drivers can be misled into adopting speeds well above the limit imposed by the traffic signs.
4. Zone 2 is the least successful in getting road users to obey the 30 km/h speed limit in effect on that section of road. The median value of E_a^*, which is 1.11, represents a "poor" level of performance. Moreover, the box is almost entirely above the line

that establishes the worst performance of the Zone in question in terms of compliance with the speed limit. If we also consider that (a) the end of the upper whisker shows how the values of the fourth quartile of E_a^* increase to a value above 1.60, (b) the first quartile is almost entirely contained in the part of the graph bounded below by the acceptance threshold; it becomes clear that the sequence of speed tables in Zone 2 is not an effective measure to ensure the dynamics "legally" required by a Zone 30.

However, from the analysis of Table 4 and Figure 8, it can be deduced that:

1. For all three Zones, there were no outliers for any test driver. This means that in no case do the traffic calming measures lead to very anomalous behavior in terms of uniformity of speed.
2. The medians for Zones 2 and 3 take similar values (0.62 and 0.72). Thus, in both Zones, users are induced to equalize their speed very close to the average value. It is also interesting to note that 50% of the R_a values are in an overall range between 0.48 and 0.83. This confirms the excellent response of the traffic calming measures in the two Zones to ensure a uniform speed. Instead, looking at the whiskers of the two boxplots, it can be seen that Zone 3 guarantees even more than Zone 2 a narrower dispersion of R_a values, actually containing all values between the minimum value of 0.36 and the maximum value of 1.03 (dispersion range 0.67). Zone 2, on the other hand, has a much wider dispersion of R_a values, equal to 1.03, although the end of the upper whisker, corresponding to $R_a = 1.27$, indicates in any case a "good" performance of the considered Zone.
3. Zone 1 influences user behavior, but to a lesser extent than the other two Zones. The median of R_a, which is 1.15, is below the threshold of 1.5 and, thus, representative of a "good" condition, but is significantly higher than the values for the other two Zones. Additionally, in this case, 50% of the R_a values (height of the box) are distributed in a range with reduced amplitude (equal to 0.37) between the extreme values of 0.93 and 1.3. However, the analysis of the whiskers of the boxplot shows how the fourth quartile of the distribution, bounded above by the value 1.66, indicates that the area in question induces a certain number of users to reach speeds well above the average value; in these cases, the Zone actually presents an "acceptable" operating condition, even if it is far from the one considered "poor".

Thus, the analysis of the distribution of the parameter E_a^* using the boxplot technique allows us to confirm that Zone 1, consisting of a sequence of five speed tables (h = 8 cm), is best able to ensure compliance with the speed limit characteristic of the Zone 30 in which it is located. On the other hand, the analogous measure consisting of the sequence of speed tables in Zone 2 (h = 4 cm) results in a significant tendency of road users to drive at speeds above the 30 km/h limit, which is insufficient overall to fulfil the task of limiting speed within the prescribed limit. These results are in agreement with numerous previous studies that have shown that the effectiveness of traffic calming measures has a strong positive correlation with the height dimension of the vertical speed control device [43–45]. Zone 3, on the other hand, provides an overall "acceptable" performance and can be considered sufficiently suitable for the role of traffic calming aimed at limiting the driving speed below the legal limit of 30 km/h. This result is also consistent with the scientific literature. In fact, a review of numerous studies on chicanes concludes that chicanes can significantly reduce the vehicle speed and traffic accident rate [46].

Moreover, the analysis of the distribution of the parameter R_a using the boxplot technique allows us to confirm that Zone 3, characterized by the chicane and the central island, is the one that most affects the uniformity of the assumed speed of users who played the role of test driver. On the other hand, the succession of speed tables in Zone 1 leads to speed variations with respect to the average value, which, although a condition for the acceptance of the speed-calming measures offered, proves the inhomogeneous behavior of the users who carried out the test in this Zone. This result does not agree with the study of Agerholm et al. [47], which shows that chicanes lead to larger speed fluctuations before reaching the chicane compared to humps. However, it is necessary to point out the small number of studies analyzing the uniformity of the speed of TCMs.

The discussion up to this point has been based on a separate consideration of the two indicators, E_a^* and R_a. In order to derive a final judgment on the overall effectiveness of the three Zones studied, the following considerations were made, based on the simultaneous analysis of the two aforementioned indicators:

- Zone 2, although not able to guarantee the basic requirement of a Zone 30, i.e., the respect of the speed limit, is characterized by the presence of TCMs suitable to influence the behavior of road users, who are induced to travel through the Zone without particular variations compared to the average speed, which is always below 40 km/h. This means that this design solution could be suitable to be installed in a context where it is necessary to respect a speed limit higher than 30 km/h (e.g., a 40 zone).
- The TCMs in Zone 2 are of the same type (vertical misalignments) as in Zone 1 and differ in that the height of the speed boards is 4 cm in Zone 2 and 8 cm in Zone 1. This difference in height speaks in favor of the effectiveness of Zone 1, which is largely acceptable in terms of compliance with the speed limit and can also ensure good uniformity of speed.
- Zone 3, similar to Zone 1, guarantees compliance with the 30 km/h speed limit, although at a lower level of acceptance. On the other hand, Zone 3 offers the best performance in terms of uniformity of speed (100% of Ra indicator values are well below the "Good" threshold).

Finally, it must be emphasized that no previous study has based the evaluation of the effectiveness of TCMs on the simultaneous analysis of the two indicators (E_a^* and R_a). This is certainly the original aspect of the SPEIR methodology proposed in this study. The authors strongly believe that the effectiveness of TCMs is not exclusively related to the value of speed reduction they produce, but also to the uniformity of speed, a parameter strongly related to the safety conditions of the site where the TCMs are installed.

5. Conclusions

The installation of TCMs results in a fluctuating speed profile along the road, and under these conditions, the road sections may not be suitable to provide a consistent travel speed for users. In this study, the authors define an analysis procedure, called the SPEIR methodology, based on the use of two effectiveness indicators. The SPEIR methodology allows the analysis of the performance of traffic calming measures that can be used wherever traffic calming measures exist or are to be installed to ensure compliance with a speed limit in a more or less extended urban infrastructure context, such as a Zone 30. These performances can be evaluated both in terms of compliance with the prescribed speed limit and in terms of their effectiveness in ensuring a consistent speed profile without too much deviation from the average.

The authors believe, also because they are reassured by the results of applying the SPEIR methodology to three survey sites, that:

(1) It is possible to compare different design solutions and better understand their effectiveness, also in terms of reproducibility within the same site or in terms of implementation in urban contexts that one wants to adapt, for example, by establishing a Zone 30;

(2) If the proposed methodology is applied internationally to evaluate the effectiveness of TCMs in different urban contexts and the results are made publicly available, an important database could be created within a reasonable period of time. Thanks to this database, the values of E_a^* and R_a corresponding to the different design configurations would be clearly known. In this way, the values of the two effectiveness indicators would be available for the different types of traffic calming measures (e.g., chicanes, speed tables, speed bumps, chokers, etc.) and their installation criteria (e.g., distance between bumps, characteristic length of the chicane, extent of narrow-

ing). These indicators could, thus, be used as reference parameters in new designs or in functional adaptation measures;

(3) Achieving the objectives mentioned in the previous two points (comparison of different design solutions and creation of a reference database) can also be carried out using design scenarios created in simulated environments. The SPEIR methodology could be tested in virtual environments and, once validated, used to characterize different simulated configurations in which traffic calming measures are present through the proposed effectiveness indicators.

Systematic applications of the SPEIR methodology can greatly assist urban road managers in optimizing design decisions during the planning and design phases of Zones 30 or other urban contexts where extensive traffic slowing measures need to be implemented. They would also benefit from a tool that would certainly be useful under the sustainable mobility goal to identify corrective actions that need to be put into practice to improve the effectiveness of TCMs in the urban areas to be managed.

In this context, it is desirable that further case studies, similar to those presented here, are conducted by researchers and brought to the attention of managers and policy makers through the usual channels of cultural and scientific dissemination.

Author Contributions: Conceptualization, N.D. and S.L.; methodology, N.D.; software, N.D. and S.L.; validation, S.L.; formal analysis, N.D. and S.L.; investigation, N.D. and S.L.; resources, N.D. and S.L.; data curation, S.L.; writing—original draft preparation, N.D. and S.L.; writing—review and editing, N.D. and S.L.; visualization, N.D.; supervision, S.L. All authors have read and agreed to the published version of the manuscript.

Funding: This research received no external funding.

Institutional Review Board Statement: The study was conducted according to the guidelines of the Declaration of Helsinki, and the protocol was approved by the DISS-Center for Road Safety of the University of Parma (Deliberation of the Steering Committee—prot. 211112/2021 of 22 February 2021).

Informed Consent Statement: Informed consent was obtained from all subjects involved in the study. Written informed consent has been obtained from the patient(s) to publish this paper.

Data Availability Statement: The data presented in this study are available on request from the corresponding author. The data are not publicly available due to confidentiality issues.

Conflicts of Interest: The authors declare no conflict of interest.

References

1. Balant, M.; Lep, M. Comprehensive Traffic Calming as a Key Element of Sustainable Urban Mobility Plans—Impacts of a Neighbourhood Redesign in Ljutomer. *Sustainability* **2020**, *12*, 8143. [CrossRef]
2. Dickerson, A.; Peirson, J.; Vickerman, R. Road Accidents and Traffic Flows: An Econometric Investigation. *Economica* **2000**, *67*, 101–121. [CrossRef]
3. Zang, K.; Batterman, S. Air pollution and health risks due to vehicle traffic. *Sci. Total Environ.* **2013**, *15*, 307–316. [CrossRef] [PubMed]
4. Fuks, K.; Moebus, S.; Hertel, S.; Viehmann, A.; Nonnemacher, M.; Dragano, N.; Möhlenkamp, S.; Jakobs, H.; Kessler, C.; Erbel, R.; et al. Long-term urban particulate air pollution, traffic noise, and arterial blood pressure. *Environ. Health Perspect.* **2011**, *119*, 1706–1711. [CrossRef]
5. Albalate, D.; Fageda, X. Congestion, Road Safety, and the Effectiveness of Public Policies in Urban Areas. *Sustainability* **2019**, *11*, 5092. [CrossRef]
6. Granà, A.; Giuffrè, T.; Guerrieri, M. Exploring Effects of Area-Wide Traffic Calming Measures on Urban Road Sustainable Safety. *J. Sustain. Dev.* **2010**, *3*, 38–49. [CrossRef]
7. Useche, S.A.; Alonso, F.; Montoro, L. Validation of the walking behavior questionnaire (WBQ): A tool for measuring risky and safe walking under a behavioral perspective. *J. Transp. Health* **2020**, *18*, 100899. [CrossRef]
8. Pulvirenti, G.; Distefano, N.; Leonardi, S. Elderly perception of critical issues of pedestrian paths. *Civ. Eng. Archit.* **2020**, *8*, 26–37. [CrossRef]
9. Distefano, N.; Pulvirenti, G.; Leonardi, S. Neighbourhood walkability: Elderly's priorities. *Res. Transp. Bus.* **2021**, *40*, 100547. [CrossRef]

10. Gálvez-Pérez, D.; Guirao, B.; Ortuño, A. Road Safety of Elderly Pedestrians in the Urban Context: An Approach Based on Infrastructure and Socioeconomic Variables. *Transp. Res. Procedia* **2021**, *58*, 254–261. [CrossRef]

11. Leonardi, S.; Distefano, N.; Pulvirenti, G. Identification of road safety measures for elderly pedestrians based on k-means clustering and hierarchical cluster analysis. *Arch. Transp.* **2020**, *56*, 107–118. [CrossRef]

12. WHO—World Health Organization. *Global Status Report on Road Safety*; World Health Organization: Geneva, Switzerland, 2018.

13. European Commission. Facts and figures—Pedestrians. In *European Road Safety Observatory; Directorate General for Transport*; European Commission: Brussels, Belgium, 2021.

14. ACI-ISTAT. Incidenti stradali in Italia (English translation: Road accidents in Italy). In *Report of ACI-ISTAT—Anno 2019*; ACI-ISTAT: Rome, Italy, 2020.

15. ACI-ISTAT. Incidenti stradali in Italia (English translation: Road accidents in Italy). In *Report of ACI-ISTAT—Anno 2020*; ACI-ISTAT: Rome, Italy, 2021.

16. Distefano, N.; Leonardi, S.; Consoli, F. Drivers' Preferences for Road Roundabouts: A Study based on Stated Preference Survey in Italy. *KSCE J. Civ. Eng.* **2019**, *23*, 4864–4874. [CrossRef]

17. Ciampa, D.; Diomedi, M.; Giglio, F.; Olita, S.; Petruccelli, U.; Restaino, C. Effectiveness of unconventional roundabouts in the design of suburban intersections. *Eur. Transp. Eur.* **2020**, *80*, 1–16. [CrossRef]

18. Gonzalo-Orden, H.; Pérez-Acebo, H.; Linares Unamunzaga, A.; Rojo Arce, M. Effects of traffic calming measures in different urban areas. *Transp. Res. Procedia* **2018**, *33*, 83–90. [CrossRef]

19. Sharpin, A.B.; Adriazola-Steil, C.; Job, S.; Obelheiro, M.; Welle, B.; Imamoglu, C.T.; Bhatt, A.; Liu, D.; Lleras, N.; Luke, N. Low-Speed Zone Guide. 2021. Available online: https://files.wri.org/d8/s3fs-public/2021-05/WRI_LowSpeedZone_web.pdf?VersionId=mMwA8aq.BSpHZ6w97Z4E2NSd5o1UPN7B (accessed on 17 January 2022). [CrossRef]

20. Kempa, J. Respecting a speed limit and its effectiveness in a traffic calming zone. *IOP Conf. Ser. Mater. Sci. Eng.* **2019**, *603*, 042068. [CrossRef]

21. Distefano, N.; Leonardi, S. Evaluation of the Benefits of Traffic Calming on Vehicle Speed Reduction. *Civ. Eng. Arch.* **2019**, *7*, 200–214. [CrossRef]

22. Vaitkus, A.; Čygas, D.; Jasiuniene, V.; Jateikiene, L.; Andriejauskas, T.; Skrodenis, D.; Ratkeviciute, K. Traffic Calming Measures: An Evaluation of the Effect on Driving Speed. *Promet-Traffic Transp.* **2017**, *29*, 275–285. [CrossRef]

23. Daniel, B.D.; Nicholson, A.; Koorey, G. Analysing speed profiles for the estimation of speed on traffic-calmed streets. *Transp. Res. Rec.* **2011**, *20*, 57–70.

24. Pérez-Acebo, H.; Ziółkowski, R.; Linares-Unamunzaga, A.; Gonzalo-Orden, H. A Series of Vertical Deflections, a Promising Traffic Calming Measure: Analysis and Recommendations for Spacing. *Appl. Sci.* **2020**, *10*, 3368. [CrossRef]

25. Gayathri, K.B. Evaluation of the Effectiveness of Traffic calming measures. In Proceedings of the International Conference on Systems, Energy & Environment (ICSEE), Singapore, 3–5 February 2021. [CrossRef]

26. Ziółkowski, R. Speed Management Efficacy on National Roads—Early Experiences of Sectional Speed System Functioning in Podlaskie Voivodship. *Transp. Probl.* **2018**, *18*, 5–12. [CrossRef]

27. Zito, R.; Taylor, M.A.P. Speed profiles and vehicle fuel consumption at LATM devices. In Proceedings of the 18th ARRB Conference, Christchurch, New Zealand, 2–6 September 1996; Volume 18, pp. 391–406.

28. Jurewicz, C.; Sobhani, A.; Woolley, J.; Dutschke, J.; Corben, B. Exploration of vehicle impact speed-injury severity relationships for application in safer road design. *Transp. Res. Procedia* **2016**, *14*, 4247–4256. [CrossRef]

29. Wann, J.P.; Poulter, D.R.; Purcell, C. Reduced sensitivity to visual looming inflates the risk posed by speeding vehicles when children try to cross the road. *Psychol. Sci.* **2011**, *22*, 429–434. [CrossRef] [PubMed]

30. Lobjois, R.; Cavallo, V. Age-related differences in street-crossing decisions: The effects of vehicle speed and time constraints on gap selection in an estimation task. *Accid. Anal. Prev.* **2007**, *39*, 934–943. [CrossRef] [PubMed]

31. Webb, E.A.; Bell, S.; Lacey, R.E.; Abell, J.G. Crossing the road in time: Inequalities in older people's walking speeds. *J. Transp. Health* **2017**, *5*, 77–83. [CrossRef]

32. 20's Plenty for Us. 20mph Limits Save Time and Improve Traffic Flow. 2012. Available online: http://www.20splentyforus.org.uk/BriefingSheets/20mph_Improves_Traffic_Flow.pdf (accessed on 10 January 2022).

33. 20's Plenty for Us. 20mph Limits Improve Air Quality Where People Live. 2010. Available online: http://www.20splentyforus.org.uk/BriefingSheets/pollutionbriefing.pdf (accessed on 10 January 2022).

34. Jones, S.J.; Brunt, H. Twenty miles per hour speed limits: A sustainable solution to public health problems in Wales. *J. Epidemiol. Community Health* **2017**, *71*, 699–706. [CrossRef]

35. Brindle, R.E. Speed-Based Design of Traffic Calming Schemes. In *Institute of Transportation Engineers (ITE) Annual Meeting 2005*; ITE: Melbourne, Australia, 2005.

36. Berloco, N.; Colonna, P.; Intini, P.; Masi, G.; Ranieri, V. Low-cost smartphone-based speed surveying methods in proximity to traffic calming devices. *Procedia Comput. Sci.* **2018**, *134*, 415–420. [CrossRef]

37. Kyriakoua, C.; Christodouloua, S.E.; Dimitrioua, L. Do Vehicles Sense, Detect and Locate Speed Bumps? *Transp. Res. Procedia* **2020**, *52*, 203–210. [CrossRef]

38. Dong, D.; Li, Z. Smartphone Sensing of Road Surface Condition and Defect Detection. *Sensors* **2021**, *21*, 5433. [CrossRef]

39. Tu, W.; Xiao, F.; Li, L.; Fu, L. Estimating traffic flow states with smart phone sensor data. *Transp. Res. Part C* **2021**, *126*, 103062. [CrossRef]

40. Polus, A.; Mattar-Habib, C. New Consistency Model for Rural Highways and Its Relationship to Safety. *J. Transp. Eng.* **2004**, *130*, 286–293. [CrossRef]
41. Moreno, A.T.; García, A. Use of speed profile as surrogate measure: Effect of traffic calming devices on crosstown road safety performance. *Accid. Anal. Prev.* **2013**, *61*, 23–32. [CrossRef] [PubMed]
42. Domenichini, L.; Branzi, V.; Meocci, M. Virtual testing of speed reduction schemes on urban collector roads. *Accid. Anal. Prev.* **2018**, *110*, 38–51. [CrossRef] [PubMed]
43. Dinasty Purnomo, A.; Dewi, H.; Syafi, I. Correlation analysis between speed bump dimensions and motorcycle speed in residential areas. *MATEC Web Conf.* **2018**, *195*, 04013. [CrossRef]
44. Antic, B.; Pesic, D.; Vujanic, M.; Lipovac, K. The influence of speed bumps heights to the decrease of the vehicle speed-Belgrade Experience. *Saf. Sci.* **2013**, *57*, 303–313. [CrossRef]
45. Bachok, K.S.; Mohamed, M.Z.; Ibrahim, M.; Hamsa, A.A.K. A theoretical overview of road hump effects on traffic speed in residential environments. *Plan. Malays. J.* **2016**, *14*, 343–352. [CrossRef]
46. Zhang, C.; Qin, S.; Yu, H.; Zheng, B.; Li, Z. A Review on Chicane Design based on Calming Theory. *J. Eng. Sci. Technol. Rev.* **2020**, *13*, 188–197. [CrossRef]
47. Agerholm, N.; Knudsen, D.; Variyeswaran, K. Speed-calming measures and their effect on driving speed—Test of a new technique measuring speeds based on GNSS data. *Transp. Res. F Traffic Psychol. Behav.* **2017**, *46*, 263–270. [CrossRef]

Article

Traffic Calming Measures and Their Slowing Effect on the Pedestrian Refuge Approach Sections

Stanisław Majer and Alicja Sołowczuk *

Department of Construction and Road Engineering, West Pomeranian University of Technology in Szczecin, 71-311 Szczecin, Poland; stanislaw.majer@zut.edu.pl
* Correspondence: alicja.solowczuk@zut.edu.pl; Tel.: +48-91-449-40-36

Abstract: The ever-increasing use of motor vehicles causes a number of traffic safety and community issues, which are particularly severe in cities, accompanied by a scarcity of parking spaces and challenges encountered in road layout alteration projects. The commonly applied solutions include the designation of through streets, the implementation of on-street parking on residential streets, and retrofitted traffic calming measures (TCMs). This article presents the results of the study conducted on a two-way street where the Metered Parking System (MPS) was implemented together with diagonal and parallel parking spaces, refuge islands, horizontal deflection, and lane narrowing by a single-sided chicane. The aim of this study was to identify those TCMs that effectively helped to reduce the island approach speed. The heuristic method was applied to assess the effect of the respective TCMs on reducing the island approach speed, and the key speed reduction determinants were defined using a cause-and-effect diagram and a Pareto chart. The determinants were evaluated with the binary system and tautological inference principles, whereby a determinant was rated as true when it was found in the field, with a simultaneous speed reduction determined in the survey. Determinants that were not confirmed in the field were rated untrue. Comparative analyses were carried out to rate the respective TCMs as effective, moderately effective, or ineffective. In this way, the following three determinants were rated as the most important for speed reduction at refuge islands: free view, visibility of a pedestrian on the right-hand side of the island, and the refuge island surroundings. Although the study was limited to a single street in Poland, the findings may hold true in other countries where similar TCMs are used.

Keywords: pedestrian refuges; refuge islands; reduce speed; traffic calming measures; TCM; horizontal deflection; free view; Pareto chart; cause-and-effect diagram

Citation: Majer, S.; Sołowczuk, A. Traffic Calming Measures and Their Slowing Effect on the Pedestrian Refuge Approach Sections. *Sustainability* **2023**, *15*, 15265. https://doi.org/10.3390/su152115265

Academic Editors: Salvatore Leonardi and Natalia Distefano

Received: 13 September 2023
Revised: 16 October 2023
Accepted: 23 October 2023
Published: 25 October 2023

1. Introduction

The ever-increasing use of motor vehicles causes more and more severe traffic issues in urban areas in particular. Various traffic management measures are applied to address these issues, including the designation of urban transit routes, implementation of traffic calming schemes, parking planning, etc. A well-planned metered parking system requires a smooth coincidence of traffic calming plans with the planned parking spaces and carefully planned pedestrian mobility improvements. The design aspects of different traffic calming measures (TCMs) are laid out in the basic design guidelines [1–5]. TCMs include raised intersections, speed tables, narrowing the carriageway by chokers or pinch points, various speed humps, and speed bumps. Horizontal deflections are also applied in the planning of parking spaces depending on the parking configuration.

Elvik [6] suggests using a meta-analysis approach in designating urban transit routes or traffic calming zones to address the relevant traffic safety issues. These should lead to defining a hierarchical road system and moving through traffic out of the residential streets, thus improving traffic safety in these residential areas. Different approaches to urban traffic safety and traffic and parking resource management scenarios in metered parking settings are presented, for example, in [7–10]. It should be noted, though, that the issues tackled in these articles concern mainly parking in urban areas. A different TCM study approach, taking into account their effect on traffic performance, traffic safety, the natural environment, public health, and the economy, was presented in articles [11–16], showing that traffic calming has some undesirable effects as well. The group of TCMs that were found to have undesirable environmental effects included speed cushions, speed bumps, speed humps, and stop signs.

1.1. Review of Studies on the Speed-Reducing Effect of Horizontal Deflections Located on the Refuge Island Approach Sections

The efficacy of various TCMs used on city streets, i.e., their slowing effect, has been studied by many researchers. In most cases, these studies analyse TCMs in relation to traffic safety improvement [17–20]. The article by Le et al. [12] is different in this respect in that it also considers the environmental and public health impacts of the analysed TCMs.—the study involved in situ tests conducted using a special test vehicle. Le et al. [12] used a comparative analysis technique to demonstrate the superiority of chicanes among the analysed TCMs, except in terms of vehicle emissions. That said, most studies are limited to analysing the efficacy of speed humps, speed tables, and chicanes in terms of speed reduction on the approach to pedestrian crossings. Some authors took into account landscape features and visibility of the pedestrian crossing and the road ahead, relating the obtained speed reduction not only to the TCMs but also to various factors of the townscape surrounding the refuge island [21,22]. For example, Balant and Lep [22] analysed the improvement in community life thanks to the implemented traffic calming scheme. Other researchers noted the slowing effect of repeating the speed humps or speed tables and the length of slowed driving [23–27].

The efficacy of various TCMs was analysed, for example, by Gonzalo-Orden et al. [28]. They compared through comparative analyses the speed reductions obtained with the applied raised crossings, lane narrowings or chokers, speed cameras, and speed camera signs. These analyses led to the conclusion that the obtained speed reductions depended on the TCM type, its geometric features, and emplacement in the street. Distefano & Leonardii [29,30] arrived at similar conclusions on the efficacy of chicanes and horizontal deflections in city streets. They compared speed profiles (85th percentile and average values) on local streets before and after installation of speed tables and up to 1 m wide chicanes on a one-way street and road narrowing treatment accompanied by a horizontal deflection on a two-way street. The before-and-after study results presented by Distefano & Leonardi [29] show the highest percentage reduction of operating speed for a single-lane chicane installed on a narrow one-way street with an on-street parallel parking configuration. The lowest percentage reduction was, in turn, noted on a two-way street with a carriageway narrowing treatment on one side, accompanied by a horizontal deflection (with parallel on-street and pavement parking). In this case, very good visibility of the road past the narrowing treatment was ensured. Kruszyna & Matczuk-Pisarek [31] arrived at different conclusions in their study on speed reduction obtained with a refuge island, speed table on the approach section, or a raised pedestrian crossing. The comparative analyses showed that raised pedestrian crossings offered the highest speed reductions. Sołowczuk [32] studied speed reductions obtained with raised pedestrian crossings in a downtown Tempo 30 zone, relating the obtained values not only to the TCM geometry and the townscape surrounding the street but also to the specific traffic volume in a given street.

Akgol et al. [33] and Aydin et al. [34] conducted a driving simulator study to investigate the effect of chicanes installed near pedestrian crossings. The factors they considered in their study included the effective lane width, the shapes of islands, and vehicle trajectories. In conclusion [33], it is stated that effective speed reduction may be obtained with a set of three chicanes located at the refuge island on streets with a 3-m effective lane width or with a more economical option of two chicanes on streets with a 2.7-m effective lane width. Hussain et al. also used a driving simulator, yet with a different approach, as presented in their article [35] investigating the effect of roadway narrowing, horizontal deflection, and various road markings and upright signs. These studies confirmed the highest efficacy of road narrowing used in combination with horizontal deflection and carriageway narrowing obtained by zigzag markings or variable message signs.

The first study that related speed reduction to the travel path deflection by a median island or chicanes was conducted in the UK by Sayer and Parry [36,37]. In the test track trials, the test vehicles navigated through artificially simulated horizontal deflection and chicanes. Experienced drivers were employed for these trials. The output of the study confirmed that the primary speed reduction factors were the stagger length, free view through the chicane, deflected path angle, and the visual obstruction type (Figure 1). In this study, the free view width "*a*" had a positive value if the median island between opposing lanes allowed the driver to see the travel lane behind it at the road surface level. If, on the contrary, the driver approaching the island could not see the whole lane width at the road surface level past the island, "*a*" acquired a negative value. The wider the median island, and thus, the less of the travel lane at the road surface level was visible to the driver, the greater the obtained speed reduction. These findings were confirmed by Zhang et. al. [38], who, in addition, investigated reductions in noise and vehicle emissions.

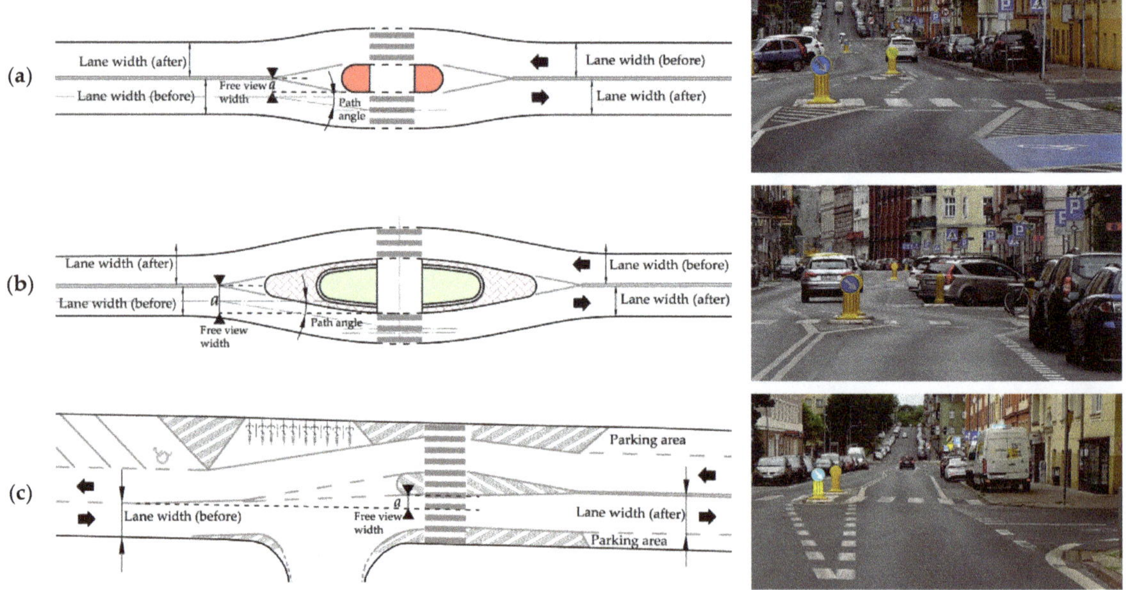

Figure 1. Free view and path angle illustration: (**a**) "*a*"—small; (**b**) "*a*"—larger; (**c**) "*a*" +. Source: own work.

The above literature review allowed us to compile in Figure 2 and compare the calculated 85th percentile and mean speeds noted just before the pedestrian crossing or chicane. Figure 2 shows a high degree of inconsistent data obtained by different researchers due to different locations (test tracks, transition zone, village centre, suburban two-lane, single-carriageway streets) and data selection. As regards the data selection, the researchers chose to analyse free traffic flow only or use the steady traffic flow data with varying hourly volumes and separately the free traffic flow data.

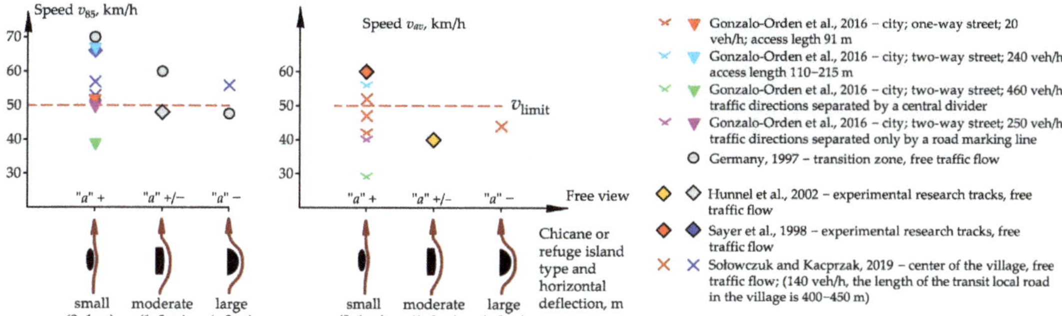

Figure 2. Comparison of v_{85} and v_{av} values ahead of a refuge island or median island in different locations. Source: Own work based on data presented in: Gonzalo-Orden et al., 2016 [28], Germany, 1997 [39], Hunnel et al., 2002 [40], Sayer et al., 1998 [37], Sołowczuk & Kacprzak, 2019 [21].

1.2. Review of Previous Before-And-After Speed Studies with the Use of the Heuristic Method

Heuristic methods are used in management analyses when dealing with complex situations and lots of information. They allow us to assess the efficacy of the analysed parameters based on the established determinants. The principles of this method were described by different scholars, including Juran (first edition in 1951) [41–43] and Deming [44] (first edition in 1982), and were elaborated by Kaoru Ishikawa, who proposed seven basic quality tools for the Total Quality Management (TQM) system [45–47] and their researchers [48–53]. Quality management principles may be used successfully for assessments of other issues, including road maintenance [54,55], road operating speed management [56–60], or very specific applications, such as analysing fluid velocity variations in medical equipment [61]. The seven tools developed by Kaoru Ishikawa [45–47] are:

a. flow chart presenting the steps of the analysis,
b. check sheet, specifically statistical tests to check speed consistency among the consecutive survey sites deployed on the street under analysis,
c. normal distribution histograms,
d. scatter diagram showing relationships,
e. control chart showing speed changes along the analysed street,
f. cause-and-effect diagram (diagram fishbone diagram or Ishikawa diagram) for defining the primary and secondary factors,
g. Pareto chart to define the final identified speed reduction determinants.

These allow the determination of factors that contributed to attaining the final effect in consideration.

In traffic speed studies, the heuristic method allows us to estimate the influence of the different determinants on the final operating speed reduced by various treatments, including TCMs. The abovementioned seven tools of the heuristic method were used in this study to assess the efficacy of different TCMs implemented in the analysed downtown street section.

The above literature review revealed that the research publications and various existing design guidelines have so far not covered the issue of the efficacy of repeated and varied TCMs before refuge islands on two-way city streets. The purpose of this study was to find the most effective TCM configuration before refuge islands located on two-way streets in urban areas. TCM effectiveness is understood as a reduction in operating speeds to improve traffic safety as a result. Section 1 of this article presents the literature review on TCM application near refuge islands and a general description of the heuristic method principles. In Section 2, the reader will find:

- information on the study site (a two-way city street with 50 km/h speed limit) and details of the respective study sections with different parking and TCM arrangements,

- traffic safety analyses before and after changes to the traffic organisation plan,
- description of the heuristic method used in the study.

Section 3 presents the results of speed change analyses for the studied street sections. Section 4 discusses the obtained results and analyses the predefined determinants that, in combination with hourly traffic volumes, may cause operating speed reduction ahead of the refuge island. These analyses were made with the use of a cause-and-effect diagram and Pareto charts. Section 5 presents conclusions that may be used by traffic engineers designing traffic calming for two-way city streets. The sequence of analyses as they appear in the article is presented in Figure 3.

Input data		Initial and final traffic measurements		Heuristic method		Dependency analyses of speed variation along the street		**Definition of the determinants**

Figure 3. Sequence of analyses conducted in this study. Source: own work.

2. Materials and Methods

2.1. Study Area

A two-way street in downtown Szczecin, Poland, was chosen as the study site. In 2015, an urban block alteration scheme was started in Szczecin in order to improve the transport network and in relation to the planned Metered Parking System MPS implementation. This required the demarcation of metered parking spaces. Tempo 30 zones were introduced on some streets, and various TCMs were implemented elsewhere in the area as part of this road system alteration scheme [62]. This study deals specifically with a two-way street, including demarcated parking spaces, refuge islands, and horizontal deflection of the travel path imposed by road markings and refuge islands (Figures 4 and 5). The street had a 50 km/h speed limit. In Poland, a 10 km/h allowance is applied in routine speed checks by means of speed cameras, as guided by relevant codes [63]. This allowance is deemed to account for measurement and driver errors, as the case may be.

Figure 4. Visualisation of the analysed two-way street before the 2014 alteration showing traffic directions. Source: own work on a satellite image background Google Earth [64]).

(a)

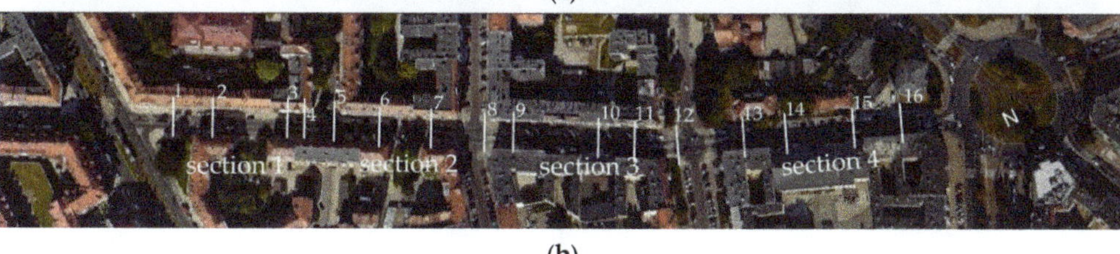

(b)

Figure 5. Visualisation—four study sections located between three junctions were selected for the analyses of this study, on which a total number of sixteen survey stations were set up: (**a**) three junctions and six pedestrian refuges; (**b**) four study sections and sixteen survey stations. Source: own work on a satellite image background Google Earth [64].

Considering the two-way traffic arrangement on the analysed street, the study sections were identified with geographical symbols and numbers (Figures 5 and 6). Thus, in the direction W→E, sections between the signal-controlled junction and the roundabout were designated WE1, WE2, and WE3. Accordingly, in the direction E→W, sections located in the same area were designated EW1, EW2, and EW3. All the study sections are shown in Figure 5. The geometrical features of the respective study sections are given in Figure A1 in Appendix A.

(**a**)

EW1

WE1

Figure 6. *Cont.*

EW2

WE2

EW3

WE3

Figure 6. Study sections: (**a**) WE1 & EW1; (**b**) WE2 & EW2; (**c**) WE3 & EW3. Source: Photo A. Sołowczuk.

2.2. Traffic Safety and Volume Count Data for the Analysed Street

The effect of the changed traffic management arrangements on the analysed street was assessed through a road incident statistical analysis carried out using the Accident and Collision Registration System SEWIK [65] software program output data. These input and output data are given in Table 1 below.

Table 1. Input data and statistical analysis output for the analysed street. Source: Own research based on data presented in [65].

Years	Traffic Accidents in General	Pedestrian Accidents
Before data: 1 January 2000–31 December 2015	27	6
After data: 1 January 2016–31 May 2023	6	1

The Chi-square significance test χ^2 was used to confirm or refute the efficacy of a given TCM and the resulting traffic safety improvement. Null hypothesis H_0: $\chi^2 = (n_1\,t_2 - n_2\,t_1)^2/(t_1\,t_2\,(n_1 + n_2))^1 \leq \chi_\alpha^2$; (no statistically significant difference exists).
Alternative hypothesis H_1: $\chi^2 > \chi_\alpha^2$; (a statistically significant difference does exist).
The following inequation should be satisfied at the same time: $n_1/t_1 > n_2/t_2$.
Critical value $\chi_\alpha^2 = 3.84$ at the significance level $\alpha = 0.05$.

$\chi^2 =$	3.0 < 3.84	1.0 < 3.84
$n_1/t_1 > n_2/t_2$	1.7 > 0.8	0.38 > 0.13

Legend: n_1—before-project road incidents/accidents, n_2—after-project road incidents/accidents, t_1—years before, t_2—years after.

The statistical test results compiled in Table 1 have not definitely confirmed the efficacy of the changed traffic arrangements, i.e., fewer road incidents and vehicle/pedestrian collisions. However, the substantial growth of traffic on the analysed street in the timespan of the study must be taken into account at this point, as it could have some bearing on the number of road incidents. The traffic volume output data are given in Figure A2 in Appendix B. The cause of a higher traffic volume in one direction of travel is the local traffic arrangement with a two-lane, one-way in the direction E→W street to the north and a two-way street to the south, the latter including a two-way tram line (Figure 4). This arrangement results in nonuniform traffic loading of the two travel directions with almost two times higher in the direction W→E traffic volume (Figure A2 in Appendix B).

2.3. Measurement and Analysis Method

In order to assess the slowing effect of the implemented modification of the traffic arrangements, round-the-clock traffic count and speed measurement surveys were carried out on site for two days, i.e., for a total of 48 h on each of the sixteen survey stations. SR4 [66] synchronised traffic detection devices were used and mounted on the existing signposts. The locations of the sixteen survey stations and deployment positions of the SR devices are shown in Figure 5.

Traffic counts and speed measurement surveys started on Friday morning and ended on Sunday. These weekend surveys lasted through May and June. The weather was dry during that time, ensuring uniform driving conditions. Four survey stations were deployed on each one-block section, positioned as follows: at the section (block) entry, just before the refuge island, and within and past the junction. This deployment allowed observation of speed variation along a given portion of the street under analysis. In total, 16,000–18,000 travel speed readings were logged at each survey station.

Considering daily variations in traffic volume (ranging from a few to dozen plus veh/h overnight to about 500 veh/h during the day), the data were subjected to a statistical analysis using the Two-sample Kolmogorov-Smirnov test and Median test in order to determine whether the hourly traffic volumes may be analysed as one group or must be treated individually. The authors conducted a two-day preliminary traffic count and speed measurement surveys at two survey stations on the analysed street before and after metered parking and TCM scheme implementation. For a majority of the results, the standard deviation was a variable statistic, and negative results were obtained in both tests. Therefore, it was required for statistical analysis purposes to split the speed data set into subsets corresponding to traffic volume intervals of 50 veh/h. The statistical tests for four of these subsets for different traffic flow directions (including two "before" and two "after" subsets) are given in Table 2.

Table 2. Results of statistical tests to check whether speed data may be analysed as a single set. Source: own work.

No.	Traffic Volume, veh/h	Traffic Flow Directions							
		Before Measurement Data				After Measurement Data			
		W→E	E→W	W→E	E→W	W→E	E→W	W→E	E→W
		Test K-S [1]		Median Test [2]		Test K–S [1]		Median Test [2]	
1	$N \leq 50$ & $50 < N \leq 100$	9.8	12.4	120.3	1392.4	14.3	15.6	2469.4	1544.9
2	$50 < N \leq 100$ & $100 < N \leq 150$	12.1	13.5	1737.1	308.2	17.9	17.4	2344.7	3113.2
3	$100 < N \leq 150$ & $150 < N \leq 200$	20.4	11.9	16,893.3	189.6	16.9	20.9	1744.0	7370.2
4	$150 < N \leq 200$ & $200 < N \leq 250$	24.6	–	7490.0	–	17.3	25.4	2735.2	16,035.3
5	$200 < N \leq 250$ & $250 < N \leq 300$	–	–	–	–	19.6	23.8	4905.4	5532.8
6	$250 < N \leq 300$ & $300 < N \leq 350$	–	–	–	–	24.2	19.3	26,238.0	2549.3
7	$300 < N \leq 350$ & $350 < N \leq 400$	–	–	–	–	33.8	–	56,117.9	–
8	$350 < N \leq 400$ & $400 < N \leq 450$	–	–	–	–	32.2	–	21,759.6	–

[1] Two-sample Kolmogorov–Smirnov test λ: H_0: $F(v^{Ni}) = F(v^{Ni+1})$ and H_1: $F(v^{Ni}) \neq F(v^{Ni+1})$, $\lambda_\alpha = 1{,}36$, $\alpha = 0.05$.
[2] Median test: H_0: $F_1(x) = F_2(x)$ and H_1: $F_1(x) \neq F_2(x)$, $\chi_\alpha^2 = 3.84$, $\alpha = 0.05$.

Note also that since the lowest hourly traffic of up to 50 veh/h was recorded only during a few hours overnight, with the actual number of only a dozen plus vehicles per hour, the collective analysis of these data with daytime speeds measured at two or even three times greater traffic volumes, would not be in line with the design of experiments (DOE) principles. However, another issue supporting the subdivision of the speed data set into hourly volume subsets was only sporadic crossing of the street or walking across or driving out of the parking spaces during nighttime.

The main objective of this study was to assess the efficacy of the applied TCMs used, understood as speed reduction, before the refuge island. To this end, the authors conducted preliminary measurements of the initial velocity at which drivers applied brakes before the pedestrian crossing and the drive-in and drive-out speeds. These speeds were measured with an SR4 synchronised traffic detector combined with a video camera during preliminary 1-h long observations on two one-block sections with diagonal on-street parking. The SR4 [66] logging chart example is shown in Figure A3 in Appendix C. These preliminary results were analysed, and the readings below 10 km/h were left out, as they were most likely associated with braking before the pedestrian crossing or driving in or out of parking. The number of occurrences of these speeds in the dataset varied from just one to several depending on the time of the day (boxed in blue in Figure A3 in Appendix C). Generally, there were not more than 2–4% of such speeds in each hourly data set. An increased frequency of their occurrence (from a few to a dozen plus records) coincided with higher traffic volumes, i.e., 250–450 veh/h in the morning between 7 a.m. and 8 a.m. and in the afternoon between 4 p.m. and 6 p.m. The final speed analysis results, with or without considering the readings below 10 km/h in each subset, were:

- 85th percentile speeds varied by up to 0.1–0.2 km/h,
- average speeds ranged from 0.5 to 1.0 km/h.

The data recorded by the SR4 traffic detectors (boxed in green in Figure A3 in Appendix C): vehicle speed in km/h, headway in meters, time intervals, measurement date and time to one-second accuracy, and all the statistical data (values boxed in brown and red in Figure A3 in Appendix C). These data allowed the carrying out of other, supplementary analyses, for example, to determine the effect of braking on the speed of the following vehicle. The results showed lower following vehicle speeds for up to 4 sec time intervals between consecutive readings (boxed in blue in Figure A3 in Appendix C), which depended on the headway to the decelerating vehicle. For time intervals greater than four seconds, the following vehicle speed readings that depended on the headway to the decelerating or parking vehicles did not depart from the relevant mean speeds of other vehicles in the street. For the sake of consistency of the data used in the speed variation analysis, the readings below 10 km/h were left out in all analyses; this is in line with the design of experiment (DOE) principles [67–69].

2.4. Research Methods

The analysed parameters were 85th percentile speed, mean speed, and speed reduction ratio, determined in the data subsets defined by hourly traffic volume ranges. As mentioned, the heuristic method was chosen for the purposes of this study.

The sequence of the flow chart analyses is shown in Figure 7. Standard statistical analyses are conducted as the first step (Figure 7), including the normality test, plotting histograms of the factors under analysis, Two-sample Kolmogorov–Smirnov test, and median test. The third tool of the heuristic method used in this study was scatter diagrams relating the vehicle speeds to the hourly traffic volumes (Figure 7). Relationships between v_{85} and v_{av} on the one hand and the hourly traffic volumes on the other were obtained in almost all cases with a correlation coefficient greater than 0.7. However, this relationship was not confirmed for entry to and exit from a signalled junction or roundabout. On all other survey stations, both these speeds were found to depend on the hourly traffic volume.

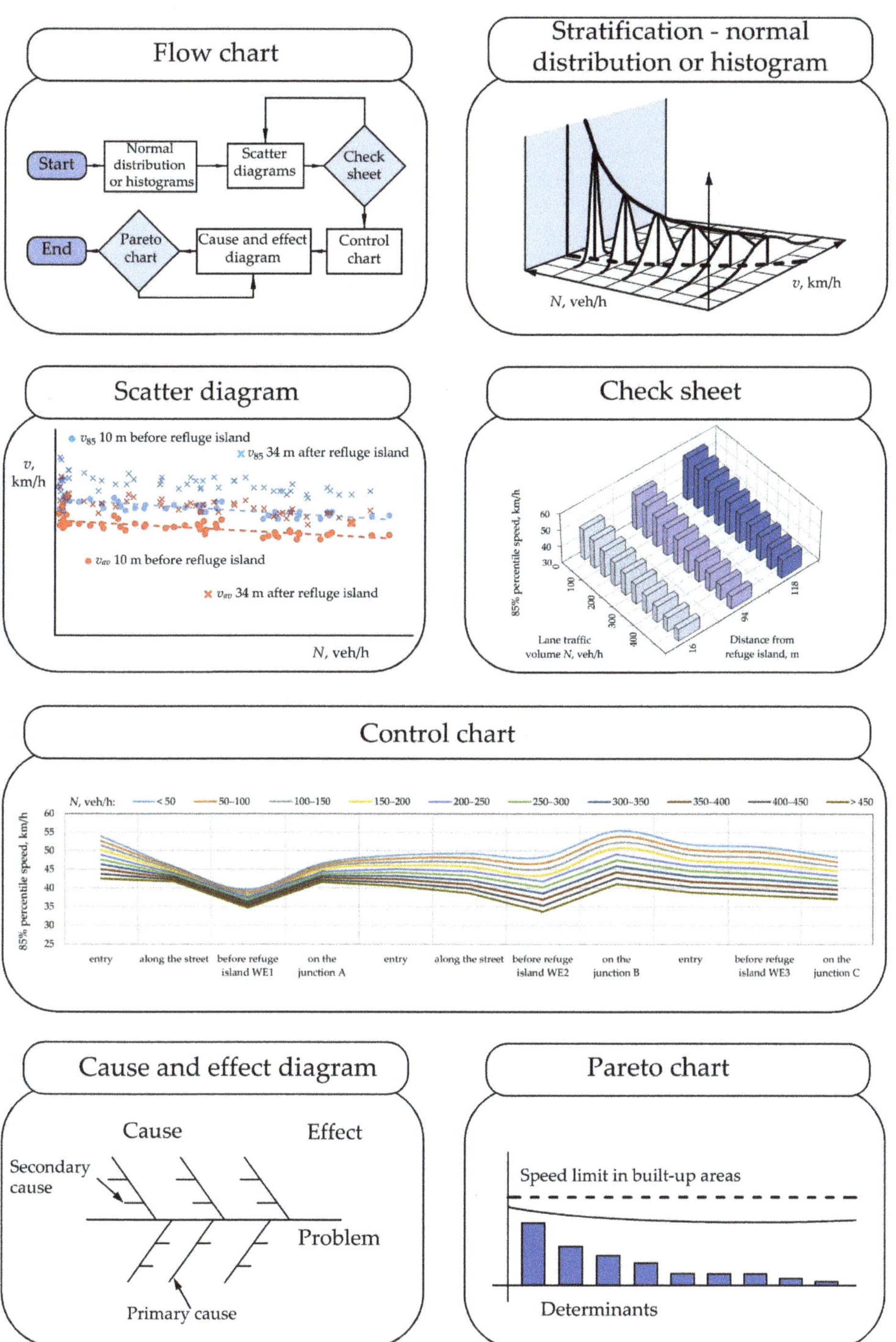

Figure 7. The seven basic tools of the heuristic method used for analysing the efficacy of speed reduction treatments before refuge islands. Source: own work.

The fourth tool of the heuristic method used in this study was 3D diagrams (Figure 7) representing speed and speed reduction ratio variations between the block entry and the refuge island. The speed and speed reduction ratio distributions turned out useful and were used to define the determinants associated with the refuge island itself and its visibility to the driver.

The fifth tool was 3D and linear control charts of speed changes and speed reduction ratios along the street under analysis (Figure 7). The analyses of the geometric and qualitative parameters, various speed distributions, and statistical test results were used to define the determinants initially. These determinants were presented in the Ishikawa cause-and-effect diagram, the sixth tool of the heuristic method applied in this study. A division into primary and secondary causes of the analysed slowing effect was made at this point. It was assumed that these determinants may be related to each other or independent. Stratification or concordance matrices are applied when dealing with a large number of determinants, most conveniently represented in the Pareto charts [47–53], the seventh tool of the heuristic method. In the Pareto chart, the determinants were rated in the order of decreasing effect, i.e., from the lowest to the highest approach speed or from the greatest to the lowest speed difference between the block entry and the refuge island station. The determinants were assessed in two ways: as a series of speed values before the refuge island and speed differences related to a given determinant or using an illustrated, summed-up number of determinants confirmed on a given study section. The adopted sequence of the heuristic method analyses allowed us to identify the refuge island approached at the lowest speeds or featuring the greatest speed difference on the approach section (Figure 7) and, as the final outcome, also identify the relevant determinants. In the summary of the conducted analyses, it will also be possible to identify the most effective among the applied TCMs. The control charts, in turn, allowed the determination of treatments having a prolonged slowing effect also past the terminal junction, i.e., in the next section of the street.

3. Results

As mentioned, the speed data set was subdivided into subsets defined by hourly traffic volume ranges. Considering the amount of data from the round-the-clock, two-day speed survey with about 16,000–18,000 readings per one SR4 detector, it was necessary to decide on the appropriate approach to be taken in the subsequent analyses. The parameters considered in previous studies [28–30,35,36] were the 85th percentile speed and the mean speed, while in traffic safety analyses, mean speed was considered [12,22,27,70–75] Kruszyna & Matczuk-Pisarek [31], in turn, used a speed reduction ratio, and Distefano et al. [29,30] expressed the speed reductions between the point in front of the refuge island and some distance earlier in percentages. Jamroz et al. [25] used solely the 85th percentile speed as this parameter is used for the purpose of speed limit analyses and the associated selection of the appropriate speed limit sign. From this wide selection of the available speed parameters, 85th percentile speed, mean speed, and speed difference (between the section entry and refuge island) were chosen for the purposes of this study, i.e., TCM efficacy assessment. Having in hand such an extensive database, it was possible also to consider in this study the hourly traffic volume effect. The analysis of 24-h speed data with the measurement time given to 1 sec accuracy (Figure A3 in Appendix C) showed that the so-far used free-flow speed may be deemed to correspond to the values obtained at an hourly traffic volume below 50 veh/h. However, one should bear in mind that these speeds concern mainly the night period when they are not influenced by pedestrian traffic (Appendix B Figure A2).

As per the adopted methodology and the statistical test results (Table 2), the speed data normality and stratification depending on the hourly traffic volume and the survey station location in relation to the analysed refuge island were checked as the first step. The obtained results of speed changes at the refuge island are shown in Figure 8. Figure 8 shows the 85th percentile speed distribution among three survey stations for all six analysed street sections along the refuge island approach section. The obtained speed changes presented in Figure 8 show a strong relationship with the hourly traffic volume and other factors, including those related to the implemented TCMs. These are most likely the determinants sought in this study.

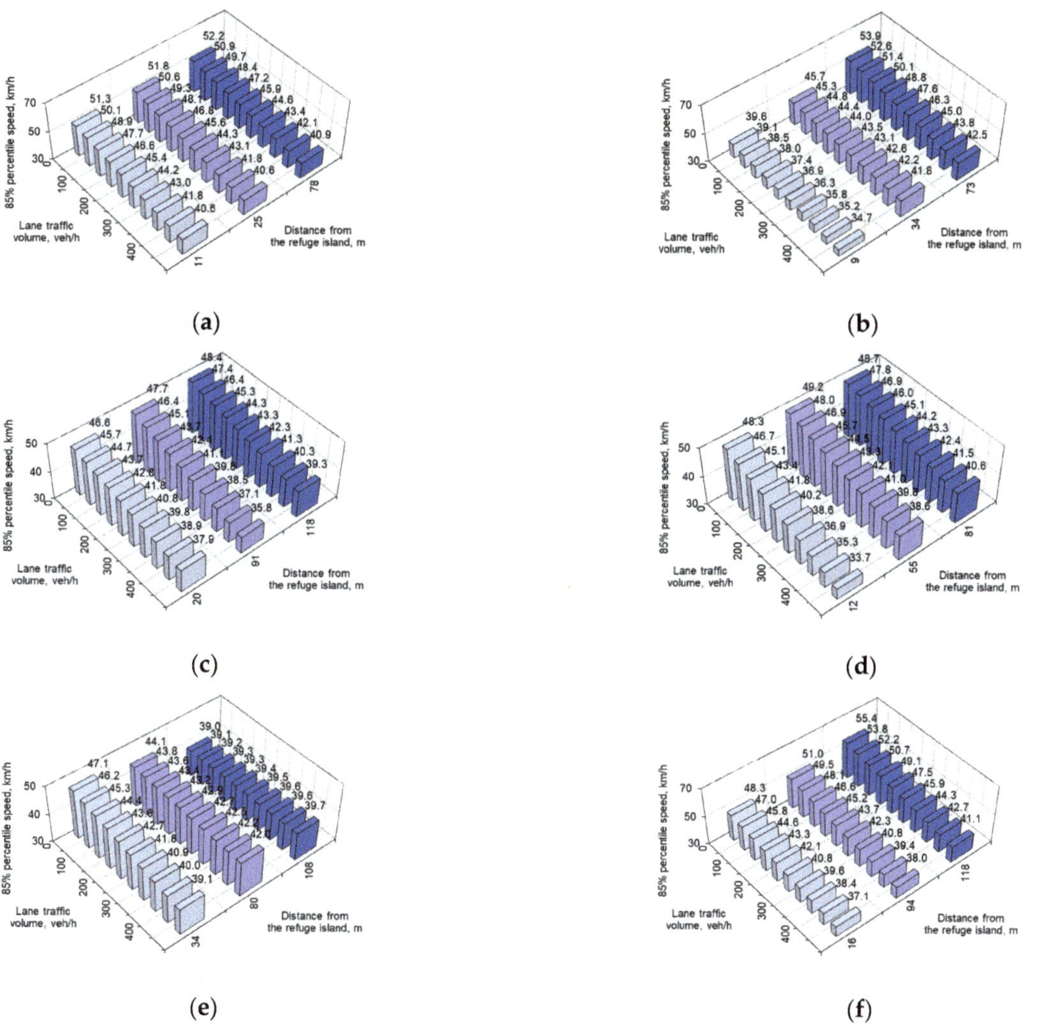

Figure 8. 85th percentile speed distribution example: (**a**) EW1; (**b**) WE1; (**c**) EW2; (**d**) WE2; (**e**) EW3; (**f**) WE3. Source: own work.

The values of v_{85} and v_{av} were calculated for all the survey stations and each survey hour. Next, the speed results data set was subdivided into hourly traffic volume subsets, and regression analyses were carried out. Appropriate regression relationships were obtained for all the results. Considering low hourly volumes overnight (not exceeding a dozen plus veh/h), larger scatters of v_{85} and v_{av} were obtained only for the up to 50 veh/h range. Illustrative speed vs. hourly traffic volume relationships are given in Figure 9.

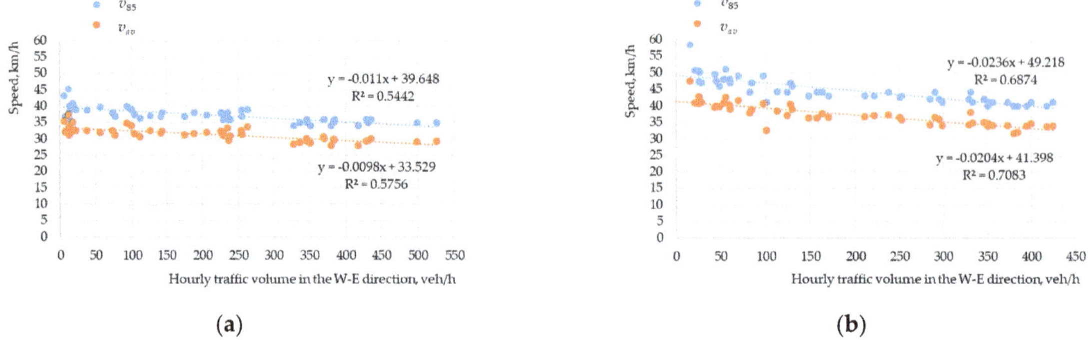

Figure 9. Regression analysis examples for different survey station locations: (**a**) at a refuge island; (**b**) on the refuge island approach section. Source: own work.

Linear control charts were the next heuristic method tool used in this study. Specifically, these were linear speed change diagrams along the analysed street sections (Figure 10). From the graph in Figure 10, it can be figured out that not all the refuge island-related TCMs should be considered effective. The greatest speed variations were noted in the direction W→E sections WE1, WE3, and WE2 (Figure 10a,b). In the direction E→W, the speed changes were only minute. Figure 10c,d also shows an increase in speed past the refuge island EW1 associated with the widening of the carriageway to two travel lanes ahead of a signalled junction and signal phases rather than the applied TCMs.

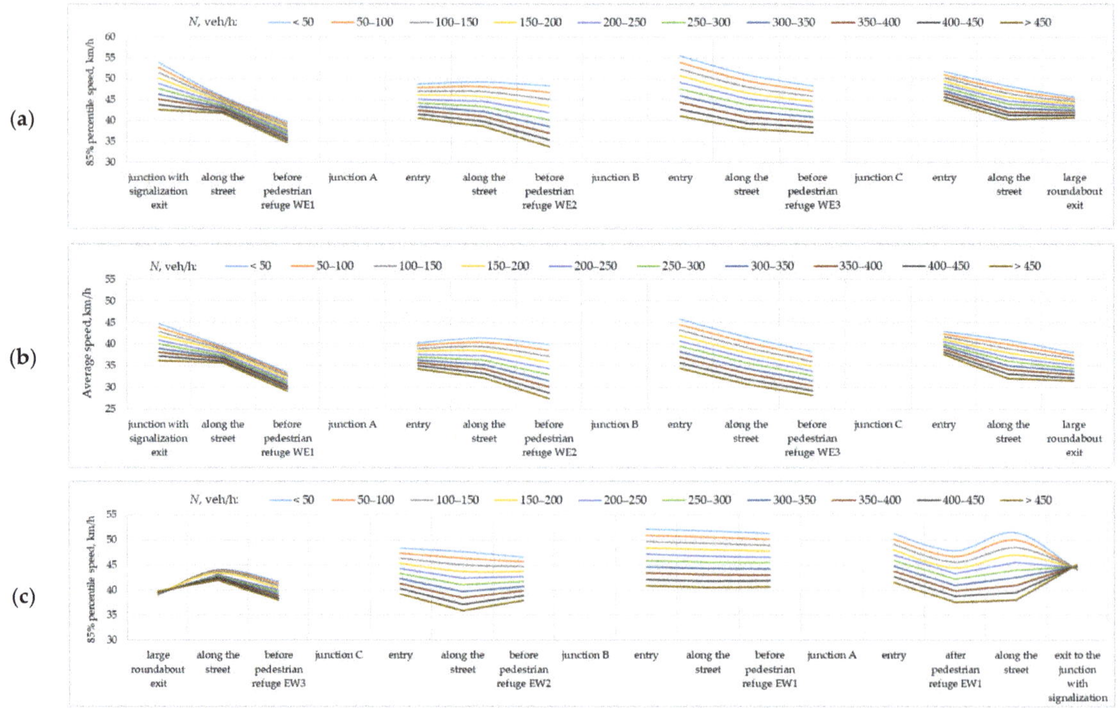

Figure 10. *Cont.*

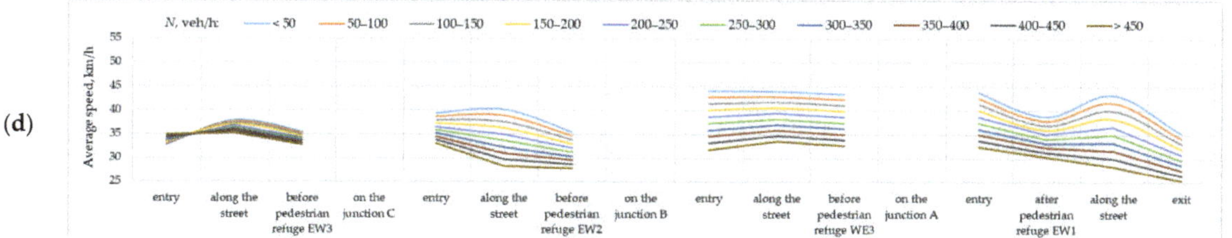

Figure 10. Speed distribution along the analysed street: (**a**) 85th percentile speed in the direction W→E; (**b**) average speed in the direction W→E; (**c**) 85th percentile in the direction E→W; (**d**) average speed in the direction E→W. Source: own work.

4. Discussion

4.1. Primary and Secondary Determinants—Cause-and-Effect Diagram

Figure 11a,c shows the driver's central vision area at different speeds. In order to reflect the 50 km/h speed limit on the analysed street (i.e., the statutory built-up area speed limit in Poland), the driver's central vision area at this speed was represented by a colour area on the images of the respective refuge islands, turned to greyscale in the fringe vision area. The latter is also a focal vision area, yet it requires the driver to move ahead and direct their eyes sideways while driving. In line with the heuristic method principles and guidelines to use the cause-and-effect diagram to identify the determinants, the probable determinants noted on the analysed refuge island are presented in Figure 11b as the next step of the mentioned analyses. The probable determinants in Figure 11b were identified initially based on the TCMs located at the refuge islands shown in Figure 11a,c.

The parameters recorded on the analysed refuge islands showed the relevance of free view width "*a*" in line with the already published findings [34–36]. The studies described in [34–36] investigated the effect of horizontal deflection treatment located before a median island on speed reduction past the island. It is a different case in this article, where we assess the slowing effect of TCMs and refuge islands on the approach section to the latter. A double horizontal deflection treatment with a 1.3-m offset to the right, followed by a 3.3-m offset to the left was found only in the study section WE1 (Figure 11c). The free view width "*a*" was large there, encompassing the whole lane past the refuge island. The greatest speed differences were noted, in this case, in all hourly traffic ranges. A 1 m offset (equal to half the refuge island width) to the right was noted at WE3 (Figure 11c). The study sections WE2 and EW2 (Figure 11a,c) featured a horizontal deflection to the right, clearly visible to the approaching driver, located before the junction with half the refuge island width offset, and a horizontal deflection to the right with the same offset on the section past the junction. The sections EW1 and EW3 had no deflections, giving small, if any speed, differences noted there (Figure 11a). Horizontal deflection and free view are intrinsically linked to visibility, as shown in Figures 11 and 12. Figure 12a,c also shows the driver's central vision area against the clear sight width at the road surface level before the first refuge island (green area) and past the second refuge island (blue area). The issue of visibility at refuge islands was dealt with by several researchers [25,76–82]. The vision field depends on the driving speed and stopping distance (braking distance plus reaction distance) [78,80,82–87]. In the abovementioned articles, vision fields varied, as besides the driving speed, they depended on different country standard reaction times and decelerations, the road surface condition, and the longitudinal profile of the carriageway.

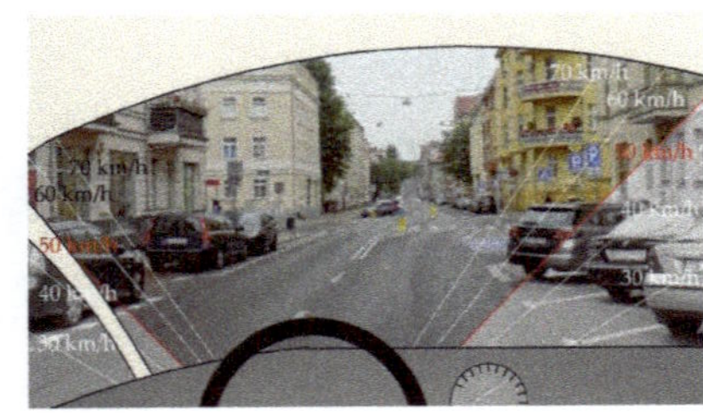

EW1

EW2

EW3

(a)

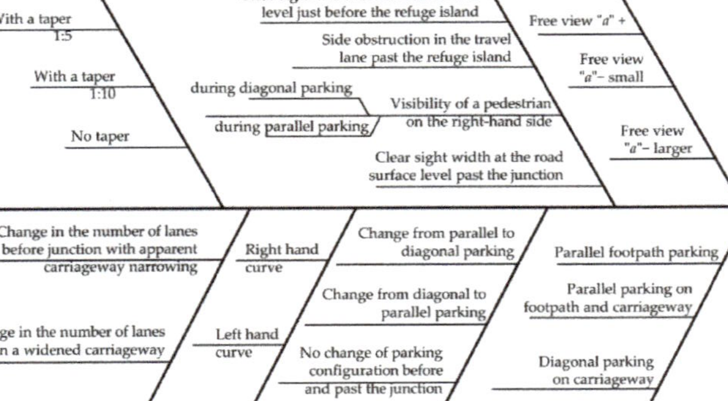

(b)

WE1

WE2

WE3

(c)

Figure 11. Cause-and-effect diagram and identification of determinants: (**a**) windscreen view in the direction E→W; (**b**) identification of primary and secondary determinants; (**c**) windscreen view in the direction W→E. Source: own work.

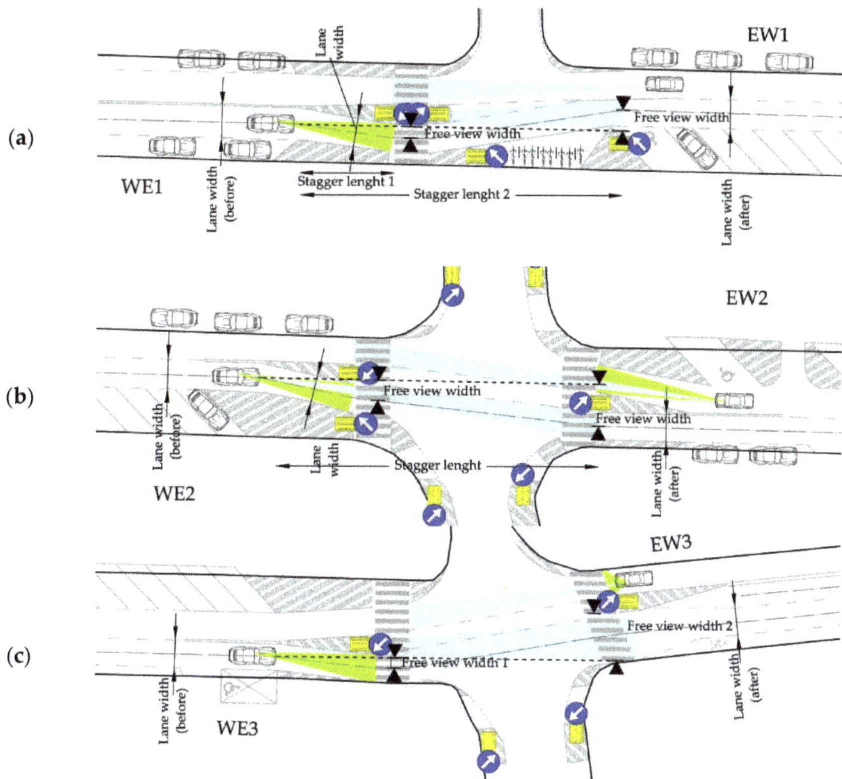

Figure 12. Free view width and clear sight width ahead of and past the refuge island: (**a**) WE1 and EW1; (**b**) WE2 and EW2; (**c**) WE3 and EW3. Source: own work.

For the purposes of this study, a 13 m clear sight width was adopted as per the guidance of [83]. The field of vision analysis presented in Figure 12 showed that in horizontal deflection layouts, the fields of vision add up, and a driver approaching the refuge island for pedestrians, sees a vehicle parked past the island as a side obstacle with a clear sight width past the island obstructed by the traffic signs located on the island (Figure 11a). The combined effect of these determinants may be considered the most likely cause of the large speed reduction in the approach to WE1. The layout of traffic signs located on the refugee islands and the horizontal and vertical curves in the C junction also partly restrict the clear sight width past the WE3 island and past the junction (Figure 12c). Horizontal deflection was also found there. However, the geometrical features, two lanes past the junction, a cycle lane, and cars parked on the footpath, were found to have less effect on the speed reduction before WE3 [3]. In the next section, WE2, the configuration of these geometrical features was found to have much less effect on the speed reduction obtained before the refuge island (Figure 12b).

On the sections located in the opposite traffic direction, the geometrical features and the TCMs were not found to have any significant slowing effect. EW1 may be an exception to that, where the view of cars parked partly on the carriageway on the right-hand side and restricted view past the island due to traffic signs positioned on the refuge island could possibly have some effect on the small, any-way speed reduction (Figures 6a and 12a).

The next of the identified determinants concerns the applied painted taper before the refuge island. It is similar to the path angle issue addressed in [35]. However, the TRL trial data [35] cannot be compared with the results of this study, as the former (speed data from over a dozen TRL track rides by experienced drivers) lacks traffic volume information. Nevertheless, the TRL trial data may still be roughly compared with the results obtained in this study at 50 veh/h traffic volume level. Note that the data given in this article relate to the actual traffic conditions in an existing street, taking into account a number of other determinants. The tapered design varied among the analysed refuge islands, with 1:5 tapers on WE1, WE2, and EW2 and 1:10 tapers on WE3. EW1 and EW3 islands had no tapers at all (Figure 12). The issue of tapers in front of refuge islands was tackled in [25]. It was concluded that a 1:5 taper should be used in city traffic conditions, possibly with one or more painted tapers before. However, it should not be used at entrances to sections that include TCMs.

A determinant that may affect the desired slowing before refuge islands could also be a variation in the number of lanes heading in one direction before and past the junction. Such variation in the number of lanes occurs in the terminal sections WE3 and EW1, shown in Figure 11 above and in Figure A1 in Appendix A. The next three determinants identified for the purposes of this study are related to the travel lane geometry (arrangement of straight sections and curves on the approach to and within the junction) and parking orientation on both sides of the street. Change of parking configuration from parallel to diagonal, or vice-versa, imposes a horizontal deflection, the use of horizontal curves, and the associated road markings (Figure 4, Figure 11, and Figure 12).

4.2. Analysis of Determinants Based on the Pareto Chart

The identified determinants, as defined in Section 4.1 above, were assessed using logical tautologies. Thus, if a determinant was confirmed in a given section on the approach to the refuge island, it received a quantification measure score of 1 as per the binary system. Otherwise, it received a 0 score. In some cases, a 0.5 score was given as an intermediate value. This includes free view "*a*"—(small) situations, in line with the conclusions of [37] (WE2 and EW2 in Table 3). Similarly, an intermediate score was given for a 1:10 painted taper; this is in line with the recommendations of [25] (Table 3—WE3). Intermediate scores were also given where a left-hand curve was found in the junction (WE3 in Table 3), as the horizontal curve configuration had a direct bearing on the visibility of the road section past the junction. A possibility of apparent carriageway narrowing past the junction was confirmed in two cases, possibly due to compromised visibility of the road surface past the junction (WE3 and EW1 in Table 3). The determinants used in the analysed sections and their quantification scores are summarised in Table 3. Table 3 includes only the determinants that were found in the sections under analysis.

Table 3. Determinants found in the analysed sections and quantification measure scores. Source: own work.

Determinants	Scores Given to the Study Sections					
	WE1	WE2	WE3	EW1	EW2	EW3
Free view	1	0.5	1	0	0.5	0
Side obstruction in the travel lane past the refuge island	1	0	0.5	0	0	0
Lack of visibility of a pedestrian on the right-hand side of the island	1	0	0	0	1	1
Lack of visibility of the road surface past the junction	0	0	0.5	0	0	0.5
Painted taper applied to the section	1	1	0.5	0	1	0
Left-hand curve in the junction	0	0	0.5	0	0	0
Parking configuration changed from diagonal to parallel	0	1	0	0	1	0
Parking configuration changed from parallel to diagonal	1	0	0	0	0	1
Apparent carriageway narrowing past the junction	0	0	0.5	0.5	0	0
Total quantification measure scores:	**5**	**2.5**	**3.5**	**0.5**	**3.5**	**2.5**

The classified quantification measures assigned to logical tautologies were separate and joint. The Pareto chart (Figure 13 left) shows the 85th percentile speeds before the refuge island and their total scores. The upper part of the Pareto chart shows stratification of the

85th percentile speeds recorded before the refuge island at different hourly traffic volumes, and the bottom part shows the confirmed determinants and total scores of the quantification measure applied to them. Having a closer look at the data presented in Figure 13 left, we see that speed variation at the refuge island depends on the combined effect of all the above determinants and the traffic volume rather than any one of them on their own. The 85th percentile speed variations on the respective sections showed their strong relationship with free view and visibility parameters. That said, the strongest relationship was found for the combination of the confirmed logical tautologies and the hourly traffic volume, especially for the volumes greater than 150–200 veh/h (Figure 13). Similar analyses are presented for the mean speed variations before refuge islands (Figure 13 right). The mean speed analysis showed that it depended even more on the hourly traffic volume above the 100–150 veh/h range and on the combined effect of the following determinants: free view, visibility, change of parking configuration, and painted taper. However, this dependency is very complex. The joint analysis of the mean speed variation, traffic volumes, and the determinants showed that the more determinants are involved (including TCMs), the greater their overall effect on the speed of vehicles approaching a refuge island. For example, without painted tapers, with travel lane visibility before the island, without visibility of pedestrians approaching the island from the right-hand side, and poor visibility of the road surface past the junction on EW3 or no confirmed determinants as was the case on EW1, we obtained a 85th percentile speed of about 50 km/h (Figure 13 left), i.e., the statutory built-up area speed limit in Poland and 40–45 km/h mean speed (Figure 13 right). Now, based on the above analyses, we can rate the analysed TCMs and other factors noted at the refuge islands as:

(a) effective (WE1)—with a change of on-street parking configuration from parallel to diagonal or vice-versa requiring the driver to change the travel path, a 1:5 taper or road and island geometry designed to get free view "*a*"—larger so that a vehicle parked in the travel lane is visible as a side obstacle and the travel lane at the road surface level past the island is not visible by the driver approaching the island, altogether resulting in lower island approach speeds;

(b) moderately effective (WE2, WE3, and EW2) with narrower free view width of "*a*"—small, a 1:5 or 1:10 painted taper, and change in parking configuration and different ways of targeting parking spaces, which in combination produce different geometry and visibility configurations offering the driver a reliable assessment of the road situation during approaching and passing the island and resulting in moderate speed reduction;

(c) ineffective (EW2 and EW3) with "*a*" + free view, no change of parking configuration, no painted tapers, no horizontal deflection, and no sight restrictions for the driver approaching and passing the island, discouraging speed reduction.

As mentioned, the speed reduction analyses conducted in this study also considered, besides the 85th percentile speed, the relative speed reduction in percentages, calculated as a ratio between the 85th percentile approach speed and the block entry speed [29] and various other speed reduction indicators [31]. However, these parameters were assessed in relation to the free flow speed. Having in hand the two-day, 24-h survey data it was also possible to also analyse the effect of the geometrical features and TCMs in combination with hourly traffic volumes on the calculated approach speed parameters v_{85} and v_{av}, estimated before refuge island. It was found that v_{85} and v_{av} variations depended on the hourly traffic ranges from free-flow conditions to the maximum of 500 veh/h at 50 veh/h intervals. The underlying cause of the two-day continuous survey was to support the determination of which of the analysed six pedestrian refuge arrangements turns out to be the most effective in real traffic conditions (with 0–500 veh/h hourly traffic volumes) rather than in a free-flow situation.

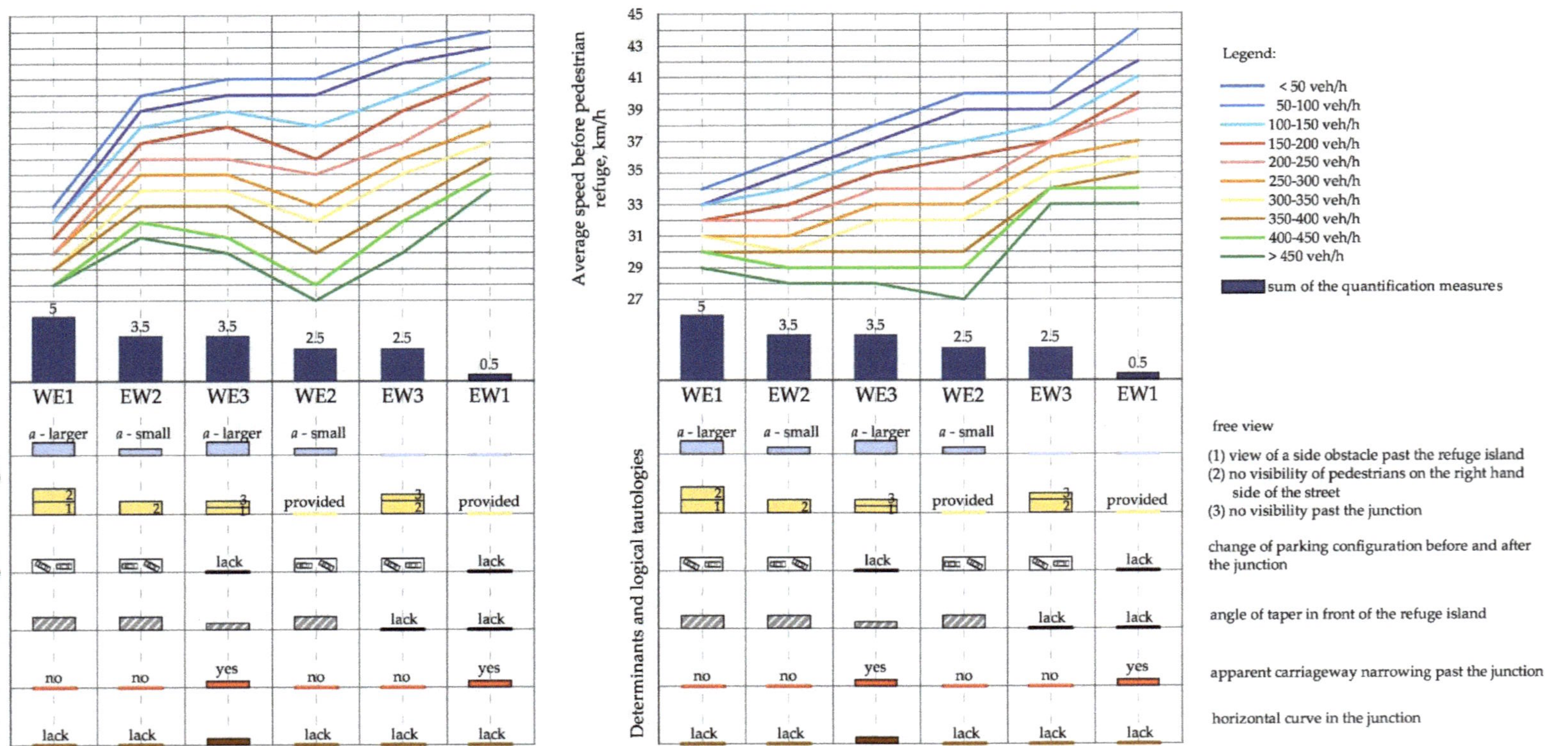

Figure 13. Pareto chart showing the variation of speed distribution parameters (v_{85} and v_{av}) and the determinants adopted for the analysed sections: the sections are listed in the growing order of v_{85} just before the refuge island; and the test sections are listed in the order v_{av} just before the refuge island. Source: own work.

Next, having in mind the speed change effect on noise, safety, and vehicle emissions, i.e., the factors intrinsically related to sustainable road construction, we compared the values of Δv_{85} and Δv_{av} noted before the refuge islands with the block entry values of v_{85} and v_{av} (Figure 14). A detailed analysis of the obtained hourly traffic volumes together with the adopted determinants on Δv_{85} and Δv_{av} the island approach speed difference depended on the identified determinants and the hourly traffic volume on a majority of the analysed sections. Noteworthy, this relationship was found to vary depending on the island geometry and the specific TCMs and traffic volume compilation. Highly relevant in this respect were the free view width and visibility of side obstacles, pedestrians, and the road surface past the junction. Where the various determinants were confirmed (regarding the free view, visibility, painted taper angle, and change of parking configuration), as the hourly traffic intensity increased, Δv_{85} and Δv_{av} were found to decrease on the approach to the island. This may be due to lower block entry speeds or higher speeds just before the refuge island. The lowest speed difference was obtained for EW1 where only free view of "*a*"—small and apparent carriageway narrowing past the junction was noted. Without a horizontal deflection or changed parking configuration, the section offers very good visibility on the approach to and past the pedestrian island, while the view of two travel lanes heading in the same direction past the junction and of the cantilevered traffic lights was found to have no slowing effect. WE2 is one exception in this analysis, in that it offered a good view of the constant geometry travel lane and the cars parked on the footpath parallel to the road past the island and past the junction, despite the free view "*a*"—small, a 1:5 painted taper, and a change of parking configuration. This being so, the obtained values of Δv_{85} and Δv_{av} most probably depended on the hourly traffic volume only. In the case of WE2 and EW3, the growing hourly traffic volume gradually increased the difference between these two speed parameters. On EW3, the travel lane does not change its geometry before the island, and the change in the parking configuration past the island has no significant effect on the analysed speed differences before the island.

The above findings are apparently consistent with the findings of the simulator study by Akgol et al. and Aydin et al. [33,34]. The difference between these studies was in single lane narrowing of the two-lane carriageway at the pedestrian crossing, which is not the case in this study, where there are two travel lanes running in opposite directions in all cases. For economic reasons, the authors of [33,34] recommended a one-sided splitter island on the right-hand side of the travel lane before the refuge island, as is the case in the WE2 and EW2 sections analysed in this study. Comparing their recommendations and the results presented in Figures 13 and 14, it can be concluded that this arrangement would not be effective for refuge islands located on junctions. The second, more expensive option recommended in [33,34] are two one-sided islands on either side of the refuge island, as in our study section WE1. The above findings are apparently consistent with recommendations for refuge islands located in junctions with horizontal deflection by three islands, resulting in a flattened U-deflected path of travel instead of a flattened "S" shape and a one-sided island on the right-hand side of the approach to the pedestrian refuge.

The findings of this study are also highly consistent with the findings presented in [28] despite different study areas and countries with different tempers and driving behaviours found there. In both cases, lane narrowing was found to be the most effective treatment at refuge islands, accompanied by horizontal deflection, additional one-sided splitter islands on both ends of the island, and painted tapers, that is the use of a few TCMs deployed within a short distance.

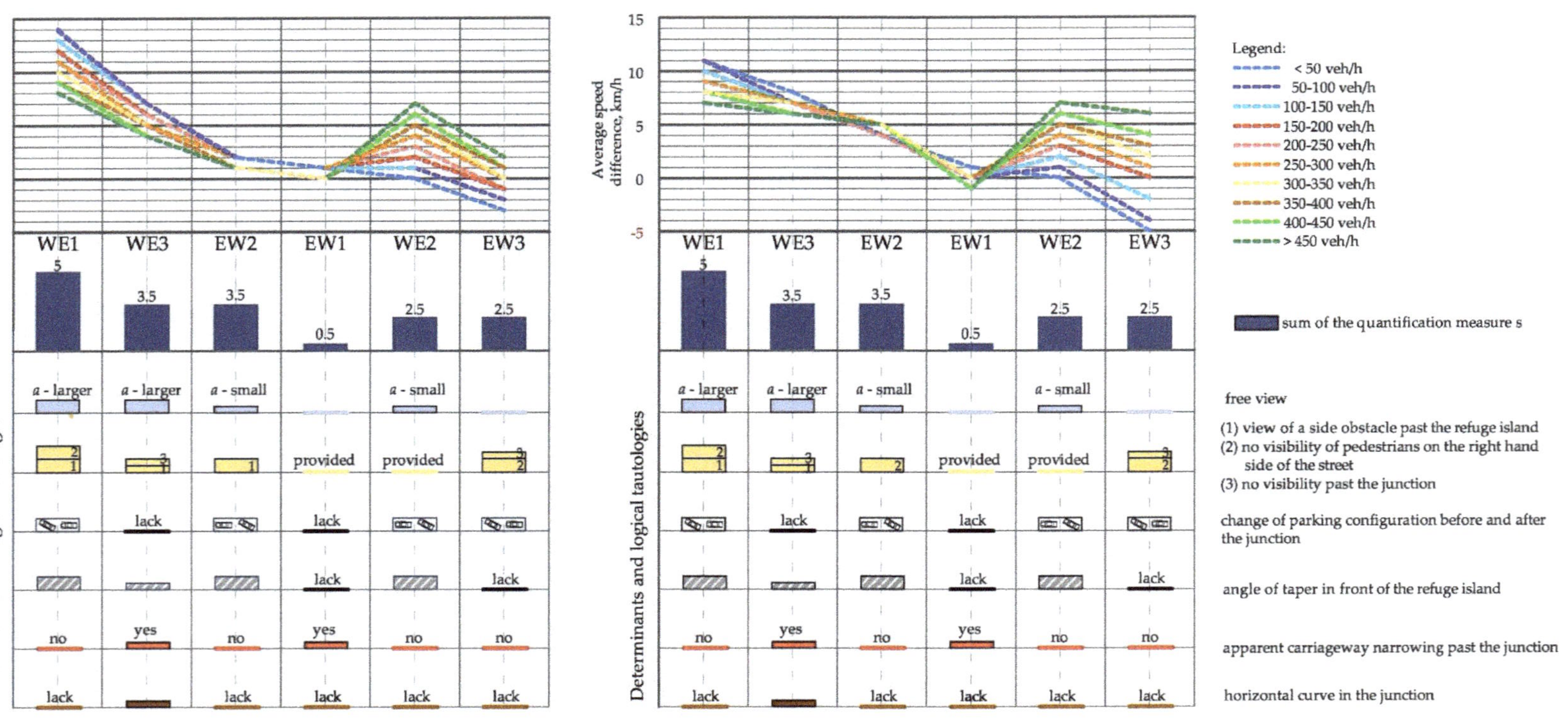

Figure 14. Pareto Chart 2—showing the values of Δv_{85} and Δv_{av} and the determinants adopted on the respective sections: sections listed in the order of Δv_{85} values calculated as $\Delta v_{85} = v_{85}$ (entry) $- v_{85}$ (before the refuge island); and sections listed in the order of Δv_{av} values calculated as $\Delta v_{av} = v_{av}$ (entry) $- v_{av}$ (before the refuge island). Source: own work.

5. Conclusions

As the literature review showed, studies on refuge island treatments date back to the late 1990s. Nevertheless, international experiences show that the issue has not been studied completely as yet, and we still do not have definite design principles at hand. The speed-reducing effect of pedestrian refuges and, more importantly, improved safety of pedestrians have been demonstrated by the study results of many researchers. The available design guidelines focus on the island width and angle of the painted taper applied in front of the island. Less attention is paid to the conditions relevant to the visibility of pedestrians on the way to the refuge island, visibility of the nearby junction at the road surface level, visibility of side obstacles, specifically cars parked at a small distance past the island, and to the effect of hourly traffic volume on the degree of slowing on the way to the island. The relevance of these determinants has been demonstrated in this article. The speed analyses conducted as part of this study in the refuge site locations confirmed high relevance:

- of free view
- visibility of pedestrians, and,
- refuge island surroundings.

Less relevant were:

- change in the parking configuration, both on the way to and past the island, and,
- taper angle.

Also highly relevant was the compilation of the above determinants, considered in combination with hourly traffic volumes.

Comparing the results of this study with the former study results we also found that the various speed parameters and ratios were so far compared in free-flow traffic conditions, thus ignoring the effect of the traffic volume. The relevance of traffic volume was demonstrated in this study, since the drivers tend to react differently when exposed to a higher number of stimuli present on a two-way road, as compared to the test track or simulator-based situations.

The determinants whose relevance has been proven in this article show a need for further analyses and studies to be conducted by other researchers to consider different drivers' tempers, engineering experiences, and cutting-edge technologies applied in various island, lighting, and structural details. A complete set of design guidelines would also be desired that would consider, besides the island width and the island taper angle, also the design parameters of the junction ahead, visibility parameters, and townscape surrounding. Only when considered all together, these recommendations will ensure the safe design of pedestrian refuges and the associated speed management in downtown streets. Steadier speeds being a probable consequence, the desired reduction of noise and vehicle emissions around refuge islands may well be expected.

Author Contributions: Conceptualization, A.S.; methodology, A.S. and S.M.; software, A.S. and S.M.; validation, A.S. and S.M.; formal analysis, A.S. and S.M.; investigation, A.S. and S.M.; data curation, S.M.; writing—original draft preparation, A.S.; writing—review and editing, A.S. and S.M.; visualization, A.S. and S.M.; supervision, A.S. All authors have read and agreed to the published version of the manuscript.

Funding: This research received no external funding.

Institutional Review Board Statement: Not applicable.

Informed Consent Statement: Not applicable.

Data Availability Statement: The data presented in this study are available in the article.

Acknowledgments: The authors would like to thank Krzysztof Bujak of the City Office of Szczecin for providing information regarding the project: "Traffic organization design along 5 Lipca Street on the section from Bohaterów Warszawy Avenue to Szarych Szeregów Place".

Conflicts of Interest: The authors declare no conflict of interest.

Appendix A

Study section parameters.

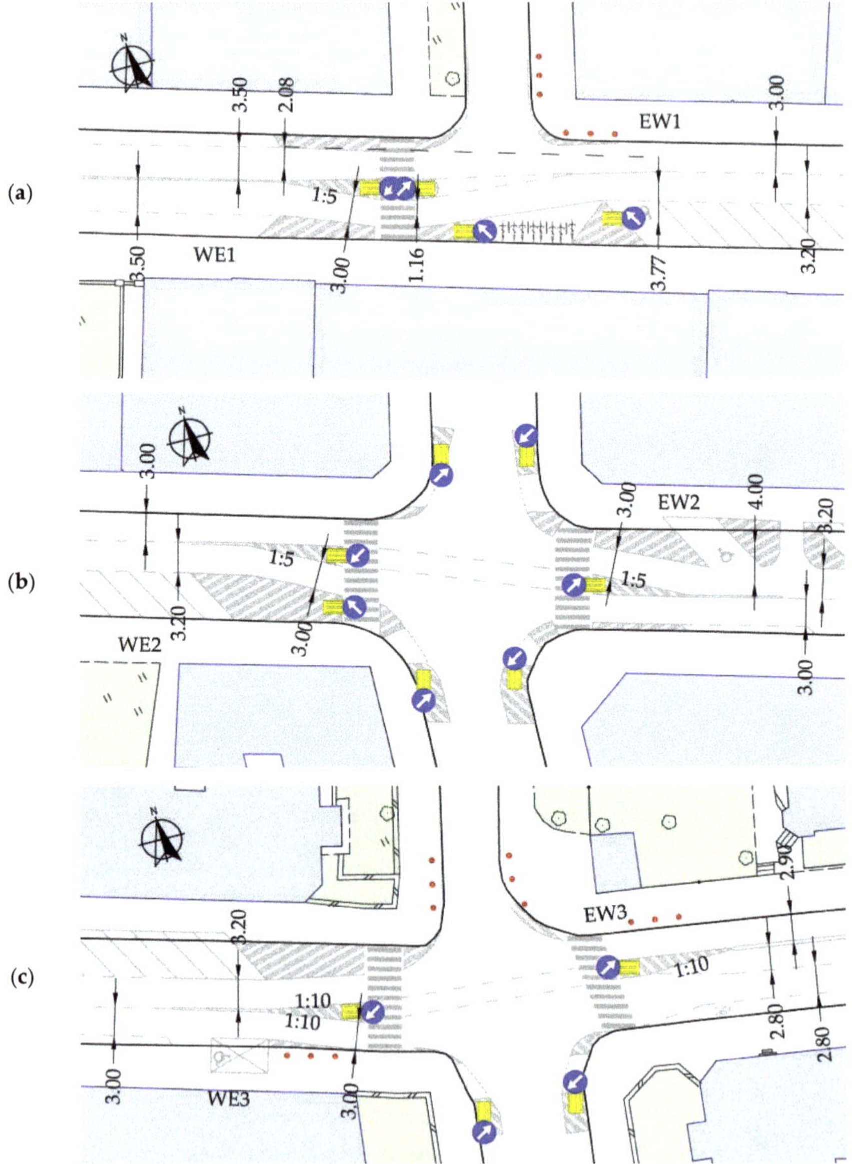

Figure A1. Geometrical features of the study sections: (**a**) WE1 and EW1; (**b**) WE2 and EW2; (**c**) WE3 and EW3. (All distances are in metres). Source: own work.

Appendix B

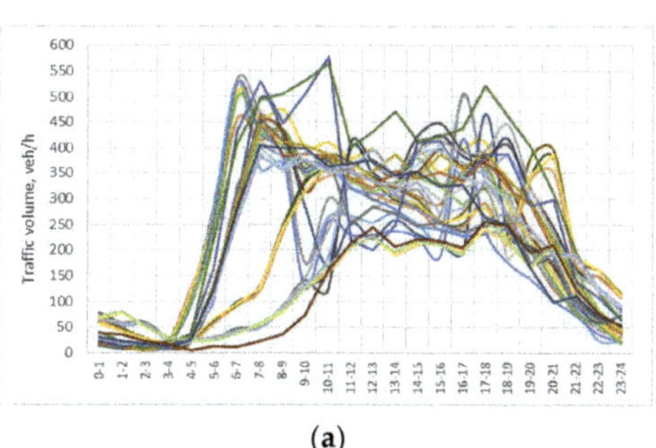 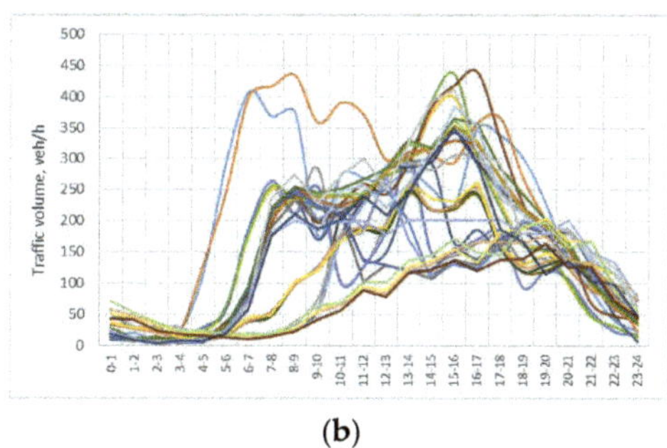

(**a**)   (**b**)

Figure A2. Hourly traffic volumes: (**a**) in the direction W→E; (**b**) in the direction E→W. Source: own work.

Appendix C

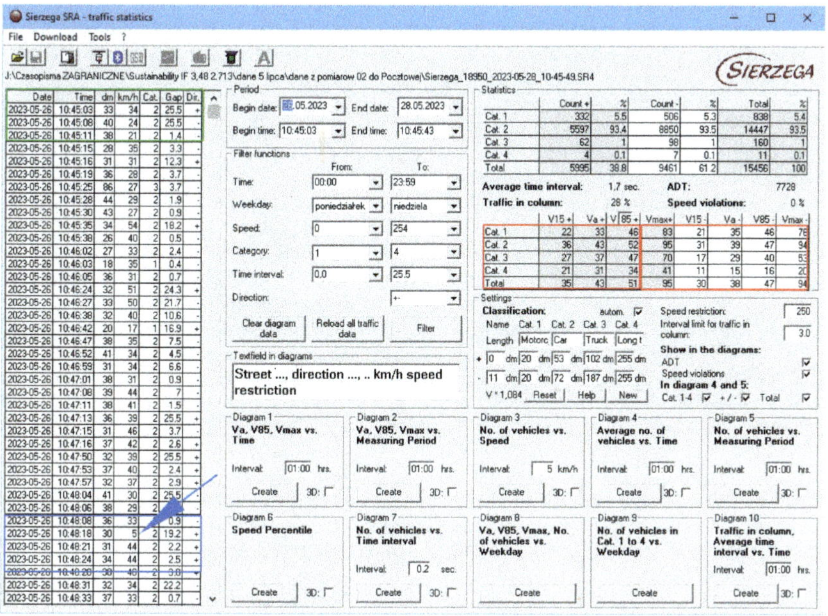

Figure A3. Example of SR4 traffic detector log (of the date: 26 May 2023). Source: own work.

References

1. National Association of City Transportation Officials. *Urban Street Design Guide*; National Association of City Transportation Officials: Washington, DC, USA, 2013.
2. Vejdirektoratet-Vejregeludvalget. *Urban Traffic Areas—Part 7—Speed Reducers*; Vejdirektoratet-Vejregeludvalget: Copenhagen, Denmark, 1991.
3. Institute of Transportation Engineers. *Traffic Calming Guidelines*; Devon County Council Engineering & Planning Department: Devon, UK, 1992.
4. East Ayrshire, Strathclyde Regional Council. *Roads Development Guide*; East Ayrshire, Strathclyde Regional Council: London, UK, 2010.
5. Working Group Highway Design. *Directives for the Design of Urban Roads RASt 06*; Working Group Highway Design FGSV: Köln, Germany, 2006.
6. Elvik, R. Area-wide urban traffic calming schemes: A meta-analysis of safety effects. *Accid. Anal. Prev.* **2001**, *33*, 327–336. [CrossRef] [PubMed]

7. Jones, P.; Hervik, A. Restraining car traffic in European cities: An emerging role for road pricing. *Transp. Res. Part A Policy Pract.* **1992**, *26*, 133–145. [CrossRef]
8. Mei, Z.Y.; Feng, C.; Kong, L.; Zhang, L.H.; Chen, J. Assessment of different parking pricing strategies: A simulation-based analysis. *Sustainability* **2020**, *12*, 2056. [CrossRef]
9. Mingardo, G.; van Wee, B.; Rye, T. Urban parking policy in Europe: A conceptualization of past and possible future trends. *Transp. Res. Part A Policy Pract.* **2015**, *74*, 268–281. [CrossRef]
10. VanHoose, K.; de Gante, A.R.; Bertolini, L.; Kinigadner, J.; Büttner, B. From temporary arrangements to permanent change: Assessing the transitional capacity of city street experiments. *J. Urban Mobil.* **2022**, *2*, 100015. [CrossRef]
11. Jazcilevich, A.; Vázquez, M.J.M.; Ramírez, P.L.; Pérez, I.R. Economic-environmental analysis of traffic-calming devices. *Transp. Res. Part D Transp. Environ.* **2015**, *36*, 86–95. [CrossRef]
12. Lee, G.; Joo, S.; Oh, C.; Choi, K. An evaluation framework for traffic calming measures in residential areas. *Transp. Res. Part D Transp. Environ.* **2013**, *25*, 68–76. [CrossRef]
13. Bellefleur, O. *Urban Traffic Calming and Air Quality: Effects and Implications for Practice*; National Collaborating Centre for Healthy Public Policy: Québec, QC, Canada, 2012.
14. Bellefleur, O.; Gagnon, F. *Urban Traffic Calming and Health: A Literature Review*; National Collaborating Centre for Healthy Public Policy: Montréal, QC, Canada, 2011.
15. Daham, B.; Andrews, G.E.; Li, H.; Partridge, M.; Bell, M.C.; Tate, J. *Quantifying the Effects of Traffic Calming on Emissions Using On-road Measurements (Report No. 2005-01-1620)*; SAE International: Warrendale, PA, USA, 2005; Available online: http://eprints.whiterose.ac.uk/2050/1/2005-01-1620SOrion_peedbump.pdf (accessed on 2 August 2023).
16. Boulter, P.G.; Hickman, A.J.; Latham, S.; Layfield, R.; Davison, P.; Whiteman, P.P. *The impact of traffic calming measures on vehicle exhaust emissions*; Report 482; Transport Research Laboratory TRL: Wokingham, Berkshire, UK, 2001.
17. Sun, M.; Sun, X.; Shan, D. Pedestrian crash analysis with latent class clustering method. *Accid. Anal. Prev.* **2019**, *124*, 50–57. [CrossRef]
18. Guéguen, N.; Meineri, S.; Eyssartier, C. A pedestrian's stare and drivers' stopping behavior: A field experiment at the pedestrian crossing. *Saf. Sci.* **2015**, *75*, 87–89. [CrossRef]
19. Balasubramanian, V.; Bhardwaj, R. Pedestrians' perception and response towards vehicles during road-crossing at nighttime. *Accid. Anal. Prev.* **2018**, *110*, 128–135. [CrossRef]
20. Li, Y.; Fan, W. Modelling severity of pedestrian-injury in pedestrian-vehicle crashes with latent class clustering and partial proportional odds model: A case study of North Carolina. *Accid. Anal. Prev.* **2019**, *131*, 284–296. [CrossRef] [PubMed]
21. Sołowczuk, A.; Kacprzak, D. Identification of determinants of the speed-reducing effect of pedestrian refuges in villages located on a chosen regional road. *Symmetry* **2019**, *11*, 597. [CrossRef]
22. Balant, M.; Lep, M. Comprehensive traffic calming as a key element of sustainable urban mobility plans-impacts of a neighbourhood redesign in Ljutomer. *Sustainability* **2020**, *12*, 8143. [CrossRef]
23. Road and Transportation Research Association. *Directives for the Design of Urban Roads*; Road and Transportation Research Association: Köln, Germany, 2012.
24. Zalewski, A.; Kempa, J. Traffic calming as a comprehensive solution improving traffic road safety. In *IOP Conference Series: Materials Science and Engineering*; IOP Publishing: New York, NY, USA, 2019; Volume 471, p. 062035. [CrossRef]
25. Jamroz, K.; Gaca, S.; Michalski, L.; Kieć, M.; Budzyński, M.; Gumińska, L.; Kustra, W.; Mackun, T.; Oskarbska, I.; Rychlewska, J.; et al. *Protection of Pedestrians. Guidelines for Pedestrian Traffic Organizers*; National Road Safety Council: Gdańsk, Poland; Warsaw, Poland; Cracow, Poland, 2014. (In Polish)
26. Rokade, S.; Kumar, R.; Rokade, K.; Dubey, S.; Vijayawargiya, V. Assessment of effectiveness of vertical deflection type traffic calming measures and development of speed prediction models in urban perspective. *Int. J. Civ. Eng. Technol.* **2017**, *8*, 1135–1146. Available online: https://iaeme.com/MasterAdmin/Journal_uploads/IJCIET/VOLUME_8_ISSUE_5/IJCIET_08_05_120.pdf (accessed on 23 July 2023).
27. Mohammadipour, A.; Archilla, A.R.; Papacostas, C.S.; Alavi, S.H. Pedestrian (RPC) influence on speed reduction. In Proceedings of the Conference: Transportation Research Board (TRB) 91st Annual Meeting, Washington, DC, USA, 22–26 January 2012; Available online: https://www.researchgate.net/publication/273687736_Pedestrian_RPC_Influence_on_Speed_Reduction (accessed on 3 August 2023).
28. Gonzalo-Orden, H.; Rojo, M.; Perez-Acebo, H.; Linares, A. Traffic calming measures and their effect on the variation of speed, Paper presented at the 12th Conference on Transport Engineering (CIT), Valencia 7–9 June, Spain. *Transp. Res. Procedia* **2016**, *18*, 349–356. [CrossRef]
29. Distefano, N.; Leonardi, S. Evaluation of the benefits of traffic calming on vehicle speed reduction. *Civ. Eng. Archit.* **2019**, *7*, 200–214. [CrossRef]
30. Distefano, N.; Leonardi, S. Effects of speed table, chicane and road narrowing on vehicle speeds in urban areas. In Proceedings of the VI International Symposium New Horizons 2017 of Transport and Comunications, Doboj, Serbia, 17–18 November 2017.
31. Kruszyna, M.; Matczuk-Pisarek, M. The Effectiveness of Selected Devices to Reduce the Speed of Vehicles on Pedestrian Crossings. *Sustainability* **2021**, *13*, 9678. [CrossRef]
32. Sołowczuk, A. Effect of Traffic Calming in a Downtown District of Szczecin, Poland. *Energies* **2021**, *14*, 5838. [CrossRef]

33. Akgol, K.; Gunay, B.; Aydin, M.M. Geometric optimisation of chicanes using driving simulator trajectory data. *Proc. Inst. Civ. Eng. Transp.* **2022**, *175*, 238–248. [CrossRef]
34. Aydin, M.M.; Akgöl, K.; Günay, B. The investigation of different chicane applications in traffic calming studies using driving symulator. *J. Fac. Eng. Archit. Gazi Univ.* **2019**, *34*, 1793–1805. [CrossRef]
35. Hussain, Q.; Alhajyaseen, W.K.M.; Kharbeche, W.; Almallah, M. Safer pedestrian crossing facilities on low-speed roads: Comparison of innovative treatments. *Accid. Anal. Prev.* **2023**, *180*, 106908. [CrossRef] [PubMed]
36. Sayer, I.A.; Parry, D.I. *Speed Control Using Chicanes—A Trial at TRL*; TRL Project Report PR 102; Transport Research Laboratory: Crowthorne, UK, 1994; Available online: https://trid.trb.org/view/425130 (accessed on 2 July 2023).
37. Sayer, I.A.; Parry, D.I.; Barker, J.K. *Traffic Calming—An Assessment of Selected On-Road Chicane Schemes*; Transport Research Laboratory TRL Report 313; Transport Research Laboratory: Crowthorne, UK, 1998.
38. Zhang, C.; Qin, S.; Yu, H.; Zheng, B.; Li, Z. A Review on Chicane Design based on Calming Theory. *J. Eng. Sci. Technol. Rev.* **2020**, *13*, 188–197. [CrossRef]
39. *Wirksamkeit Geschwindigkeitsdämpfender Maßnahmen Außerorts*; Hessisches Landesamt für Straßen- und Verkehrswesen: Hessen, Germany, 1997. Available online: https://docplayer.org/57666796-Wirksamkeit-geschwindigkeitsdaempfender-massnahmen.html (accessed on 2 August 2020).
40. Hunnel, T.; Mackie, A.; Wells, P. *Traffic Calming Measures in Built-Up Areas. Literature Review*; Unpublished Project Report PR/SE/622/02 Vägverket TR80 2002:15779; Swedish National Road Administration: Borlänge, Sweden, 2002.
41. Juran, J.M.; Gryna, F.M. *Juran's Quality Control Handbook*, 6th ed.; McGraw-Hill: New York, NY, USA, 2010.
42. Stephens, K.S. *Juran, Quality, and a Century of Improvement*, 15 ed.; ASQ: Milwaukee, WI, USA, 2004.
43. Feigenbaum, A.V. *Quality Control: Principles, Practice, and Administration*; McGraw Hill Book Company Inc.: New York, NY, USA, 1951.
44. Deming, W.E. *Out of the Crisis*, 5th ed.; The MIT Press: Cambridge, UK, 2000.
45. Ishikawa, K.; Loftus, J.H. *Introduction to Quality Control*, 3rd ed.; 3A Corporation: Tokyo, Japan, 1990.
46. Ishikawa, K. *Introduction to Quality Control*; Taylor & Francis: Philadelphia, PA, USA, 1990.
47. Ishikawa, K. *What Is Total Quality Control? The Japanese Way*; Prentice-Hall Direct: University Michigan: Michigan, MI, USA, 1985.
48. Juran, J.M. The Non-Pareto Principle; Mea Culpa. In *Juran on Quality by Design*; Juran, J., Ed.; Free Press: New York, NY, USA, 1992; pp. 68–71.
49. SAS/QC®14.1. *User's Guide: The Pareto Procedure*; SAS Institute Inc.: Cary, NC, USA, 2015.
50. Iyer, R.G. *The Art of Creating Pareto Analysis: A Complete End-to-End Guide to Understand Pareto Charts and Easily Create them in Excel*; Advanced Innovation Group Pro Excellence: Birmingham, UK, 2021.
51. Koch, R. *The 80/20 Principle: The Secret to Achieving More with Less*; Independently Published: Chicago, IL, USA, 2021.
52. Mydlarz, A. *7 Sztuczek Najlepszych Audytorów Jakości*; Sztuka Jakości Artur Mydlarz: Rzeszów, Poland, 2022. (In Polish)
53. Mydlarz, A. *Siedem Złotych Narzędzi Jakości. Praktyczny Poradnik Inżyniera*; Sztuka Jakości Artur Mydlarz: Rzeszów, Poland, 2022. (In Polish)
54. Sołowczuk, A. Metod TQM v zadačah ocenki tehničeskogo sostoâniâ dorog. *Transp. Stroit.* **2005**, *7*, 11–12.
55. Sołowczuk, A. Naučnye osnowy ocenki tehničeskogo sostoâniâ dorog po potrebitel'skim svojstvam. *Dorogi i Mosty* **2005**, *13*, 165–175.
56. Evans, S.R.; Norbacka, J.P.P. An Heuristic Method for Solving Time-Sensitive Routing Problems. *J. Oper. Res. Soc.* **1984**, *35*, 407–414. [CrossRef]
57. Wang, Z.; Wang, Z. A novel two-phase heuristic method for vehicle routing problem with backhauls. *Comput. Math. Appl.* **2009**, *57*, 1923–1928. [CrossRef]
58. Ichoua, S.; Gendreau, M.; Potvin, J.-Y. Vehicle dispatching with time-dependent travel Times. *Eur. J. Oper. Res.* **2003**, *144*, 379–396. [CrossRef]
59. Malandraki, C.; Daskin, M.S. Time dependent vehicle routing problems: Formulations, properties and heuristic algorithms. *Transp. Sci.* **1992**, *26*, 185–200. [CrossRef]
60. Hill, A.C.; Benton, W.C. Modelling intra-city time-dependent travel speeds for vehicle scheduling problems. *J. Oper. Res. Soc.* **1992**, *43*, 343–351. [CrossRef]
61. Kim, J.-S.; Jeong, J. Development of the Dripping Speed Measurement System of Medical Liquid using Heuristic. *J. Korean Inst. Intell. Syst.* **2014**, *24*, 542–547. [CrossRef]
62. Bujak, K. *Traffic Organization Design Along 5 Lipca Street on the Section from Bohaterów Warszawy Avenue to Szarych Szeregów Place*; City Office of Szczecin: Szczecin, Poland, 2015. (In Polish)
63. Ustawa—Prawo o Ruchu Drogowym, *Dziennik Ustaw* z dnia 20 Czerwca 1997 Nr 98 poz. 602, z późn. zm. Available online: https://isap.sejm.gov.pl/isap.nsf/DocDetails.xsp?id=wdu19970980602 (accessed on 2 July 2023). (In Polish)
64. Google Earth. Available online: https://earth.google.com (accessed on 2 July 2023).
65. Accident and Collision Registration System SEWIK. Available online: https://sewik.pl/ (accessed on 2 July 2023).
66. Vitronic. *Speed Displays Traffic Detection, Radar, Detection, Software*; Vitronic: Kędzierzyn Koźle, Poland, 2015.
67. Sobczyńska, D. Art of Experimental Research, habilitation monograph. In *Scientific Papers of the Philosophy and Logic*; series No. 71; Scientific Publishing House UAM: Poznań, Poland, 1993.
68. Roess, R.P.; Prassas, E.S.; McShane, W.R. *Traffic Engineering*, 4th ed.; Pearson Higher Education: London, UK, 2018.
69. Taylor, J.R. *An Introduction to Error Analysis*, 6th ed.; Scientific Publishing House: Warszawa, Poland, 2022.

70. Ziółkowski, R. Zachowania Kierowców Pojazdów w Otoczeniu Środków Uspokojenia Ruchu w Warunkach Miejskich. Ph.D. Dissertation, Politechnika Białostocka, Białystok, Poland, 2022. Available online: https://pb.edu.pl/oficyna-wydawnicza/wp-content/uploads/sites/4/2022/10/Zachowania-kierowcow-pojazdow-w-otoczeniu-srodkow-uspokojenia-ruchu-w-warunkach-miejskich.pdf (accessed on 2 August 2023). (In Polish).
71. Daniel, B.D.; Nicholson, A.; Koorey, G. Investigating Speed Patterns and Estimating Speed on Traffic-Calmed Streets; IPENZ Transportation Group Conference, Auckland, 27 March 2011. Available online: https://www.researchgate.net/publication/2673 70096_INVESTIGATING_SPEED_PATTERNS_AND_ESTIMATING_SPEED_ON_TRAFFIC-CALMED_STREETS (accessed on 30 November 2022).
72. Alavi, S.H. Analyzing Raised Crosswalks Dimensions Influence on Speed Reduction in Urban Streets. In Proceedings of the 3rd Urban Street Symposium, Seattle, WA, USA, 24–27 June 2007; Available online: https://www.researchgate.net/publication/2736 87736_Pedestrian_RPC_Influence_on_Speed_Reduction (accessed on 1 December 2015).
73. Domenichini, L.; Branzi, V.; Meocci, M. Virtual testing of speed reduction schemes on urban collector roads. *Accid. Anal. Prev.* **2018**, *110*, 38–51. [CrossRef] [PubMed]
74. Jorgensen, M.; Agerholm, N.; Lahrmann, H.; Araghi, B. *Driving Speed on Throughfares in Minor. Towns in Denmark*; Aalborg University: Aalborg Ost, Denmark, 2014; Available online: https://www.ictct.net/wp-content/uploads/26-Maribor-2013/26-Agerholm-Full-paper.pdf (accessed on 19 November 2015).
75. Kang, B. Identifying street design elements associated with vehicle-to-pedestrian collision reduction at intersections in New York City. *Accid. Anal. Prev.* **2019**, *122*, 308–317. [CrossRef] [PubMed]
76. Jamroz, K.; Gumińska, L.; Mackun, T.; Rychlewska, J. Widoczność na przejściach dla pieszych. *Drogownictwo* **2015**, *4–5*, 142–149. Available online: https://www.researchgate.net/publication/352121417_Widocznosc_na_przejsciach_dla_pieszych (accessed on 2 August 2023). (In Polish).
77. Jamroz, K.; Kempa, J.; Rychlewska, J.; Mackun, T. Metoda wyznaczania obszaru dobrej widoczności na przejściach dla pieszych w Polsce. *Transport Miejski i Regionalny* **2015**, *4*, 10–21. Available online: http://yadda.icm.edu.pl/baztech/element/bwmeta1 .element.baztech-42c6a790-a9b5-4ff0-9171-13d47b93c747 (accessed on 18 August 2023).
78. Das Land Kärnten: *Abteilung 7. Kompetenzzentrum Wirtschaftsrecht und Infrastruktur Richtlinie*. Grundlagen für die Anordnung eines Schutzweges. 2013. Available online: https://docplayer.org/64254869-Richtlinie-grundlagen-fuer-die-anordnung-eines-schutzweges.html (accessed on 18 August 2023).
79. Antoine, D.; Janssens, I.; Chevalier, N.; Dullaert, I.; Trussart, S.; Blanquet, D.; Delbart, E. Visibilité et Sécurité des abords d'écoles, Dirk DE SMET, Directeur général DGO1, Namur, Belgium. 2011. Available online: https://rue-avenir.ch/wp-content/uploads/files/resources/Securite-abords-ecoles-Wallonie.pdf (accessed on 18 August 2023).
80. Liu, C.; Wang, W. Integrating Visibility, Parking Restriction, and Driver's Field View for Enhancing Pedestrian Crossing Safety. *Int. J. Transp. Sci. Technol.* **2013**, *2*, 351–356. [CrossRef]
81. KfV, Leitfaden für die Anlage von Schutzwegen und Sonstigen Fußgängerquerungsstellen, Kuratorium für Verkehrs, LF/Schutzweg/V02, Land Oberösterreich 2006. Available online: https://www.tirol.gv.at/fileadmin/themen/verkehr/service/downloads/lf_schutzweg_v02end.pdf (accessed on 2 August 2023).
82. Høye, A.; Mosslemi, M. Fartsdempende Tiltak i Gangfelt—Eksempler og Erfaringer, TØI rapport 1033/2009, Transportøkonomisk Institutt, Oslo 2009. Available online: https://www.toi.no/getfile.php?mmfileid=16942 (accessed on 2 August 2023).
83. Fressancourt, M. Une Limitation de la Vitesse Pour un Meilleur Partage de la Route. . ., AQTr 21 June 2011. Available online: https://aqtr.com/association/actualites/limitation-vitesse-meilleur-partage-route (accessed on 3 August 2023).
84. How to Calculate Stopping Distances! Available online: https://www.drivingcrawley.co.uk/blog/how-to-calculate-stopping-distances (accessed on 3 August 2023).
85. Stopping Distances and The Theory Test 2018. Available online: https://driving-test-success.myshopify.com/pages/stopping-distances-and-the-theory-test (accessed on 3 August 2023).
86. What Is Stopping Distance? Available online: https://www.reddrivingschool.com/stopping-distances/ (accessed on 3 August 2023).
87. Everything You Need to Know about Vehicle Stopping Distances. Available online: https://www.unicominsurance.com/insurance-news/everything-you-need-to-know-about-vehicle-stopping-distances (accessed on 3 August 2023).

sustainability

Article

Towards a New Design Methodology for Vertical Traffic Calming Devices

Mauro D'Apuzzo [1,*], Azzurra Evangelisti [1], Daniela Santilli [1], Sofia Nardoianni [1], Giuseppe Cappelli [1,2] and Vittorio Nicolosi [3]

[1] Department of Civil and Mechanical Engineering, University of Cassino and Southern Lazio, 03043 Cassino, Italy; aevangelisti.ing@gmail.com (A.E.); daniela.san1992@libero.it (D.S.); sofia.nardoianni@unicas.it (S.N.); giuseppe.cappelli1@unicas.it (G.C.)
[2] University School for Advanced Studies, IUSS, Piazza della Vittoria 15, 27100 Pavia, Italy
[3] Department of Enterprise Engineering "Mario Lucertini", University of Rome "Tor Vergata", Via del Politecnico 1, 00133 Rome, Italy; nicolosi@uniroma2.it
* Correspondence: dapuzzo@unicas.it

Abstract: With increasing emphasis on new soft mobility in urban areas, it becomes more and more important to provide effective speed control measures for vehicular traffic. Among those, the ones based on vehicle vertical deflection, namely vertical traffic calming measures, are historically the most widespread. However, since the basic operating principle of these devices is related to the vertical dynamic response due to the interaction between the moving vehicle and the road profile, different vehicles may exhibit different speed behaviors when traveling on a specific road profile shape. As a matter of fact, recent research provided evidence of this connection, and therefore it has been worth investigating the dynamics underlying the phenomenon in order to develop a new approach to the design of vertical traffic calming devices. In this paper, following an initial state-of-the-art review, an in-depth study on the dynamic interaction between vehicle and road profile has been presented by means of an ad hoc-developed mathematical model. The proposed simulation model has been used to evaluate the root mean square acceleration value associated with each vehicle/traffic calming device/crossing speed. Following the outcomes provided by numerical simulations, an experimental investigation has been designed and carried out on a vertical traffic calming device. Speed profiles of different vehicles have been acquired, and preliminary results seem to provide evidence of a different dynamic response for each vehicle type, yielding the basis to reconsider the design approach of such devices in order to control urban traffic speed.

Keywords: speed control; traffic safety; vehicle dynamics; speed humps

Citation: D'Apuzzo, M.; Evangelisti, A.; Santilli, D.; Nardoianni, S.; Cappelli, G.; Nicolosi, V. Towards a New Design Methodology for Vertical Traffic Calming Devices. *Sustainability* **2023**, *15*, 13381. https://doi.org/10.3390/su151813381

Academic Editors: Salvatore Leonardi and Natalia Distefano

Received: 27 July 2023
Revised: 1 September 2023
Accepted: 3 September 2023
Published: 6 September 2023

1. Introduction

Speed management has become a major issue over the past century. For this reason, more and more often, horizontal and vertical traffic calming devices are installed (for example, speed humps, gateways, speed bumps, speed cushions, etc.). They force the drivers to maintain their speed level according to the type of road. However, traffic calming does not guarantee that speeds that comply with the imposed limits will be maintained. This has encouraged researchers to evaluate the effects of these devices and determine their effectiveness based on their size (height, length, gradient, and radius) and shape. Efficiency is evaluated by analyzing their effectiveness in speed reduction or accident rate reduction.

The first studies have been conducted since 1980 [1,2]. From that period on, the focus has been on the different devices from various points of view: efficacy studies, standards, guidelines, etc.

In the current literature review, devices with a vertical orientation were mainly analyzed, considering experimental and theoretical studies.

In the experimental studies, before-and-after analyses are carried out, in which the sites before and after the installation of the devices are compared in order to evaluate the variation in speed and traffic flows, thus determining the effectiveness of the installed measurement. Vertical traffic calming devices may have advantages and disadvantages (noise, vibrations, pollution); therefore, researchers [3–7] have been trying to determine the geometric parameters, spacing, etc. that make them more effective.

Many studies have shown that, by applying these mitigation measures at sites, various types of road issues [7–9], such as speed reduction [10–13], reduction of traffic volumes [14], reduction of road accidents [15–17], and safety concerns for pedestrians [18] due to the lack of sidewalks or narrow streets, can be effectively tackled.

Depending on the type of device and its geometry [19–24], different speed reductions are generated [25–30], also due to the different area of influence in which the vehicle deceleration occurs [31]. This suggests that, to obtain significant speed reductions, the devices could be installed in series [17]. Therefore, in this way, a speed profile rather than a spot speed value can be evaluated [32–36]. To obtain the desired speed reductions, it is necessary to correctly space the devices; in fact, for a shorter distance, lower speeds are expected since, between a bump or a hump and the next vertical traffic calming device, it would not be possible to reach high speeds [4,37–40].

Another approach used to evaluate the issue of speed reduction due to traffic calming does not use experimentation but a theoretical approach that requires complex mathematical or soft computing tools. With these approaches, it is possible to develop multiple degrees-of-freedom models with the aim of evaluating kinematic quantities such as displacements, accelerations [41,42], or rotation of the vehicle [43] and/or driver [44].

Other, more elaborated models allow for the evaluation of the vibrations transmitted to the ground [45–47] and their relationship with the road profile [48,49].

Sometimes, to calibrate the theoretical models, the results of the field campaigns are taken into consideration and then compared with those derived from ad hoc experimentations [50,51]; this allows to derive the empirical speeds [52,53] and the characteristics of the vertical traffic calming device and the road [54,55]; thus, by knowing only the layout of the devices, it is possible to somehow predict the speeds without going through experimental observations [56–59], in order to reduce deaths and serious injuries [60].

However, interactions between the different road users seem to play a fundamental role in safety conditions, and studying driver and pedestrian behaviors is essential to designing measures to reduce the risk of crashes [61]. In this paper, particular attention is given to raised pedestrian crossings (RPCs), which seem to have great efficiency in terms of speed reduction and safety for pedestrians compared to other traffic calming devices [62]. However, according to Loprencipe et al. [63], they require a high value of vertical acceleration to induce a crossing speed reduction. The aim, as well as the novelty aspect, is to study the behavior of drivers in close proximity to RPCs in order to take actions with the aim of increasing pedestrian safety and, in turn, fostering sustainable mobility.

However, summarizing these results, it appears that there is still a need to deepen the study, such as the vehicle's and the road profile's dynamic interactions, with the aim to better characterize the vibration level in time and frequency domains and to derive new design criteria based on vehicle mechanical proprieties, on the one hand, and on geometrical traffic calming layout, on the other.

To this purpose, a new insight comprising the development and calibration of numerical interaction models, helpful in analyzing and understanding real speed profile data collected through an experimental campaign, is presented and discussed in the following sections.

2. Methodology

The method proposed is summarized as follows in the figure below (see Figure 1). The next paragraphs will explain the main steps that need to be followed, as described in this schematic flow chart, to better understand the method behind and the peculiar aspects introduced in this work.

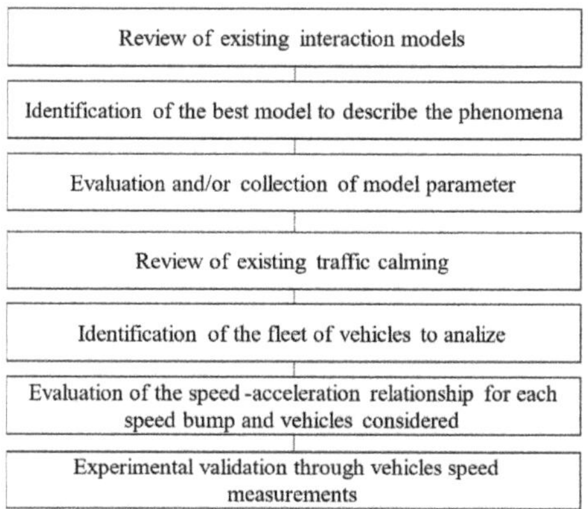

Figure 1. Description of the methodology proposed.

2.1. Vibrations' Effect on Human Body

During motion in a vehicle, the body is subject to vibrations. Vibration is a mechanical movement around a fixed point propagated as a mechanical wave by means of a mechanical medium (such as vehicles and people).

Waves can be divided into:

- deterministic, when future oscillations can be predicted from the knowledge of the previous waves;
- stochastic, commonly called random waves, where only some significant statistical properties can be evaluated due to uncertainties on several factors affecting the vibration phenomenon.

Deterministic waves can be divided into periodic and nonperiodic; the former, in turn, are divided into simple harmonic or multi-harmonic waves, which have the characteristic of being able to be represented analytically in a closed form. The latter, the nonperiodic ones, on the other hand, can in turn be divided into transient waves or shocks.

In most cases, the waves to which the human body is exposed during work and leisure are stationary and nonstationary random waves [64].

However, in the real world, it is difficult to encounter vibrations described by a simple harmonic wave, so more complex wave descriptions are required. These complex waves could be processed by superimposing multiple harmonic waves with different amplitudes, frequencies, and phases. This principle is the basis of the Fourier transform, in which, given as input a complex signal, the Fourier transform returns as output data a frequency response spectrum, which has frequency as an independent variable and amplitude as a dependent variable. It allows the writing of a time-dependent function in the frequency domain; to do this, it decomposes the function into the basis of exponential functions with a scalar product.

$$X(f) = \int_{-\infty}^{+\infty} x(t)e^{-j2\pi ft}dt \tag{1}$$

This Fourier transform can only be used for finite signals; in short, the signal must be summable, therefore the following equation must be considered:

$$lim_{T \to \infty} \int_{-\frac{T}{2}}^{\frac{T}{2}} x(t)dt < \infty \tag{2}$$

Vibration can be synthetically described by means of three physical quantities: displacement, velocity, and acceleration.

The magnitude of the acceleration of a vibration can be synthetically expressed in terms of peak-to-peak acceleration (acceleration calculated as the difference between the amplitude of the positive peak and the amplitude of the lower peak) and in terms of peak acceleration (acceleration calculated by measuring just a peak). The acceleration measure used most often in engineering is the RMS (root mean square) value, which represents the standard deviation of a digital signal with a zero mean. This quantity can therefore be calculated as the square root of the mean of the squares of the acceleration values, that is, the square root of the square mean. For a simple sinusoidal signal, the peak-to-peak acceleration is equal to twice the amplitude, while the peak amplitude is equal to the amplitude itself, where the RMS value is equal to the amplitude divided by the square root of 2.

However, it has to be highlighted that the acceleration measurement values described so far do not take into account the duration, or rather, the exposure time. With the same peak-to-peak value, peak amplitude, or RMS, it is easy to understand that the response of the human body varies according to the exposure time during which the body is subjected to this vibration. More recent guidance seems to suggest alternative measures such as the vibration dose value, in which the vibration descriptor takes exposure time into account [31].

2.2. Simulation Models

Vertical traffic calming devices cause a strong vertical acceleration in the transiting vehicle due to the excitation induced by travel on a localized road profile irregularity. This acceleration causes a strong disturbance for the driver, who is thus forced to reduce speed [65,66]. It is evident from the aforementioned scientific literature that in order to operate an effective and continuous control of the speed of the vehicles over a defined road section, it is necessary to establish in series a certain number of such devices according to a specific spacing. However, regardless of the spacing layout to be selected, the evaluation of the vibration level induced by a vertical traffic calming device is of paramount importance.

For this purpose, it is necessary to evaluate the vertical dynamic response of a typical vehicle traveling on a rough road surface.

It has to be underlined that this approach is currently adopted in mechanical engineering in order to study vehicle dynamics. Vehicles can be represented as rigid bodies and/or point masses interconnected with springs and dashpots. The most used model is the "Quarter Car Model" (Figure 2), a two-degree-of-freedom model where the sprung mass (one fourth of the suspended mass of the vehicle) is linked to the unsprung mass (that is, the half axle) with springs and dashpots assembled in parallel (describing car suspensions). The unsprung mass is also connected to the road profile with a spring coupled with a dashpot in order to simulate the tire's vertical stiffness.

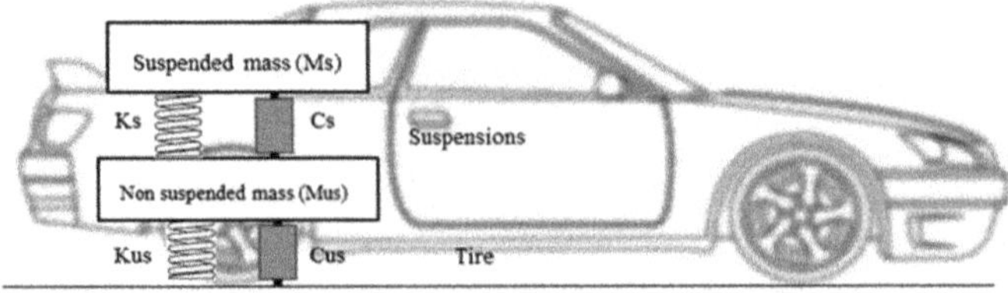

Figure 2. 2 degree-of-freedom (*2 dof*) vehicle model (quarter car).

However, it has been observed that if the vertical response of a vehicle traveling on a vertical traffic calming device is concerned, it is not correct to describe this phenomenon with such models, so it appears more suitable to use a more detailed model, such as the "Half Car Model" (Figure 3), a four-degree-of-freedom model in which both front and rear axles are represented. The reason why a "Half Car Model" is considered is related to the "wheel base filtering" problem: if a vehicle is moving with a certain speed on an irregular surface, vibrations are detected by the front axle and then, with a time lag related to vehicle speed and vehicle wheel base, by the rear axle. This phenomenon dramatically changes the way the vehicle is vertically excited when moving on a rough road surface and, for certain speed and wheel base values, could induce some unexpected resonant frequencies due to a change in the shape of the transfer function between the road excitation and the vertical load transmitted by the vehicle [45].

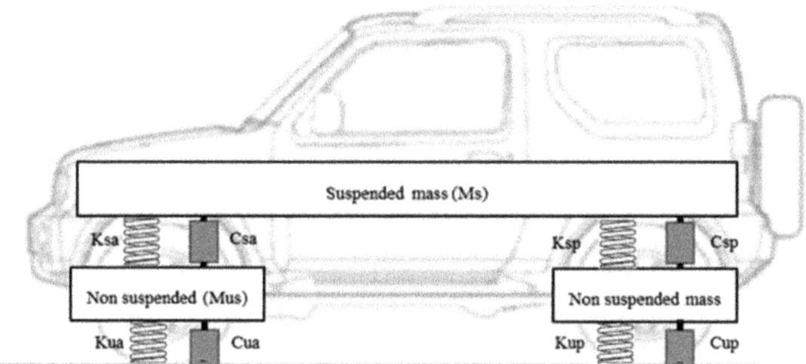

Figure 3. 4 degree-of-freedom (*4 dof*) vehicle model (half car).

A further insight into riding comfort may consider the insertion of the driver, which, in turn, can be represented by a lumped mass system (Figure 4). However, it is worth highlighting that, in this latter case, more information related to the inertial and mechanical properties of seat suspensions and the human body itself needs to be collected that is not easily available.

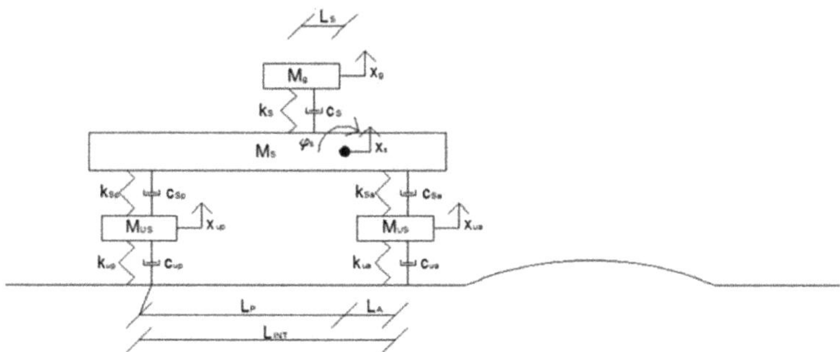

Figure 4. 5 degree-of-freedom (*5 dof*) vehicle model (half car plus driver insertion).

On these premises, it seemed more appropriate to consider a Half Car Model in evaluating vertical acceleration. Within this model, the vehicle chassis is considered a rigid body with its own mass and its own specific rotational inertia. The chassis is connected to both axles (rear and front) by means of springs coupled with dashpots describing the suspension system, whereas the axle (described by a point mass) is connected to the road surface by another spring–dashpot system accounting for tire mechanical properties.

On a computational point of view, for a specific vehicle model, the vehicle motion on an irregular surface can be easily described through a classical set of second-order linear differential equations:

$$[M]\ddot{X} + [C]\dot{X} + [K]X = F \tag{3}$$

where $[M]$ is the mass matrix containing the inertial properties of the vehicle, $[C]$ is the damping matrix containing the damping properties of the dashpots, $[K]$ is the stiffness matrix containing the stiffness properties of the springs, X is the vector of displacements pertaining to each degree-of-freedom characterizing the model, and F is the vector of the external forces acting on the vehicle (in this case the vertical excitation of the profile).

By making use of the complex notation to describe the displacement of each degree-of-freedom, the aforementioned set of differential equations can be transformed into an algebraic set of equations in the frequency domain that can be solved in order to derive the complex transfer functions (or frequency response function, FRF) between the road profile vertical excitation and the kinematic physical entities associated with the degree-of-freedom of the dynamic system itself (i.e., translational and rotational displacement and their first and second order derivatives).

Once all the relevant parameters (such as the road surface characteristics, the mechanical and inertial properties, and the speed of heavy vehicles) are known, time histories of the vertical displacement and of rotation sensed in the traveling vehicle are computed by convoluting the FRF of the vehicle model with the spectral representation of the vertical road profile and by performing an inverse Fourier transform with the aim of evaluating the time histories.

2.3. Input Data

By analyzing the accelerogram in the time domain, it may be necessary to evaluate the acceleration in the frequency domain by making a change of variables through the use of the DFT (fast discrete Fourier transform) in order to evaluate the weighted root mean square (RMS) of vertical acceleration according to the procedure based on the exposure threshold for human vibration reported in the frequency domain by ISO Standards [67–69].

Once the most suitable vehicle model had been identified, the travel of four different types of vehicles, representative of the entire vehicle fleet that can be encountered in an urban environment, was simulated on several vertical traffic calming devices. The four macrocategories of vehicles can be summarized as follows:

- Small cars are represented with the designation A;
- Medium-sized cars are represented with the designation C;
- Sedans are represented with the designation D;
- Off-road/sport utility vehicles are represented with the designation E.

The main characteristics of the vehicles considered are shown in Table 1. These quantities were collected and rearranged from the technical data sheets [70].

Table 1. Main characteristics of the vehicles considered in this analysis.

Vehicle	A	C	D	E
Elastic stiffness of the front suspension (N/m)	12,400	12,800	12,950	13,400
Elastic stiffness of the rear suspension (N/m)	10,600	13,800	14,720	17,000
Linear coefficient of kinematic viscosity of the front suspension (N/m)	1550	1600	1618	1675
Linear coefficient of kinematic viscosity of the rear suspension (N/m)	1325	1725	1840	2125
Suspended mass (kg)	893	963	1220	1700
Front unsprung mass (kg)	80	92	105	115
Rear unsprung mass (kg)	72	120	110	130
Elastic stiffness of the front tire (N/m)	165,000	170,000	180,000	200,000
Elastic stiffness of the rear tire (N/m)	150,000	150,000	170,000	200,000
Moment of inertia	1018.76	1284	2313.21	3264
Wheel base length (m)	2.35	2.47	2.69	2.85
Total length (m)	3.70	4.0	4.77	4.80
Tire imprint length (m)	0.12	0.14	0.16	0.2
Length of the distance between the center of gravity of the car and the driver (m)	1	1.2	1.3	1.6

2.4. Implementation of Speed Humps

The initial objective of this work is to evaluate, for a given profile shape and for a given vehicle, a relationship that links acceleration level sensed by the driver to actuated vehicle speed. It is intuitively possible to imagine that acceleration increases with increasing speed, the height of the hump, and decreasing vehicle characteristics. A careful analysis must therefore be carried out to understand how the input parameters can vary the acceleration intended as a disturbance index. In this study, a fleet of private vehicles (described above in Table 1) representing the traffic spectrum was considered.

A total of 5 types of bumps have been implemented in the model, 4 of which are parabolic and 1 trapezoidal. Below are the geometric characteristics of each bump:

1. Parabolic speed hump (length 3.65 m; height 0.0635 m);
2. Parabolic speed hump (length 3.65 m; height 0.0762 m);
3. Parabolic speed hump (length 6.70 m; height 0.0762 m);
4. Parabolic speed hump (length 6.70 m; height 0.0889 m);
5. Trapezoidal speed hump (length 6.70 m; height 0.0762 m).

The analyses were conducted in a speed range between 0 km/h and 40 km/h, with a speed step of 5 km/h for each vehicle on each bump.

Thus, diagrams have been obtained in which acceleration is placed in terms of RMS.

The results for each vehicle on each type of bump are shown below in graphical form (Figure 5).

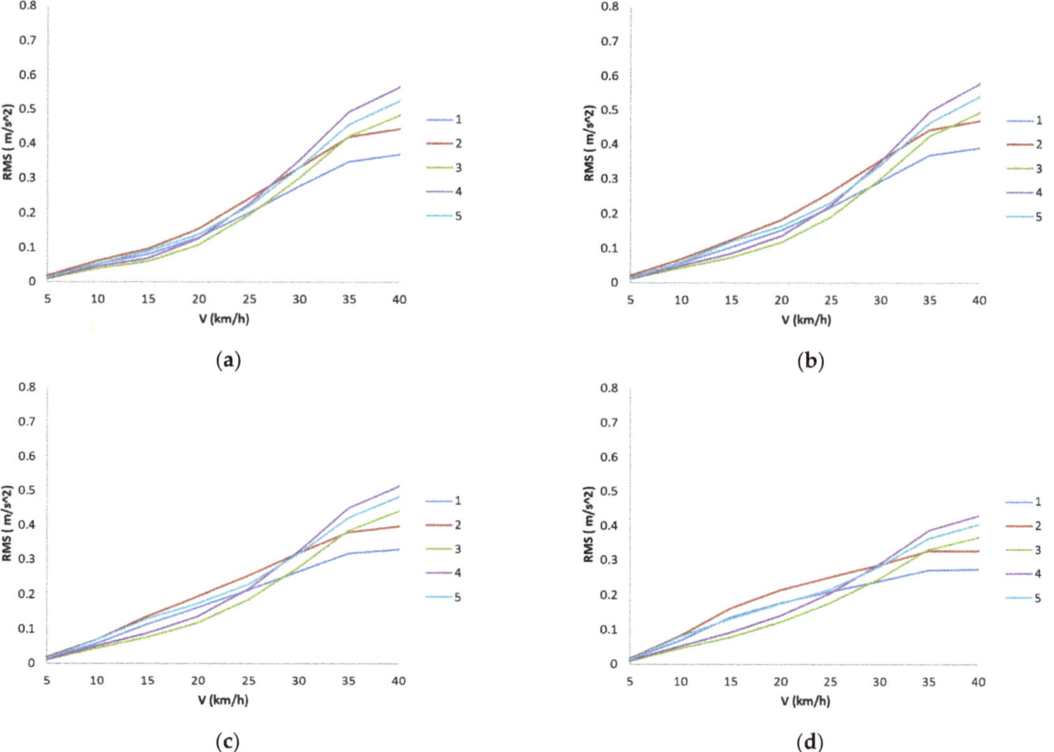

Figure 5. (**a**) Acceleration in terms of RMS on several speed humps (1 to 5) as a function of speed for the vehicle category A; (**b**) for the vehicle category C; (**c**) for the vehicle category D; and (**d**) for the vehicle category E.

From the graphs, it is possible to deduce how the response in terms of RMS varies with the variation of the geometric characteristics of the device (height and length).

It should be noted that at the same speed, the device that causes greater acceleration for all vehicles, in terms of RMS, is the fourth device, with a length of 6.70 m and a height of 8.89 cm.

It can be similarly interesting to evaluate the acceleration level caused by the same vertical traffic calming device on different vehicles. To this purpose, RMS diagrams of vertical acceleration versus vehicle speed have been conveniently derived and reported in the following figures (see Figure 6).

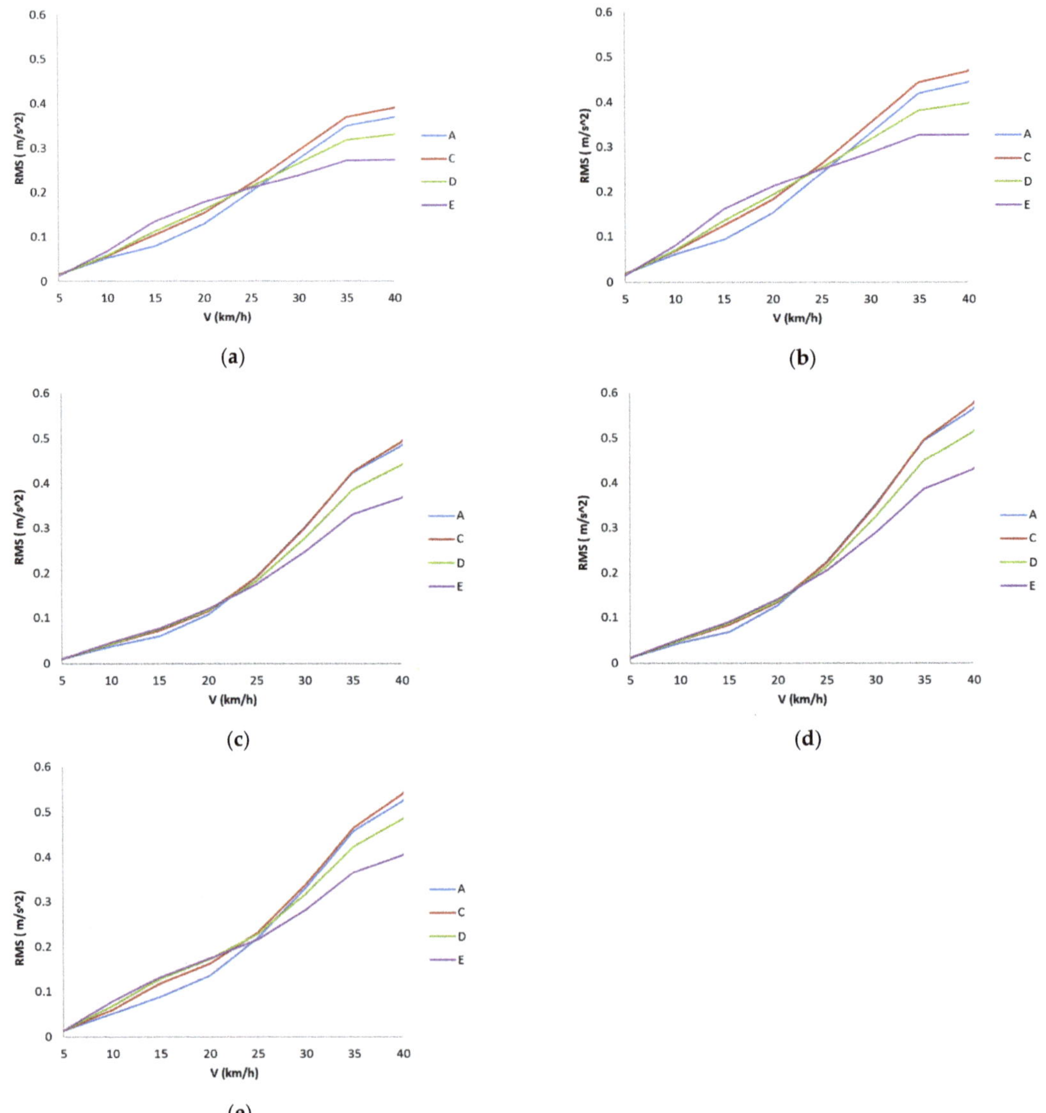

Figure 6. (a) Acceleration in terms of RMS on different classes of vehicles (A to E) caused by the speed hump 1; (**b**) by the speed hump 2; (**c**) by the speed hump 3; (**d**) by the speed hump 4; and (**e**) by the speed hump 5.

It can be observed from these latter diagrams that the vibration-induced disturbance is very similar for speed values ranging from 22 to 25 km/h, depending on the specific vertical traffic calming device traveled. For higher speed values, it can be seen that vehicles belonging to category E and category D receive less "vibrational stress" than vehicles belonging to category A and category C.

These results seem to provide some theoretical basis to investigate the possibility of highlighting different crossing speed behaviors according to the specific vehicle category traveling onto a defined vertical traffic calming device.

To this purpose, a preliminary experimental campaign, which is detailed in the following, has been carried out on a real vertical traffic calming device located in an urban environment.

3. Case Study

An application of this research has been developed within a city in Central Italy.

A raised pedestrian crossing (RPC) located close to a university facility was selected since it was easier to monitor. The vertical shape of the traffic calming device is trapezoidal, with an overall length of 3.75 m. The location of the investigated vertical traffic calming device is reported in Figure 7, and in Figure 8, a closer view of the RPC is shown.

Figure 7. Areal view and location of the monitored raised pedestrian crossing (RPC) with the indication of the directions "a" in entrance to the city of Cassino and "b" in exit from the city of Cassino.

Figure 8. Closer view of the investigated RPC.

A digital camera located at an elevated point of view has been used to monitor vehicular traffic at different hours of the day and on different typical weekdays. In order to avoid errors caused by the experimental setting, the same camera placement and investigator were employed throughout the experimentation.

In order to derive consistent vehicle speed profiles, it was necessary to define several equally spaced fixed targets in the video frames (every 5 m) departing from the monitored RPC along the upstream and downstream directions (see Figure 9).

Figure 9. Postprocessing video editing to derive speed profiles (red lines represents spaced fixed targets in the video frames every 5 m).

It was therefore possible to measure spot speeds of vehicles belonging to monitored traffic flows in both directions of travel, defined in Figure 7 with the letters "a" and "b".

As far as significant speed profiles are concerned, only "isolated" vehicles were considered, i.e., only vehicles with headway higher than 15 s with respect to the preceding vehicle or that were not affected in their run by any side or frontal obstacle or pedestrian's crossing. An overall amount of 252 and 229 speed profiles were collected and evaluated along the "a" and "b" directions, respectively, and speed values were manually derived from acquired videos.

A sample of such vehicle speed profiles derived from video recordings is conveniently depicted as follows (see Figures 10 and 11).

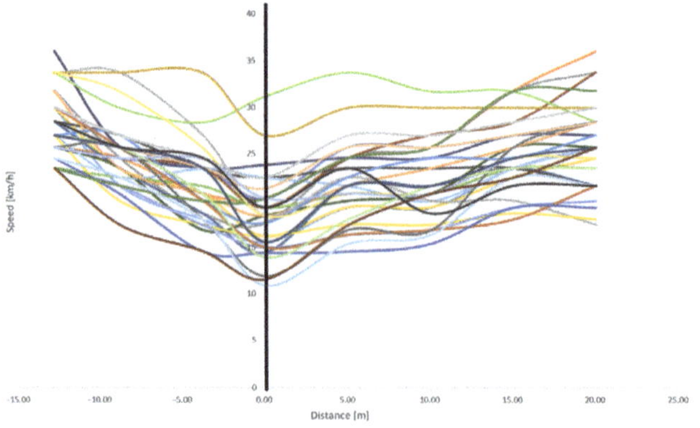

Figure 10. Speed profiles, represented in several colors in the graph, of a typical working daytime slot in the forward direction "a". Zero distance identifies the RPC location.

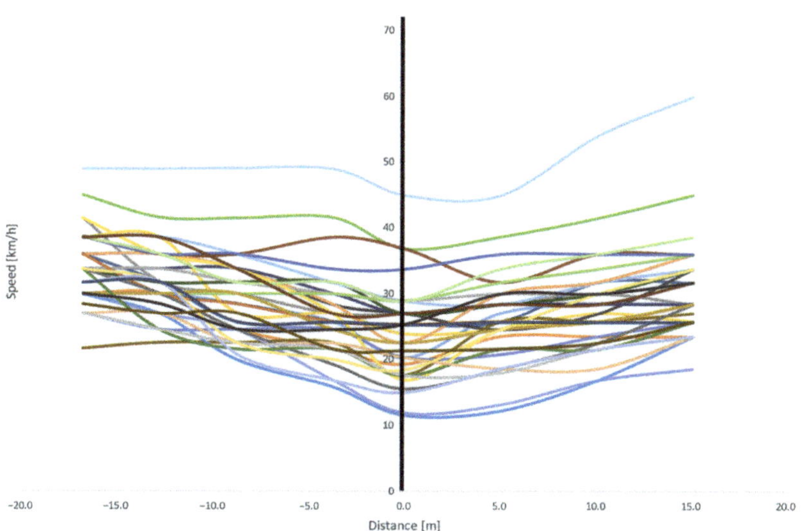

Figure 11. Speed profiles, represented in several colors in the graph, of a typical working daytime slot in the forward direction "b". Zero distance identifies the RPC location.

4. Discussion

In accordance with what has been seen in the literature in the previous paragraphs, it appeared worth analyzing if and how the trend of the speed on the RPC varies according to different types of vehicles.

In order to highlight the significant difference between crossing speeds of different vehicle categories, a statistical ANOVA test has been carried out with the aim of determining whether groups differ from the others or not, i.e., whether the null hypothesis (H_1) of the same speed crossing value should be rejected (if the likelihood of the observed data under the null hypotheses is low) or not. In other words, ANOVA makes it possible to estimate the probability of obtaining a difference between the values of the two averages at least as large as that observed in the sample when the null hypothesis is true. This probability is called the *p*-value.

If the p-value is low (*p*-value < alpha), it can be concluded that the observed difference between the averages of the groups is statistically significant, whereas when this probability is high, it can be concluded that the observed difference between the averages of the groups is not statistically significant. This situation occurs when the *p*-value is large (*p*-value > alpha).

It is essential to remember that as input data, the spot speeds recorded when they are traveling on the monitored RPC are used (dependent value) to assess how the average of them was different in order to reject or confirm the theoretical hypothesis that the different types of vehicles (independent value) produce different accelerations and vibrations at the passage of the bump and consequently also different crossing speeds.

The following Table 2 reports the values of the *p*-value (and also other relevant statistics) between the different vehicles.

Table 2. ANOVA of the main statistics for directions "a" and "b", respectively.

	F	***p***	**F$_{crit}$**
Direction "a"	0.671	0.570	2.641
Direction "b"	1.785	0.151	2.645

As can be observed from the results reported in the aforementioned tables, the results of the tests do not clearly show the difference between the crossing speeds when traveling on the RPC between the various types of examined vehicles. In detail, along the "a" direction, the null hypothesis seems to hold, whereas in the opposite direction ("b"), it is more questionable.

This may be due to a limitation given by a preliminary, nonexhaustive data collection that should, in a second phase, be extended to confirm and/or refute the results thus obtained.

However, having obtained the average spot speed value on the examined RPC, it was possible to estimate the corresponding RMS parameter (see Figure 12a,b) by making use of the RMS–Speed relationship derived for the most similar road hump type among those aforementioned examined (see Figure 6).

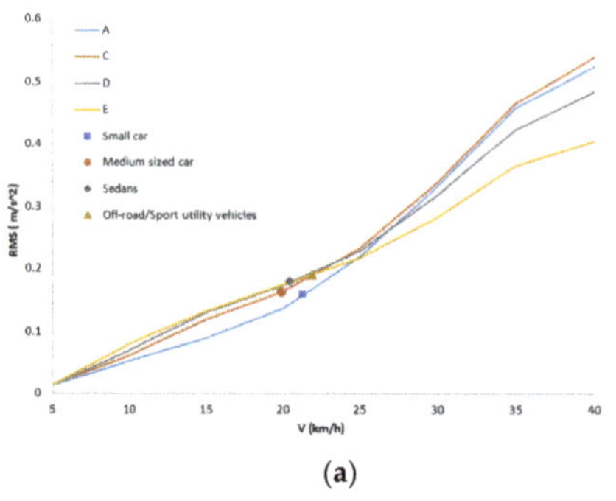

(a)

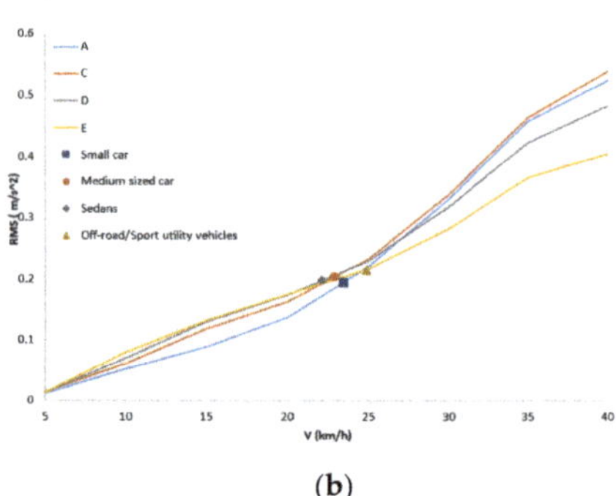

(b)

Figure 12. (**a**) Acceleration level in terms of weighted RMS as a function of average crossing speed (trapezoidal speed hump 5) measured for the specific vehicle category along the "a" direction; (**b**) Acceleration level in terms of weighted RMS as a function of average crossing speed (trapezoidal speed hump 5) measured for the specific vehicle category along the "b" direction.

These RMS values are shown in tabular form (see Tables 3 and 4).

Table 3. Estimated weighted RMS values of direction "a" based on vehicle average speed.

Type of Vehicles	Small Car	Medium-Sized Car	Sedans	Off-Road/Sport Utility Vehicles
RMS [m/s^2]	0.159	0.161	0.179	0.190
V [km/h]	21.27	19.93	20.43	21.89

Table 4. Estimated weighted RMS values of direction "b" based on vehicle average speed.

Type of Vehicles	Small Car	Medium-Sized Car	Sedans	Off-Road/Sport Utility Vehicles
RMS [m/s^2]	0.193	0.203	0.198	0.215
V [km/h]	23.48	22.93	22.14	24.85

The weighted RMS standard deviation for both directions has therefore been evaluated and reported in the following table (Table 5):

Table 5. Standard deviation of RMS in directions "a" and "b".

	Standard Deviation
Direction "a"	0.0863
Direction "b"	0.0466

From the values obtained from the standard deviation, it can be highlighted that although the dispersion of the estimated weighted RMS based on the average measured crossing vehicle speed is rather small, the weighted RMS may not be the optimal parameter to estimate the relationship between vibration levels sensed by vehicle occupants and actuated driver speed when traveling on the RPC. Nevertheless, it is worth noting that the lowest standard deviation is obtained along the direction (direction "b"), according to which previous statistical tests provided some evidence towards the rejection of the null hypothesis (similar crossing speeds).

However, the evaluation of perceived discomfort remains a very complex problem since several literature studies have shown that a very important factor in characterizing the effects of vibrations on the human body is their frequency. With the same amplitude of accelerations, the human body behaves differently since there is a different perception threshold depending on the frequency itself. This highlighted that stopping to evaluate only the acceleration as a function of time was reductive; for this reason, the acceleration was represented as a function of the frequencies, subsequently identifying a single parameter (the acceleration expressed in terms of weighted RMS).

Numerical simulations obtained taking into account different vehicle categories and different vertical traffic calming devices highlighted a range of vehicle speeds yielding a similar acceleration level. As a matter of fact, it is worth mentioning that in the literature, there are few works that evaluate the speeds according to the category of vehicle that crosses them, but according to this limited evidence, the crossing speeds on bumps seem to remain unchanged regardless of the type of vehicle [35,37].

According to these preliminary results, the null hypothesis of similar crossing speeds cannot be rejected or accepted, and therefore more data will need to be collected in the future from the same site and from different sites to validate or refute the developed theoretical approach. In addition, new vibration descriptors such as the vibration dose value, in which vibration level is also correlated with the duration of exposure as reported in the ISO 2631 standard [67–69], are worth investigating.

5. Conclusions

Nowadays, with the increasing attention paid to social and environmental sustainability, there is a great demand both to reduce vehicles' speeds and to achieve an acceptable level of safety in urban areas. With this aim in mind, vertical traffic calming devices seem to meet these two requirements. Great efforts are made by researchers all over the world to study the main interactions, which is why, in this paper, as an initial step to frame the topic, an in-depth analysis of the current scientific literature review is conducted. A side analysis of the literature review is also conducted on mathematical models that are able to better describe the phenomenon; based on these scientific evidences, a four-degree-of-freedom model (also known as the "Half Car Model"), considering the right compromise between simplicity in implementation and accuracy of outputs, is implemented. With this model, the weighted root mean square acceleration (RMS), associated with each vehicle category/traffic calming devices/crossing speed, is evaluated, since the aim of this paper is to understand the behavior of drivers in close proximity to traffic calming devices.

Attention was paid to vertical traffic calming devices (raised pedestrian crossings and speed tables) that exploit vertical acceleration; this causes discomfort in the driver, who tends to moderate the speed below a certain threshold to reduce the vibrational disturbance. A numerical relationship was then identified between vehicle crossing speed and vertical acceleration by using a four-degree-of-freedom lumped mass vehicle model to evaluate the dynamic interaction between the vehicle itself and the vertical traffic calming device.

The final goal was to define acceleration thresholds (that can be expressed in the time or frequency domain) beyond which the driver feels discomfort. Therefore, a complex numerical model able to simulate the vibration level sensed by vehicle occupants when traveling on a specific vertical traffic calming device was developed and calibrated.

In the following step, a deep dive into a real-life case study, in particular a raised pedestrian crossing (RPC), is presented in this paper. This vertical traffic calming device allows for reducing crash risks and speed and increasing pedestrian safety levels in crossing sections in urban contexts, and for this reason it is essential to study drivers' behavior.

As shown from the limited literature on this specific topic, at speeds within the 22–25 km/h range, the induced accelerations seem not to vary with the type of vehicle and type of vertical traffic calming device [35,37]. To gain experimental evidence for such behavior, a real vertical traffic calming device (a raised pedestrian crossing) has been monitored, and vehicle speed profiles have been collected. Statistical ANOVA tests have been performed in order to highlight the differences in crossing spot speeds according to different vehicle categories, but preliminary results seem somehow contradictory, highlighting differences between the crossing speeds on the RPC only along one direction.

It is believed that this is due to the limited amount of experimental data collected so far. It is therefore planned to expand the experimental campaign in order to increase data collection at the aforementioned site and to extend the investigation to other vertical traffic calming devices other than the ones monitored.

In addition, it can be argued that weighted RMS may not be the most suitable vibration descriptor to capture the drivers' behavior in terms of vibrational nuisance. As a matter of fact, another candidate descriptor could be the vibration dose value, in which the vibration level is also correlated with the duration of exposure as reported in the ISO 2631 standard [67–69].

However, it has to be underlined that the developed theoretical approach, partially corroborated by experimental data, appears to be promising in developing and refining a true human vibration exposure threshold to be used for the design of new vertical traffic calming devices other than those currently proposed by national standards.

Based on the different dynamic responses due to different vehicle categories, it will also be possible to estimate, in advance and in a more precise and reliable way, the crossing vehicle speed distribution according to a specific vehicle fleet characterizing the site of intervention.

Author Contributions: Conceptualization, M.D., A.E. and D.S.; methodology, M.D., A.E. and D.S.; software, D.S., S.N. and G.C.; validation, D.S., S.N. and G.C.; formal analysis, D.S., S.N. and G.C.; investigation, D.S., S.N. and G.C.; resources, M.D. and V.N.; data curation, S.N. and G.C.; writing—original draft preparation, S.N. and G.C.; writing—review and editing, S.N. and G.C.; visualization, D.S., S.N. and G.C.; supervision, M.D. and V.N.; project administration, M.D.; funding acquisition, M.D. All authors have read and agreed to the published version of the manuscript.

Funding: This study was carried out within the MOST—Sustainable Mobility Center and received funding from the European Union Next-GenerationEU (PIANO NAZIONALE DI RIPRESA E RESILIENZA (PNRR)—MISSIONE 4 COMPONENTE 2, INVESTIMENTO 1.4—D.D. 1033 17/06/2022, CN00000023). The research leading to these results has also received funding from Project "Ecosistema dell'innovazione Rome Technopole" financed by the EU in the NextGenerationEU plan through MUR Decree n. 1051 23.06.2022—CUP H33C22000420001. This manuscript reflects only the authors' views and opinions; neither the European Union nor the European Commission can be considered responsible for them.

Data Availability Statement: The data used in this study appear in the submitted article.

Acknowledgments: The help of Eng. Franco Renzullo and Eng. Gianluca Pilla in collecting and analyzing vehicle speed data is gratefully acknowledged.

Conflicts of Interest: The authors declare no conflict of interest. The funders had no role in the design of the study; in the collection, analysis, or interpretation of data; in the writing of the manuscript; or in the decision to publish the results.

References

1. Waltz, F.; Hoefliger, M.; Fehlmann, W. Speed Limit Reduction from 60 to 50 km/h and Pedestrian Injuries. In Proceedings of the Twenty-Seventh Step Car Crash Conference; International Research Council on Biokinetics of Impacts (IRCOBI), Warrendale, PA, USA, 17–19 October 1983.
2. Van Houten, R.; Van Huten, F. The Effects of a Specific Prompting Sign on Speed Reduction. *Accid. Anal. Prev.* **1987**, *19*, 115–117. [CrossRef] [PubMed]
3. Johansson, C.; Rosader, P.; Leden, L. Distance between speed humps and pedestrian crossing does it matter. *Accid. Anal. Prev.* **2011**, *43*, 1846–1851. [CrossRef] [PubMed]
4. García, A.; Torres, A.; Romero, M.; Moreno, A. Traffic Microsimulation Study to Evaluate the Effect of Type and Spacing of Traffic Calming Devices on Capacity. *Procedia-Soc. Behav. Sci.* **2011**, *16*, 270–281. [CrossRef]
5. Yannis, G.; Kondyli, A.; Georgopoulou, X. Investigation of the impact of low cost traffic engineering measures on road safety in urban area. *Int. J. Inj. Control Saf. Promot.* **2014**, *21*, 181–189. [CrossRef]
6. Jasiūnienė, V.; Pociūtė, G.; Vaitkus, A.; Ratkevičiūtė, K.; Pakalnis, A. Analysis and evaluation of trapezoidal speed humps and their impact on the driver. *Balt. J. Road Bridge Eng.* **2018**, *13*, 104–109. [CrossRef]
7. County Surveyors' Society. *Village Speed Control Working Group*; Final Report; Department for Transport: London, UK, 1994.
8. County Surveyors' Society. *Traffic Calming in Practice*; Landor Publishing: London, UK, 1994.
9. Abate, D.; Dell-Acqua, G.; Lamberti, R.; Coraggio, G. Use of Traffic Calming Devices Along Major Roads thru Small Communities in Italy. In Proceedings of the 88th Annual Meeting of the Transportation Research Board TRB, Washington, DC, USA, 11–15 January 2009.
10. Antić, B.; Pešić, D.; Vujanić, M.; Lipovac, K. The Influence of Speed Bumps Heights to the Decrease of the Vehicle Speed–Belgrade Experience. *Saf. Sci.* **2013**, *57*, 303–312. [CrossRef]
11. Arnold, E.; Lantz, K. *Evaluation of Best Practices in Traffic Operations and Safety: Phase I: Flashing LED Stop Sign and Optical Speed Bars*; Final Report; Virginia Transportation Research Council: Charlottesville, VA, USA, 2007.
12. Pau, M. Speed bumps may induce improper drivers' behavior: Case study in Italy. *J. Transp. Eng.* **2002**, *128*, 472–478. [CrossRef]
13. Wayne, D.C.; Naree, K.; Peter, T.; Martin, J.; Perrin, J. Effectiveness of traffic management in Salt Lake City, Utah. *J. Safety Res.* **2006**, *37*, 27–41.
14. Herrstedt, L.; Kjemtrup, K.; Borges, P.; Andersen, P. *An Improved Traffic Environment—A Catalogue of Ideas*; Report 106; Road Data Laboratory, Road Standards Division, Road Directorate, Denmark Ministry of Transport: Roskilde, Denmark, 1993.
15. Chen, L.; Chen, C.; Ewing, R.; McKnight, E.; Srinivasan, R.; Roe, M. Safety Countermeasures and Crash Reduction in New York City—Experience and Lesson Learned. *Accid. Anal. Prev.* **2013**, *50*, 312–322. [CrossRef]
16. Forbes, G. *Traffic Management in Rural Settlements*; Final Report; Centre for the Assessment of Road Safety: Burlington, ON, Canada, 2006.
17. van der Horst, A.R.A.; Thierry, M.C.; Vet, J.M.; Rahman, A.F. An evaluation of speed management measures in Bangladesh based upon alternative accident recording, speed measurements, and DOCTOR traffic conflict observations. *Transp. Res. Part F Traffic Psychol. Behav.* **2017**, *46*, 390–403. [CrossRef]
18. Rahman, F.; Kojima, A.; Kubota, H. Investigation on North American Traffic Calming Devices Selection Practices. *IATSS Res.* **2009**, *33*, 105–119. [CrossRef]
19. Kirkpatrick, S. *Tentative Guidelines on the Provision of Speed Breakers for Control of Vehicular Speeds on Minor City*; The Indian Roads Congress: New Delhi, India, 1996.
20. Klyne, M. The effect of edge lining on vehicle speeds through roundabouts and mid-block blisters. *Proc. Inst. Civ. Eng.-Munic. Eng.* **1998**, *127*, 170–173. [CrossRef]
21. Marek, J. Mid-block speed control: Chicanes and speed humps. *ITE J.* **1998**, *68*, 16.
22. Parkhill, M.; Sooklall, R.; Bahar, G. *Updated Guidelines for the Design and Application of Speed Humps*; Annual Meeting and Exhibit: Pittsburgh, PA, USA, 2007.
23. Traffic Engineering South Carolina Department of Transportation. *Traffic Calming Guidelines. . .Reducing the Speed and Volume of Traffic to Acceptable Levels*; Traffic Engineering South Carolina Department of Transportation: Columbia, SC, USA, 2006.
24. Webster, D.; Layfield, R. *Traffic Calming-Sinusoidal H and S Humps*; TRL Report 377; Transport Research Laboratory: Wokingham, UK, 1998.
25. Chadda, H.S.; Asce, M.; Cross, E. Speed (road) bumps: Issue and opinions. *J. Transp. Eng.* **1985**, *111*, 410–418. [CrossRef]
26. Russell, E.; Godavarthy, R. *Mitigating Crashes at High-Risk Rural Intersections with Two-Way Stop Control*; Report No. K-TRAN: KSU-06-4; Bureau of Materials and Research, Kansas Department of Transportation: Topeka, KS, USA, 2010.
27. Gifford, R. *Road Humps Should Be Dug Up: The Case against Parliamentary Advisory Council for Transport Safety*; Parliamentary Advisory Council on Transport Safety: London, UK, 2004.
28. Moreno, A.; Garcia, A.; Romero, M. Speed table evaluation and speed modelling for low-volume crosstown roads. *Transp. Res. Rec.* **2011**, *2203*, 85–93. [CrossRef]
29. Shauna, H.; Keith, K.; Gary, T.; Smith, D. *Temporary Speed Hump Impact Evaluation*; Final Report; Center For Transportation Research and Education: Ames, IA, USA, 2002.
30. Smith, D.J.; Knapp, K.; Hallmark, S. Speed impacts of temporary speed humps in small Iowa cities. In Proceedings of the ITE 2002 Annual Meeting and Exhibit, Philadelphia, PA, USA, 4–7 August 2002.

31. Basil, D. The Impacts of Neighbourhood Traffic Management. Ph.D. Thesis, University of Canterbury, Canterbury, UK, 2012. Available online: http://eprints.uthm.edu.my/2536/ (accessed on 26 July 2023).
32. Gunwoo, L.; Shinhye, J.; Cheol, O.; Keechoo, C. An evaluation framework for traffic claming measures in residential areas. *Transp. Res. Part D Transp. Environ.* **2013**, *25*, 68–76.
33. Heloisa, M.; Barbosa, M.R. A model of speed profiles for traffic calmed roads. *Transp. Res. Part A Policy Pract.* **2000**, *34*, 103–123.
34. Johnson, L.; Nedzesky, A.J. A comparative study of speed humps, speed slots and speed cushions. In Proceedings of the ITE Annual Meeting and Exhibit, Lake Buena Vista, FL, USA, 1–4 August 2004.
35. Namee, S.; Witchayangkoon, B. Crossroads vertical speed control devices: Suggestion from observation. *Int. Trans. J. Eng. Manag. Appl. Sci. Technol.* **2011**, *2*, 161–171.
36. Sayer, I.; Parry, D. *Speed Control Using Chicanes*; Project Report 102; Transport Research Laboratory: Crowthorne, UK, 1994.
37. Aya, K.; Hisashi, K. Effectiveness of speed humps ranged at different intervals considering roadside environment including vehicle speed, noise and vibration. *J. East. Asia Soc. Transp. Stud.* **2011**, *9*, 1913–1924.
38. Brindle, R. Speed-based design of traffic calming schemes. In Proceedings of the 75th ITE Annual Metting and Exhibit, Melbourne, VIC, Australia, 7–10 August 2005.
39. Distefano, N.; Leonardi, S. Evaluation of the Benefits of Traffic Calming on Vehicle Speed Reduction. *Civ. Eng. Arch.* **2019**, *7*, 200–214. [CrossRef]
40. Siti, S.; Hamsa, A.K. A study on the effects of road humps in reducing speed along local roads in residential areas. *Plan. Malays.* **2016**, *XIV*, 55–66.
41. Al-Nassr, E.K. Dynamic consideration of speed control humps. *Transp. Res. Part B Methodol.* **1982**, *16*, 291–302.
42. Bahram, H.; Masoud, S.; Mansour, N. Multiobjective shape optimization of speed humps. *Struct. Multidiscip. Optim.* **2007**, *37*, 203–214. [CrossRef]
43. Hessling, J.; Zhu, P. Analysis of vehicle rotation during passage over speed control road humps. In Proceedings of the International Conference on Intelligent Computation Technology and Automation, Changsha, China, 20–22 October 2008.
44. Khorshid, E.; Alfares, M. Model refinement and experimental evaluation for optimal design of speed humps. *Int. J. Veh. Syst. Model. Test.* **2007**, *2*, 80–99. [CrossRef]
45. D'Apuzzo, M. Some remarks on the prediction of road traffic induced ground-borne vibrations. In Proceedings of the 4th International SIIV Congress, Palermo, Italy, 12–14 September 2007.
46. Mhanna, M.; Sadek, M.; Shahrour, I. Prediction and mitigation of traffic induced ground vibrations in an urban zone. *WIT Trans.* **2011**, *116*, 701–711.
47. Toplak, S.; Sever, D.; Ivanič, A.; Mohorić, M.; Lubej, S. Effect of traffic calming measures on ground-borne vibrations in residential areas—The Slovenian experiences. In Proceedings of the 7th AAAA Congress on Sound and Vibration, Ljubljana, Slovenia, 22–23 September 2016.
48. Krylov, V. Traffic calming and associated ground vibrations. *Proc.-Inst. Acoust.* **1998**, *20*, 41–48.
49. Oke, S.; Salauz, T.; Adeyefaj, O.; Akanbi, O.; Oyawale, F. Mathematical Modelling of the Road Bumps Using Laplace Transformation. *Int. J. Sci. Technol.* **2007**, *2*, 129–141.
50. Fwa, T.F.; Liaw, C.Y. Rational approach for geometric design of speed-control road humps. *Transp. Res. Rec.* **1992**, *1356*, 66–72.
51. Hassen, D.; Miladi, M.; Slim Abbes, M.; Caglar Baslamisli, S.; Chaari, F.; Haddar, M. Road profile estimation using the dynamic responses of the full vehicle model. *Appl. Acoust.* **2019**, *147*, 87–99. [CrossRef]
52. Barbosa, H.M. Impacts of Traffic Calming Measures on Speeds on Urban Roads. Ph.D. Thesis, University of Leeds, Leeds, UK, 1995.
53. Vlahogianni, E. Some empirical relations between travel speed, traffic volume and traffic composition in urban arterials. *Transportation* **2006**, *31*, 110–119. [CrossRef]
54. Pau, M.; Angius, S. Do speed bumps really decrease traffic speed? An Italian experience. *Accid. Anal. Prev.* **2001**, *33*, 585–597. [CrossRef]
55. Riviera, P.; Bassani, M.; Dalmazzo, D. Field investigation on the effects on operating speed caused by trapezoidal humps. In Proceedings of the 90th TRB Annual Meeting, Washington, DC, USA, 15 November 2010.
56. Amirarsalan, M.; Ali, A. Optimization of speed hump profiles based on vehicle dynamic performance modelling. *J. Transp. Eng.* **2014**, *140*, 04014035.
57. Moreno, A.T.; García, A. Use of Speed Profile as Surrogate Measure: Effect of Traffic Calming Devices on Crosstown Road Safety Performance. *Accid. Anal. Prev.* **2013**, *61*, 23–32. [CrossRef]
58. Philip, A.; Weber, P. Toward a north American geometric design standard for speed humps. *ITE J.* **1998**, *70*, 30–39.
59. Sahoo, P. Geometric design of speed control humps in Bhubaneswar City. *Int. J. Civ. Eng. Technol.* **2009**, *1*, 70–73. [CrossRef]
60. Babak, M.; Zohn, R.; Muennig, P. The impact of urban speed reduction programmes on health system cost and utilities. *Inj. Prev.* **2018**, *24*, 262–266.
61. Loprencipe, G.; Moretti, L.; Pantuso, A.; Banfi, E. Raised Pedestrian Crossings: Analysis of Their Characteristics on a Road Network and Geometric Sizing Proposal. *Appl. Sci.* **2019**, *9*, 2844. [CrossRef]
62. Kruszyna, M.; Matczuk-Pisarek, M. The Effectiveness of Selected Devices to Reduce the Speed of Vehicles on Pedestrian Crossings. *Sustainability* **2021**, *13*, 9678. [CrossRef]
63. Orsini, F.; Batista, M.; Friedrich, B.; Gastaldi, M.; Rossi, R. Before-after safety analysis of a shared space implementation. *Case Stud. Transp. Policy* **2023**, *13*, 101021. [CrossRef]

64. Newland, D.E. *An Introduction to Random Vibrations, Spectral & Wavelet Analysis*, 3rd ed.; Dover Civil and Mechanical Engineering, Dover Publications Inc.: Mineola, NY, USA, 2005; pp. 1–477.
65. Wong, J.Y. *Theory of Ground Vehicles*; John Wiley & Sons, Inc.: Hoboken, NJ, USA, 2022; pp. 1–560. [CrossRef]
66. Cebon, D. *Handbook of Vehicle-Road Interaction*; Taylor & Francis, Ed.; CRC Press: Boca Raton, FL, USA, 2000; pp. 1–616. ISBN 9026515545.
67. *ISO 2631-1:1985*; Evaluation of Human Exposure to Whole-Body Vibration—Part 1. International Organization for Standardization: Geneva, Switzerland, 1985.
68. *ISO 2631-3:1985*; Evaluation of Human Exposure to Whole-Body Vibration—Part 3. International Organization for Standardization: Geneva, Switzerland, 1985.
69. *ISO 2631-2:1989*; Evaluation of Human Exposure to Whole-Body Vibration—Part 2. International Organization for Standardization: Geneva, Switzerland, 1989.
70. Dixon, J. *Tires, Suspension, and Handling*; SAE International: Warrendale, PA, USA, 1996.

 *sustainability*

Article

Eight Traffic Calming "Easy Pieces" to Shape the Everyday Pedestrian Realm

Giuseppe Cantisani *, Maria Vittoria Corazza, Paola Di Mascio and Laura Moretti

Department of Civil, Constructional and Environmental Engineering, Sapienza University of Rome, Via Eudossiana 18, 00184 Rome, Italy; mariavittoria.corazza@uniroma1.it (M.V.C.); paola.dimascio@uniroma1.it (P.D.M.); laura.moretti@uniroma1.it (L.M.)
* Correspondence: giuseppe.cantisani@uniroma1.it

Abstract: The need for safe pedestrian movement implies subtracting and modifying space dedicated to vehicles, especially in urban areas. Traffic control measures aim to reduce or modify the width of the carriageway and force the correct use of the space by pedestrians through two approaches: the former is hard and includes physical barriers and the latter is soft and induces psychological fashion effects on the drivers. This paper presents vertical and horizontal devices integrated by landscaping, planting, or other similar works to slow motor vehicle speed, narrow traffic lanes, and/or create smaller distances for pedestrian crossings. Mobility and boundary issues are considered to discuss their warrants and potential impacts. Indeed, the effects of speed or volume treatments should be investigated through a comprehensive multicriteria analysis without overlooking pedestrian level of service, access and connectivity to residents and emergency vehicles, drainage and snow issues, loss of on-street parking lots, and environmental goals in terms of noise and emissions to air reduction.

Keywords: traffic calming; horizontal devices; vertical devices; pedestrian mobility; walkability

check for updates

Citation: Cantisani, G.; Corazza, M.V.; Di Mascio, P.; Moretti, L. Eight Traffic Calming "Easy Pieces" to Shape the Everyday Pedestrian Realm. *Sustainability* **2023**, *15*, 7880. https://doi.org/10.3390/su15107880

Academic Editors: Salvatore Leonardi and Natalia Distefano

Received: 13 April 2023
Revised: 8 May 2023
Accepted: 9 May 2023
Published: 11 May 2023

1. Introduction

Road safety is a major health burden at the international level because it causes 1.35 million deaths [1] and more than 50 million physical injuries and disabilities [2] every year. Road accidents pose a serious issue, especially in urban areas where urbanisation and mobility demand have grown and still heighten the interaction between traffic and the surrounding environment [3]. It implies increasing health costs due to crashes whose risk for road users is higher than citizens' exposure to natural or anthropogenic events [4], even at the district scale [5]. On the other hand, social consequences in terms of the liveability of urban spaces [6], noise [7,8], air pollution [9,10], urban heat island [11], soil contamination and imperviousness [12,13], congestion [14,15], and neighbourhood unsatisfaction [16] cannot be overlooked. Due to increasing environmental attention, a new consciousness is changing how citizens move across the city [17], and slow and light mobility plays a pivotal role in this cultural change promoted after the COVID-19 outbreak [18]. The increasing demand for alternative transportation modes forces radical changes in urban layout [19] from car-centred to proximity-based cities [20]. Proximity among city areas, especially to green spaces and services, shortens spaces and time, and favours the neighbourhood relationship [21]. The transport demand modification impacts urban dynamics and requires transport infrastructures' appropriate layout to ensure comfortable and safe movements for new road users [22]. More specifically, pedestrians and bicycles need proper infrastructures designed for everyday activities in car-centred cities [23,24], with full requirement-meeting when it comes not only to safety, but also comfort, security, and attractiveness, which calls for specific surveys starting from the users' perception of the urban environment they move within [25]. On the road, vulnerable users (i.e., pedestrians, bicyclists, and two-wheeler riders) conflict with vehicles with larger dimensions and masses. Therefore, they are the "fragile" component in the event of a crash [26], although social costs are often underestimated [27]. Mobility of "vulnerable users" needs design criteria that safely places motorised and non-motorised flows (either by physically separating them or having them sharing the same space) [28] and manages conflict points thanks to vehicular speed reduction because the faster the speed of cars, the higher crash risk and severity in crashes. In particular, the difference in mass between pedestrians and motor vehicles causes the former the most severe injury. Indeed, fatality incidence rises from 5% at 20 mph (32 km/h) to 45% at 30 mph (48 km/h), and 85% at 40 mph (64 km/h) [29]. Geometry and kinematics of traffic movements significantly impact the risk management of urban intersections when vulnerable users or light vehicles interact with heavy vehicles [30,31]. Likewise, it is important to associate the geometry of roads with the appropriate vehicular flows. The concept of traffic calming moves from the above and summarises different technical solutions to slow motor vehicles through commercial and residential areas and increase safety for vulnerable users [32]. Traffic is not eliminated but is regulated to reach just origins or destinations and avoid through-traffic (especially due to cruising for parking). The speed drops to less than 30 km/h and reduces the negative impacts of cars on the urban environment [33] and vulnerable users, such as students and elderly people [25]; the ultimate goal is to create pedestrian realms by designing a continuous network of pedestrian paths, with reduced conflict points, made safer by giving priority to non-motorised modes. All of the above is conducive to great benefits, especially from a sustainability point of view since traffic calming not only meets vulnerable users' safety requirements but also: (i) contributes to mitigate noise and air pollution levels (mainly achieved through low-speed levels and slowing down traffic flows which prevent queues, sudden gear changes, accelerations, and decelerations, which cause the emission of harmful exhaust gases and noise); (ii) saves energy (by optimising transport resources and prioritising walking); and (iii) reduces surface consumption by increasing the space for pedestrians at the expenses of that occupied by cars.

Traffic calming is achieved through changes in road design, which involve many different geometrical and functional transformations [34]. In particular, the most common measures include vertical or horizontal deflections to maintain slow motor vehicle speed [35,36], narrowed traffic lanes to create smaller distances for pedestrian crossings [37], blocked or restricted access to avoid conflict points or signals to control the traffic flow [38]. Therefore, they include chicanes, mid-block neck-downs, altered horizontal and vertical street geometry, 45-degree or 90-degree angle parking, widened footpaths, provision of cycle lanes, and lane narrowing.

Pioneered by the *woonerven*, current residential areas with low-vehicular flows (less than 50 vehicles per hour) are virtually all eligible to become such pedestrian realms, provided to be regulated as Zone 30s or similar, depending on the local highway code standards to enforce traffic calming and reduced speed limits.

This study investigates eight traffic calming scenarios to demonstrate the flexibility of typical urban environments where traffic calming measures can reshape four- and three-arm intersections and mid-block links. Each layout is a possible solution for safety problems that solves punctual conflict points (CP), typical crossing areas, and belongs to the urban system in different locations [39]. For each scenario, both technical and functional issues are considered. Different pavement materials have been proposed to differentiate and highlight conflict points for pedestrians and motorised vehicles. The road pavement scenarios assume existing asphalt surfaces, like most Italian roads. Moreover, the analyses focus on the needs of pedestrians, the problems of rainwater disposal, and the infrastructure capacity according to Grava [40]. Each scenario is conceived to be easily replicable at CPs or mid-blocks at the district level and is specifically designed to reduce vehicular speed and ensure smooth driving style, thus reducing air and noise pollution, saving energy, and reshaping the streetscape to increase surface availability for pedestrians.

1.1. Building the Knowledge

Traffic calming initially consolidated in northern Europe and America and soon propagated elsewhere, with several case studies available in the grey and scientific literature; most notably, a series of handbooks and manuals in the 1990s guided how to design and implement traffic calming schemes. Many of these were issued by governmental bodies to provide directions and examples to replicate at the national level, as in the case of France [41], Denmark [42], Spain [43], or in some other cases, were aimed at introducing the novelty of traffic calming by transferring the best practice from abroad [44].

This dissemination process went hand in hand with more study fields, from the development of technical criteria to design traffic calming [45], to the attempt to frame it within a general redesign of the urban built environment [46], or to use traffic calming as one of the tools to improve walkability [47] and foster the sentiment of belonging among communities [48,49]. However, some prior seminal studies on the role of the urban environment in creating liveability by enhancing walking developed within urban planning theories [50–53] affected a large part of the 1990s literature on traffic calming, thus giving rise to more studies focused on pedestrians as main characters in the streetscape and as a modal priority in everyday mobility behaviours [54,55]. Concurrently, the (at that time) rising environmental concerns generated more studies on the problems associated with vehicular traffic (i.e., noise and air pollution and surface consumption) to which traffic calming and non-motorised modes could provide solutions, starting from empirical evidence [45,56,57].

These pioneering studies from the 2000s generated more case studies and theories, also thanks to the consolidation of several regulatory tools (Zone 30s, environmental islands, green zones, etc.) enforcing traffic calming schemes in cities. Moving from the implementation of speed devices [58,59] as an effective control to increase pedestrians' safety and as a prerequisite to increase liveability, traffic calming has evolved into more design concepts like "road diet" (reducing lanes' number or width to accommodate bike lanes or larger sidewalks) [60], "complete streets" (converting mono-modal roads into the multimodal and multipurpose environment) [61], and "superblocks" (redirecting traffic away from intersections to convert these into civic spaces, as in Barcelona, Spain, also with the help of nature-based solutions) [62]. These concepts' common trait lies in the awareness that traffic calming, when associated with design for all, nature-based solutions, place-making theories, and practice, can be not only a solution to safeguard non-motorised modes but an actual catalyst for urban repurposing. Examples abound worldwide, with Europe initially leading [63], and evidence that traffic calming is no longer just a road safety technique but a major requirement in urban design.

At the same time, studies demonstrating benefits in terms of improved air quality and reduced noise are still thriving, based on several case studies [64,65], as well as those stressing improved liveability [66,67], with great potential in the quality of life of the most vulnerable users [66,68].

2. Methods

The eight traffic calming scenarios, which are further elaborated, are built considering the lessons from the literature reported above and can be considered "easy pieces" to be easily adaptable to and designed for vehicular low-flow areas, mostly residential or with moderate mixed land use, typical of many European consolidated urban areas. All of them can be combined, complementing each other, and creating a continuous pedestrian path, enhancing all the benefits of walking and the environmental potential in generating a pedestrian realm. More specifically, the traffic calming "easy pieces" described next are:

- Four-arm intersection with build-outs.
- Road closure with cul-de-sac and mini roundabout.
- Road closure with narrow U-turn cul-de-sac.
- Three-arm intersection with raised crossing.
- Median opening to create a pedestrian refuge.

- Chicane.
- Crossing area with speed humps.
- Four-arm intersection with diverter.

2.1. Four-Arm Intersection with Build-Outs

A four-arm intersection with build-outs is specifically designed to reduce conflict areas and improve safety by creating sidewalk protrusions or "build-outs", thus enabling pedestrians "to see and be seen" when crossing (Figure 1). Build-outs are designed to allow turning manoeuvres according to cars' travel directions. Figure 1 describes the most common symmetric layout with the following design criteria.

A   Limit of the existing sidewalk the protrusion is built upon (Figure 2 as an example).

B   Curb extension (build-out) with bollards to facilitate turning manoeuvres.

C   3.50 m-wide one-way carriageway with protrusions shaped to create a parking lane.

D   Bollards to direct pedestrians toward the safest point for crossing and to avoid jaywalking.

E   Curb reshaped to avoid illegal turning, equipped with low vegetation (h < 50 cm), as in Figure 3; low vegetation can be used, as in B or D and substitute bollards.

F   The crossing area is designed to meet universal design requirements and standards. Sidewalks protrude and restrict the carriageway area so that two drivers have difficulty passing through simultaneously. The ramp slope is designed to enable wheelchair users to cross; likewise, the sidewalk grade is maintained for visually challenged and blind users to avoid walking on ramps. The grade and the zebras are equipped with detectable warning surfaces (tactile tiles and metallic plates or "bubbles" on the zebra flat). Low lamps (pencil-like) complete the public lighting system (cut-off lamps to enable pedestrians to see and be seen).

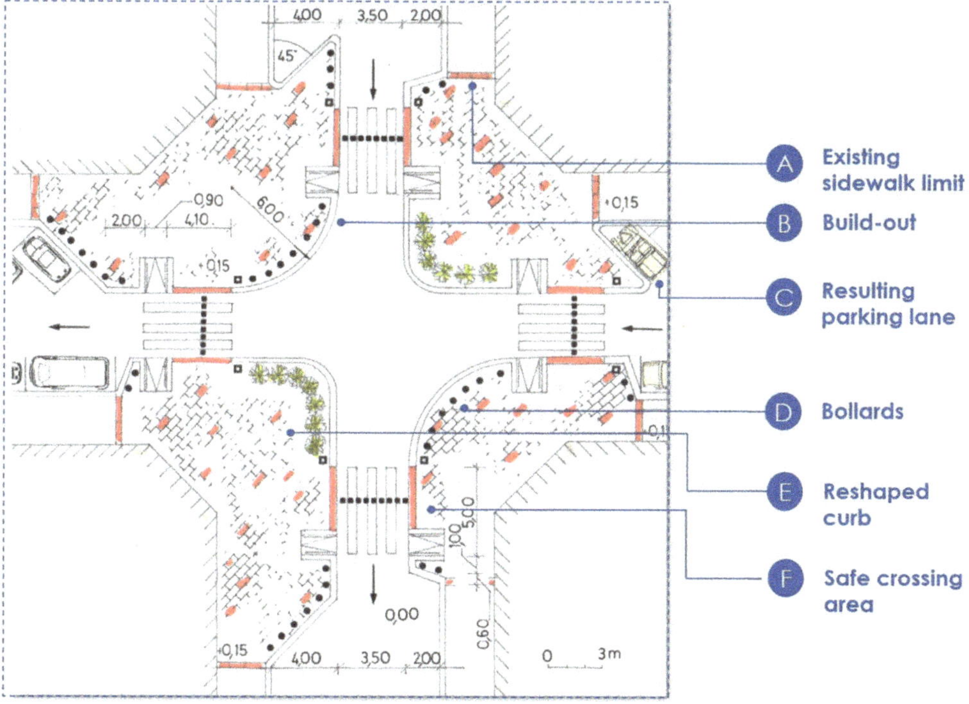

Figure 1. Four-arm intersection with sidewalk advancement (units in m).

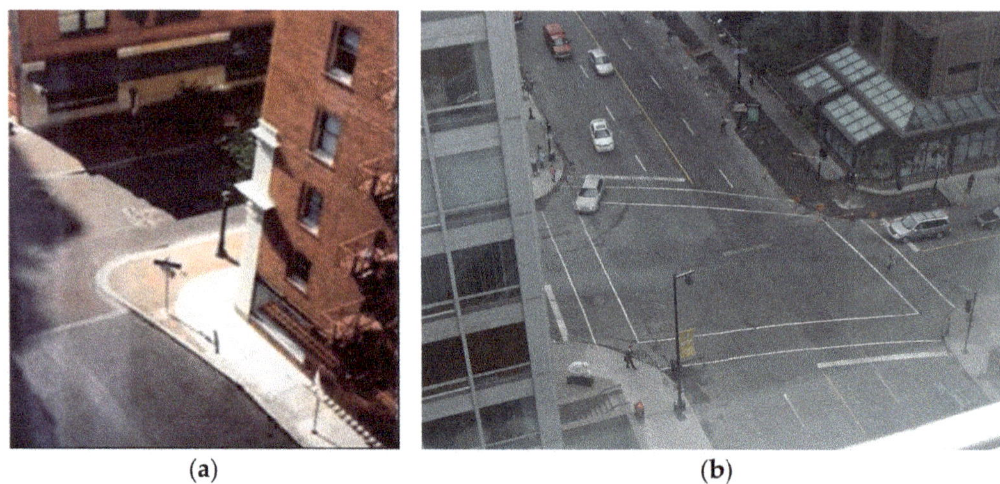

(a)   (b)

Figure 2. Protrusion from existing sidewalks (**a**) with different surface textures and colours, San Antonio, TX, USA; (**b**) under construction, Montreal, Canada.

Figure 3. Low vegetation to shape the edge of the sidewalk, Madrid, Spain.

2.2. Road Closure with Cul-de-sac and Mini Roundabout

The closure of a multi-lane road creates a safe environment for pedestrians, connects two opposite sidewalks via a raised crossing, and compels passenger cars to turn around, thus mitigating through-traffic at the district-level (Figure 4).

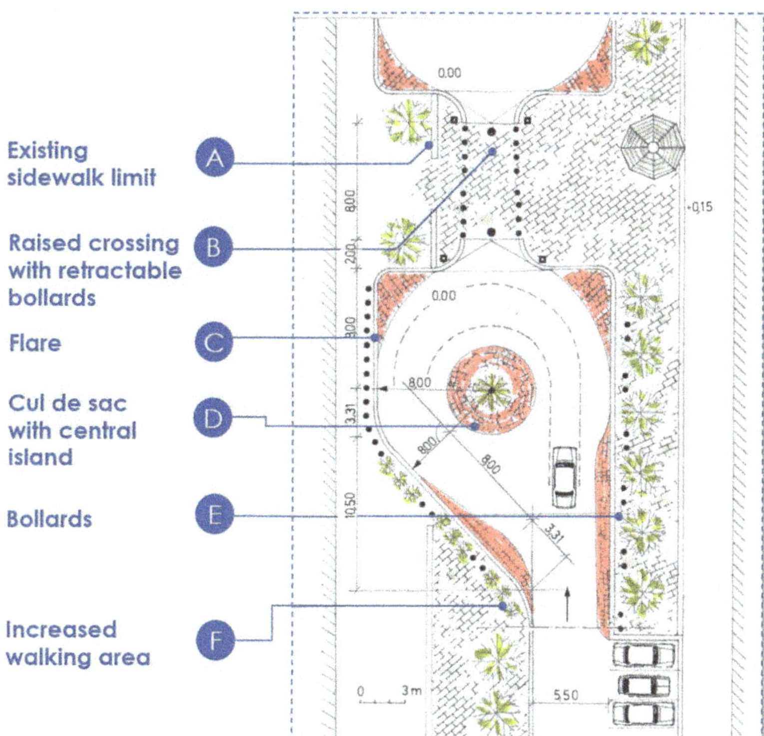

Figure 4. Road closure with cul-de-sac and mini roundabout (units in m).

This layout includes the following:

A   Curb of the existing sidewalk.

B   This raised crossing environment is designed by creating a single raised area (+0.15 m above the carriageway) connecting the two sidewalks [69]. A retractable bollard restricts vehicular access (Figure 5) except for emergency and utility vehicles, which approach the area via a 7%-slope ramp. Emergency vehicles should be considered when vertical devices are used to avoid impediments or unsafe journeys. The texture and the raised area's colours differ from those of the existing sidewalks and carriageways. High vegetation (also potted) to narrow the driver's field of vision, low "pencil-like" lamps spotting the walking area, and cut-off lamps for the whole approaching and raised crossing areas complete the environment [70]. One unique raised area facilitates physically challenged users' crossing operations and reduces the implementation of detectable warning surfaces and equipment.

C   Flares enable large vehicles to turn on narrow carriageways (see D) and prevent irregular parking; textures and colours differ from those of the sidewalks and carriageway.

D   Cul-de-sac with a non-accessible central island (Figure 6). The approach lane is 6.00 m-wide to allow both parallel parking and low-speed travel in both directions (but the width can be reduced to 5.50 m to prevent illegal parking). A turnaround signal turns around the central island (according to reference design criteria for passenger cars provided by the Swiss Standard SN640271a); emergency vehicles can drive through thanks to the retractable bollards in the middle crossing environment.

E   Bollards or vegetation to direct pedestrians to the raised area and avoid illegal parking.

F   Increased walking surfaces by decreasing vehicular lanes (see D).

Figure 5. Retractable bollard, Brussels, Belgium.

Figure 6. Cul-de-sac with a mini roundabout, Vienna, Austria.

2.3. Road Closure with Narrow U-Turn Cul-de-sac

A road closure with a cul-de-sac and mini roundabout (Figure 7) is a simplified version of the previous one. It fits very narrow local roads to reduce through-traffic in a residential area. A two-lane road is closed by simply merging the sidewalks, thus creating a cul-de-sac with the following elements.

A   Limit of the existing sidewalk.

B   Creation of one single raised pedestrian area, as in 2.2 B.

C   Bollards or vegetation to direct pedestrians to the raised area and avoid illegal parking, as in 2.2 E.

D   Cul-de-sac with a narrow U-turn. Two configurations are proposed (i.e., T- and gamma-layouts). The approach carriageway is 5.50 m-wide to allow parallel parking and travel in both directions. Sidewalks can be enlarged thanks to the restricted car lanes (Figure 8). Emergency vehicles can drive through thanks to retractable bollards placed in B.

Figure 7. Road closure with a simple cul-de-sac (units in m).

Figure 8. Enlarged sidewalk, London, United Kingdom.

2.4. Three-Arm Intersection with Raised Crossing

A three-arm intersection with a raised area where pedestrians and vehicles share the road provides benefits in terms of safety and functionality (Figure 9).

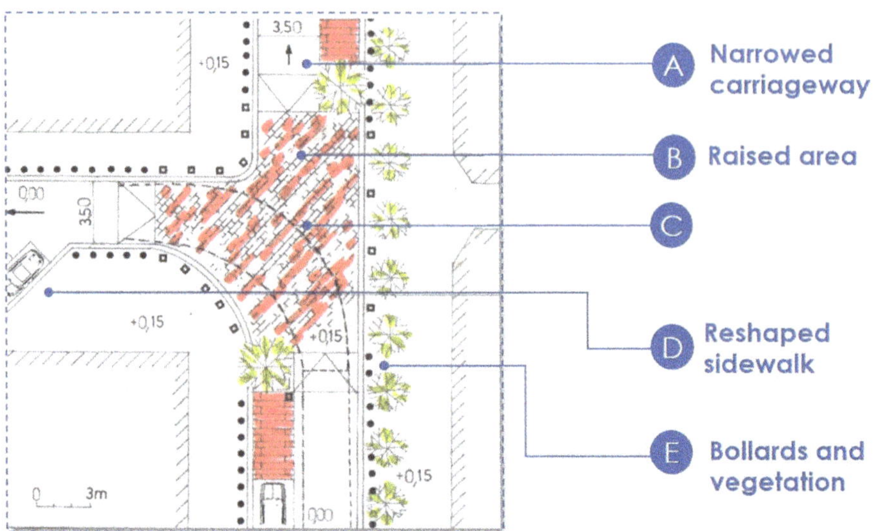

Figure 9. Three-arm intersection with raised crossing (units in m).

This layout can be designed by including the following elements:

A   Narrowed carriageway for cars approaching the raised area. Stone elements (cobblestones, bollards, etc.) narrow the carriageway to 3.5 m and prevent irregular parking (Figure 10) and high vegetation (i.e., trees, also potted) warns drivers that they are approaching a conflict point. The intervention (choker or neckdown) narrows the mouth of the intersection, causing motorists to slow and encouraging pedestrians to cross at the correct location. However, it requires bicyclists to merge with traffic.

B   Raised area creating a shared space among motorised and non-motorised modes, designed as in 2.2 B. Here, pedestrian traffic has the right of way and cars are "guests". To direct visually impaired pedestrians, detectable warning surfaces can be installed; in any case, colours and textures differ from those of the existing infrastructure.

C   Area for large vehicle (emergency or utility ones) turning manoeuvres, designed according to the Swiss Standard SN640271a.

D   Sidewalks reshaped to narrow the carriageways and parking lanes. As in the previous examples, low lighting for pedestrians and general public lighting, equipped with cut-off lamps, complement the layout.

E   Bollards, vegetation, and low-lighting fixtures to direct pedestrians to cross at the safest point and avoid irregular parking (Figure 11).

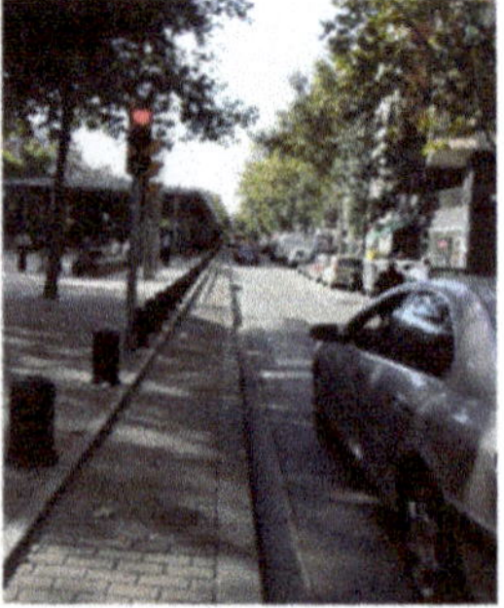

Figure 10. Stone elements restricting the carriageway and/or preventing illegal parking, Barcelona, Spain.

Figure 11. Bollards and vegetation restricting the carriageway and/or preventing illegal parking, Barcelona, Spain.

2.5. Median Opening to Create a Pedestrian Refuge

A central refuge island on a zebra crossing, created by a median opening (Figure 12), reduces the car–pedestrian conflict areas. Moreover, it forces vehicles to reduce speed and breaks the crossing distance for pedestrians into two shorter legs.

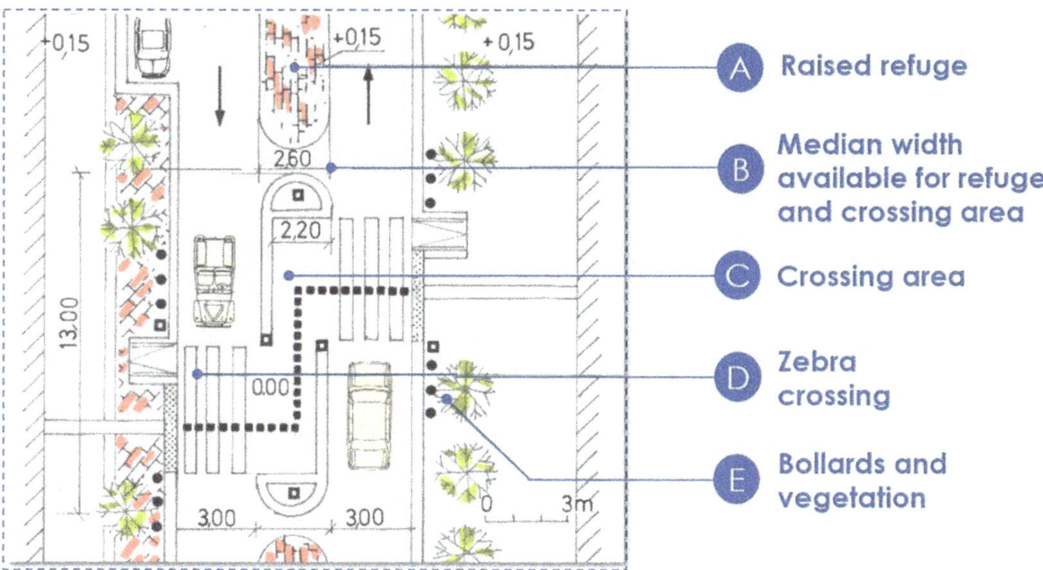

Figure 12. Median opening to create a pedestrian refuge (units in m).

This layout's main elements are:

A   Central refuge for pedestrians. It divides the carriageway into two one-way lanes and enables both pedestrians and bicyclists to wait in a safe area.

B   Median width available to create the crossing point with a refuge (Figure 13) and sidewalk protrusion to reduce the conflict area.

C   Crossing area designed to meet universal design requirements and standards, as in 2.1 F. The refuge enables pedestrians and riders to stop and resume crossing. Drivers are alerted via warning light signals. The whole crossing area is equipped with a cut-off lighting system. For increasing pedestrians and vehicles volumes, zebra crossing can be regulated by traffic lights from the pelican crossing (i.e., the standard pedestrian crossing with traffic lights) to the toucan crossing (i.e., a crossing shared by pedestrians and cyclists with traffic lights, which modifies the crossing phase depending on the presence of the people crossing), to puffin crossing (i.e., pedestrian crossing with traffic lights on call and phase regulated by an infrared system).

D   Zebra crossing equipped with a ramp for physically challenged pedestrians (typically, wheelchair users) and a step with detectable warning surfaces (i.e., the curb of the sidewalk) for the visually challenged ones. Metallic flat bollards or "bubbles" determine the centre line of the available space. The crossing approaching areas of the sidewalks are also equipped with detectable warning surfaces.

E   Bollards to direct pedestrians towards the designated crossing point and to avoid jaywalking, as D in 2.1.

Figure 13. Crossing area with a pedestrian refuge, London, United Kingdom.

2.6. Chicane

By designing a sharp double bend on the carriageway (Figure 14), it is not only possible to force drivers to slow down but to create a safer crossing point and enlarge the available surface for pedestrians. The geometric layout limits bicyclists' use of the road and implies the potential loss of on-street parking lots.

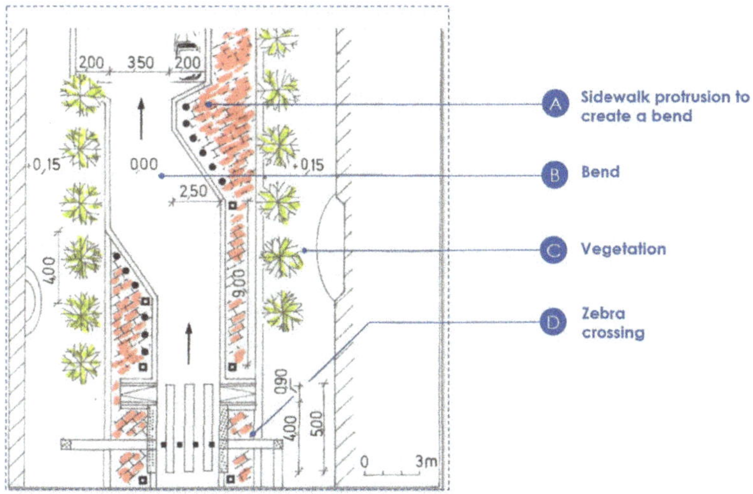

Figure 14. A chicane (units in m).

Many of the elements have already been cited in the previous examples and are resumed here as follows:

A   Sidewalk protrusions to create the bends and the parking lane and reduce the crossing distance. Shaped as in 2.3D and Figure 8 with colour and texture differing from the existing infrastructures.

B   Bends to narrow the carriageway and accommodate both passenger and large vehicles according to Swiss Norm SN640284. Bollards are placed to direct pedestrians towards the crossing area (Figure 15) and to avoid jaywalking and illegal parking. The crossing area can be placed at the end of the bends.

C   Vegetation to shape the enlarged sidewalks into two approaching areas for crossing or to shape the bends (Figure 16).

D   Zebra crossing designed as D in 2.5.

Figure 15. A chicane, London, United Kingdom.

Figure 16. A chicane with flowerbeds, San Francisco, CA, USA.

2.7. Crossing Area with Speed Humps

Simple zebras can be shaped into safer crossing areas by introducing speed humps to slow vehicular traffic (Figure 17). Speed humps are 10 cm-high and 5 m-long on average.

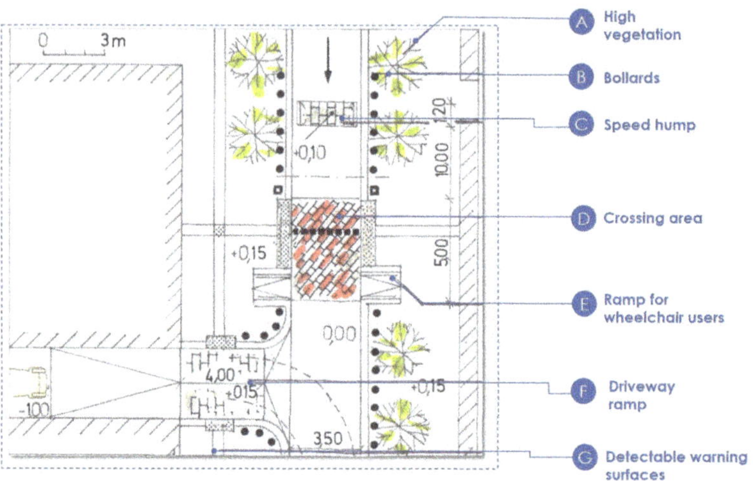

Figure 17. Crossing area with speed humps (units in m).

The main elements are:

A   High vegetation to alert pedestrians and drivers of the pedestrian area. The existing vegetation can be integrated with different shapes, foliage texture, and colours of the new greenery. Typically, trees with edges or flower beds are combined, mixing perennials with deciduous trees in line with the local climate and natural conditions.

B   Bollards to direct pedestrians towards the crossing area (as in 2.3 E, for example).

C   Speed hump (Figure 18) with possible differing colour and texture from the carriageway to slow down vehicular traffic ahead of the crossing area. As an alternative to speed humps, Berlin speed cushions are vertical devices that reduce passenger car speeds without interfering with mopeds and motorcycles [71], allowing large vehicles to pass unaffected. On the other hand, speed tables are longer than speed humps and flat-topped, with a height of 7–12 cm and a length of 6–7 m; they are suitable for collector streets and transit or emergency response routes.

D   Crossing is designed according to the universal design criteria, as in 2.1 F.

E   Ramp for wheelchair users designed according to Italian Norm 506/93.

F   Driveway ramp (slope < 10%) to create a raised area and avoid grades for pedestrians (Figure 19). A change of surface colour and texture can alert pedestrians that they are approaching a conflict point. Detectable by drivers, warning surfaces and bollards (Figure 18) increase the safety level and avoid illegal parking or jaywalking.

G   Detectable warning surfaces to provide path guidance.

Figure 18. Crossing area with speed humps, London, United Kingdom.

Figure 19. Driveway ramp with bollards, Brussels, Belgium.

2.8. Four-Arm Intersection with Diverter

A central raised area turns a four-arm intersection into two turning bends for vehicles with a fully walkable strip for pedestrians, which also substitutes zebras for crossing (Figure 20 represents a diagonal diverter). Traffic volume and speed are reduced, through-traffic is avoided, and the surface for pedestrians is enlarged. The elements included in this layout are:

A One-way, 3.50 m-narrowed car lane; the carriageway's remainder area is turned into sidewalks and the parking lane. Bollards and vegetation narrow the carriageway.

B Raised area (+0.15 m) shared by vehicles and pedestrians and designed as in 2.2 B.

C High vegetation (also potted) to alert drivers to the turning areas.

D Central diverter to direct vehicular traffic (Figure 21), designed according to Swiss Standard SN 640282. The diverter also avoids illegal parking near the intersection. It is equipped with retractable bollards to enable emergency and utility vehicles to pass through if need be. Different islands or curbed closures can be designed to prevent through or turning movements (e.g., diagonal, star, truncated, and forced turn measures).

E As shown in the previous layouts, a lighting system with low lamps for pedestrians and cut-off lamps for the area.

F Area for large vehicle (emergency or utility ones) turning manoeuvres, designed according to the Swiss Standard SN640271a, as in 2.4 C; note that in case of construction drawings and detailed design phases, it is recommended to use a software simulator.

Figure 20. Raised crossing with a diverter (units in m).

Figure 21. Diverter, Vancouver, B.C., Canada.

3. Results and Discussions

Traffic calming devices should be part of an urban system uniquely identifiable by drivers and pedestrians, repeatable, and integrated with similar systems in other city areas. Under such conditions, a set of traffic calming elements can ensure its effectiveness if properly located. Table 1 summarises traffic calming warrants in regard to mobility issues. They are useful for choosing the best option, although it depends on the study area [72] and the initial vehicle and pedestrian volume [73].

Table 1. Traffic calming requirements—mobility issues.

Road Type	Pre-Scenario Motor-Vehicle Volume (veh/h/lane)	Pedestrian Volume (vol/4 h)	Before Speed Limit of Approaching Traffic (km/h)	Emergency Path	Suggested Traffic Calming Measures
Pure residential roads	up to 350	<50	50		Four-arm intersection with build-outs
	up to 150	>500	30		Road closure with cul-de-sac and mini roundabout
	up to 150	>500	30		Road closure with simple cul-de-sac
	up to 350	>500	40		Four-arm intersection with diverter
	up to 500	250–400	30		Chicane
Moderately mixedland use roads	over than 400	250–400	50		Three-arm intersection with raised crossing
	up to 550	<50	50		Median opening to create a pedestrian refuge
Collector roads	up to 550	250–400	30		Crossing area with speed humps
Key to colours	No adaptation needed	Carriageway to be adapted	Carriageway and sidewalks to be redesigned		

Vertical devices are not suitable before driveways, on approaches to intersections, on more than 8% slope branches, or where traffic calming measures would not be visible. Berlin cushions can substitute speed humps to balance the reducing speed goal and the needs of emergency vehicles. In regard to the locations type, pedestrian refuge islands and speed humps are suitable to spot location measures; diverters, speed humps, and raised tables are suited for intersections, while horizontal measures (e.g., reducing the number of lanes, carriageway narrowing, partial- or full-street closure) are suitable for roadways. With variable boundary conditions, all the tools are appropriate for streets within a district. Nevertheless, narrow-lane roads should be avoided on primary emergency vehicle routes and no-parking areas next to fire hydrants should be designed to allow fire trucks' operation.

Table 2 lists the potential impacts of traffic calming tools, focusing on the road and urban functions; the last column on noise and air pollution mitigation is assessed by considering potential benefits from improved driving styles associated with low vehicular speed (fewer accelerations and decelerations). Green boxes refer to positive or absent interaction between the measure and the investigated variable, yellow boxes refer to critical or to-be-investigated interactions between the tool and the boundary parameter, and red boxes refer to negative impacts.

Table 2. Traffic calming impacts on urban functions and the environment.

Traffic Calming Measure	Run-Off Water Management	Loss of Parking	Snow Removal Operations	Presence of Schools or Playgrounds	Noise and Air Pollution Mitigation
Four-arm intersection with build-outs	Green	Green	Green	Green	Orange
Road closure with cul de sac and mini roundabout	Yellow	Yellow	Yellow	Green	Green
Road closure with simple cul de sac	Yellow	Yellow	Yellow	Green	Green
Three-arm intersection with raised crossing	Orange	Green	Green	Yellow	Yellow
Median opening to create a pedestrian refuge	Green	Green	Green	Orange	Orange
Chicane	Green	Orange	Yellow	Yellow	Yellow
Crossing area with speed humps	Yellow	Yellow	Orange	Yellow	Orange
Four-arm intersection with diverter	Orange	Green	Yellow	Yellow	Yellow

Key to colours:
- Green — No interaction or interaction positively impacting
- Yellow — Interaction requiring improvements or adaptation
- Orange — Interaction negatively impacting

Vertical devices (i.e., speed humps or cushions) are permanent measures mainly used to reduce vehicle speed and traffic volume and to increase the users' safety. They are generally made of hot rolled asphalt that requires attention during construction because their shapes depend on manual activities. On the other hand, temporary solutions made of rubber models can be removed for winter operations. Whatever the material, Berlin speed cushions ensure good road drainage and make it easier for cyclists to pass.

Horizontal devices require more building efforts to be implemented than vertical ones and are often limited by pre-existing geometries and adjacent land uses. They differ in traffic volume and speed reduction; horizontal markings can increase their effectiveness [74]. Chicanes imply a reduction in road volume because they discourage shortcutting and through-traffic; they are not recommended on bike routes and are ineffective on low-volume roads. Curb extensions do not affect the traffic volume, increase pedestrian visibility, and prevent irregular parking at intersections. As for pedestrian refuges, their narrower section reduces crossing distance for pedestrians and causes a potential loss of parking. Barriers across the intersection (e.g., diverters and closures) force traffic to turn on adjacent streets and lengthen trips. Environmental-energy indicators highlight the negative results of traffic calming devices on high-traffic roads [75]. Both air emissions, particularly PM_{10} from diesel vehicles, and energy consumption increase; noise is a related defect in speed humping [76]. Therefore, such measures can provide environmental benefits only if they force a reduction in traffic activity, such as partial or total road closure [77]. It is difficult to value construction and maintenance costs because they depend on materials and landscaping.

Table 3 lists the geometrical and functional properties of the pedestrian infrastructure after implementing the investigated traffic calming measures. In particular, the net width of the sidewalk refers to the actual walkable space for pedestrians in Figures 1, 4, 7, 9, 12, 14, 17 and 20, and it recognises restrictions and protrusions due to curbs, urban furniture, and buildings, according to [40]. On the other hand, the pedestrian flow has been obtained according to [78], and the level of service has been obtained from the pedestrian flow characteristics in [79].

Table 3. Geometrical and functional properties of traffic calming strategies.

Traffic Calming Measure	Walking Area Net Width (m)	Flow (ped/min/m)	Level of Service
Four-arm intersection with build-outs	1.5	32	D
Road closure with cul de sac and mini roundabout	2	32	D
Road closure with simple cul de sac	2	32	D
Three-arm intersection with raised crossing	4.5	40	D
Median opening to create a pedestrian refuge	3	32	D
Chicane	1.5	16	B
Crossing area with speed humps	2	23	C
Four-arm intersection with diverter	9.5	32	D

Finally, Table 4 summarises the direct (D) and indirect (I) effects of the investigated measures.

Table 4. Direct and indirect effects of the investigated traffic calming measures.

Traffic Calming Measure	Effects					
	Reduction in Through-Traffic	Speed Reduction	Clarity of the Residential Function	Pedestrians' Safety	Environmental Benefits	Appeal to Correctness
Speed hump	I	D			D	D
Berlin cushion	I	D	D	D	D	D
Raised crossing	I	D	D		D	D
Sidewalk protrusion	I	D		D	D	D
Choker	I	D		D	D	D
Chicane	I	D			D	D
Narrowed carriageway	I	D	D	D	D	D
Diverter	D	D	D		D	D
Cul de sac	D	D	D	D	D	D
Zebra crossing		D	D	D	U	D
Toucan crossing		D		D	I	D
Pelican crossing		D		D	I	D
Puffin crossing		D		D	I	D
Pedestrian refuge				D		
Shared area	I	D	D		D	D
Raised crossing	I	D	D	D	D	D

Note: I = indirect; D = direct.

To conclude, the eight scenarios comply with the design criteria reported in the literature sources cited in Section 1.1 [41–45], and the use of materials, vegetation, urban furniture, and public lighting is conceived to go beyond the mere meeting of the safety requirements and create fully liveable environments (i.e., pedestrian realms with what was postulated in [50–55]). Being "easy pieces", they can be enforced in any regulatory tools [58,59] and smoothly introduced in the most advanced design concepts for urban repurposing [60–62], eventually deploying their higher potential to enhance the quality of life of the communities they are designed for [66–68].

4. Concluding Remarks

Traffic calming measures offer solutions for traffic concerns and ensure a safe environment for vulnerable users (e.g., motorists, bicyclists, pedestrians, and residents on neighbourhood streets). Several strategies are currently used, providing examples of physical obstacles that divert traffic horizontally or vertically, reduce lane or carriageway width, or reduce or avoid conflict points between motor vehicles and light ones or vulnerable users.

Nevertheless, in consolidated areas, such measures can conflict with surrounding constraints that cannot be overlooked. Therefore, a multicriteria approach should consider the input data about road type, current traffic volume and speed, and traffic composition without overlooking emergency vehicles, snowploughing, and stormwater drainage. The before–after analysis should predict the transportation effects and the environmental consequences of increased noise and emissions, construction and maintenance works and costs, aesthetics, and boundary constraints. The ongoing process to convert the transportation systems to electric vehicles gives opportunities to air quality improvement even in traffic calming areas. Indeed, environmental disadvantages from vertical tools (e.g., increased noise and pollution) can be prevented.

A further issue to consider is the regulatory support that enables the enforcement of traffic calming. Although, as already observed, grey and scientific literature abound, the crucial element in appropriate design and enforcing traffic calming is the availability of specifications and standards and their framing within sustainable mobility policies. Although standards are available in many European countries, they are still missing in some others, e.g., Italy. This slackens the pace of full-scale enforcement of traffic calming as shared and replicated practice at the urban level, even when regulatory tools that include it among the active tools to improve life quality are in place, typically the Sustainable Urban Mobility Plans. This discrepancy calls for a supranational regulation to overcome differences, pave the way for its full enforcement, and create consolidated practice everywhere in Europe.

Author Contributions: Conceptualization, G.C., M.V.C. and P.D.M.; methodology, P.D.M. and M.V.C.; formal analysis, L.M. and M.V.C.; investigation, P.D.M. and M.V.C.; data curation, L.M. and M.V.C.; writing—original draft preparation, L.M. and M.V.C.; writing—review and editing, G.C., M.V.C. and L.M.; visualization, M.V.C.; supervision, G.C. All authors have read and agreed to the published version of the manuscript.

Funding: This research received no external funding.

Institutional Review Board Statement: Not applicable.

Informed Consent Statement: Not applicable.

Data Availability Statement: The data presented in this study are available on request from the corresponding author.

Conflicts of Interest: The authors declare no conflict of interest.

References

1. Jafari-Khounigh, A.; Iranzad Khiyabani, I.; Nouri, M.; Mahmoodi, J. Epidemiological Pattern of Deaths Caused by Traffic Accidents in East Azerbaijan, Iran. *J. Inj. Violence Res.* **2019**, *11*, 136. [CrossRef]
2. World Health Organization. *Global Status Report on Road Safety 2018*; CC BYNC-SA 3.0 IGO; WHO: Geneva, Switzerland, 2018. Available online: https://www.who.int/violence_injury_prevention/road_safety_status/2018/en/ (accessed on 25 March 2023).
3. Horn, B.E.; Jansson, A.H.H. Traffic safety and environment: Conflict or integration. *IATSS Res.* **2000**, *24*, 21–29. [CrossRef]
4. Coppola, D.P. *Introduction to International Disaster Management*; Elsevier: Abingdon, UK; Butterworth-Heineman: Oxford, UK, 2015.
5. Ragnoli, A.; Corazza, M.V.; Di Mascio, P. Safety ranking definition for infrastructures with high PTW flow. *J. Traffic Transp. Eng.* **2018**, *5*, 406–416. [CrossRef]
6. Al-Mashaykhi, A.A.; Hammam, R.A.; Wahab, M.H.; Rani, W.N.; Jasimin, T.H. Shared Street as A Means of Liveable Urban Space. *IOP Conf. Ser. Earth Environ. Sci.* **2020**, *409*, 0120442020. [CrossRef]
7. Correia, C.R.; Roseland, M. Addressing Negative Externalities of Urban Development: Toward a More Sustainable Approach. *Urban Sci.* **2022**, *6*, 38. [CrossRef]
8. D'Alessandro, F.; Di Mascio, P.; Lombardi, L.; Ridolfi, B. Methodology for the identification of economic, environmental and health criteria for road noise mitigation. *Noise Mapp.* **2022**, *9*, 10–22. [CrossRef]
9. Badach, J. The Potential of Improving Air Quality by Urban Mobility Management: Policy Guidelines and a Case Study. In Proceedings of the 10th International Conference on Future Environment and Energy, Kyoto, Japan, 7–9 January 2020; Volume 581.
10. Schweitzer, L.; Zhou, J. Neighborhood air quality, respiratory health, and vulnerable populations in compact and sprawled regions. *J. Am. Plan. Assoc.* **2010**, *76*, 363–371. [CrossRef]

11. Moretti, L.; Cantisani, G.; Carpiceci, M.; D'andrea, A.; Del Serrone, G.; Di Mascio, P.; Peluso, P.; Loprencipe, G. Investigation of parking lot pavements to counteract urban heat islands. *Sustainability* **2022**, *14*, 7273. [CrossRef]

12. Adobati, F.; Garda, E. Soil releasing as key to rethink water spaces in urban planning. *City Territ. Archit.* **2020**, *7*, 9. [CrossRef]

13. Asabere, S.B.; Zeppenfeld, T.; Nketia, K.A.; Sauer, D. Urbanization Leads to Increases in pH, Carbonate, and Soil Organic Matter Stocks of Arable Soils of Kumasi, Ghana (West Africa). *Front. Environ. Sci.* **2018**, *6*, 119. [CrossRef]

14. Bao, Z.; Ou, Y.; Chen, S.; Wang, T. Land Use Impacts on Traffic Congestion Patterns: A Tale of a Northwestern Chinese City. *Land* **2022**, *11*, 2295. [CrossRef]

15. Wang, M.; Debbage, N. Urban morphology and traffic congestion: Longitudinal evidence from US cities. *Comput. Environ. Urban Syst.* **2021**, *89*, 101676. [CrossRef]

16. Howley, P.; Scott, M.; Redmond, D. Sustainability versus liveability: An investigation of neighbourhood satisfaction. *J. Environ. Plann. Man.* **2009**, *52*, 847–864. [CrossRef]

17. Çolak, S.; Lima, A.; González, M.C. Understanding congested travel in urban areas. *Nat. Commun.* **2016**, *15*, 10793. [CrossRef] [PubMed]

18. De Palma, A.; Vosough, S.; Liao, F. Na overview of effects of COVID-19 on mobility and lifestyle: 18 months since the outbreak. *Transp. Res. Part A Policy Pract.* **2022**, *159*, 372–397. [CrossRef] [PubMed]

19. Corazza, M.V.; Moretti, L.; Forestieri, G.; Galiano, G. Chronicles from the new normal: Urban planning, mobility and land-use management in the face of the COVID-19 crisis. *Transp. Res. Interdiscip. Perspect.* **2021**, *12*, 100503. [CrossRef]

20. Rodríguez-Hernández, K.L.; Narezo-Balzaretti, J.; Gaxiola-Beltrán, A.L.; Ramírez-Moreno, M.A.; Pérez-Henríquez, B.L.; Ramírez-Mendoza, R.A.; Krajzewicz, D.; Lozoya-Santos, J.d.-J. The Importance of Robust Datasets to Assess Urban Accessibility: A Comparable Study in the Distrito Tec, Monterrey, Mexico, and the Stanford District, San Francisco Bay Area, USA. *Appl. Sci.* **2022**, *12*, 12267. [CrossRef]

21. Jansson, M. Green space in compact cities: The benefits and values of urban ecosystem services in planning. *Nord. J. Archit. Res.* **2014**, *2*, 139–160.

22. Klanjčić, M.; Gauvin, L.; Tizzoni, M.; Szell, M. Identifying urban features for vulnerable road user safety in Europe. *EPJ Data Sci.* **2022**, *11*, 27. [CrossRef]

23. Callari, T.; Tirello, G.; Occelli, C.; Re, A. Infrastructure redesign to improve vulnerable road users' safety. In *Proceedings of the Human Factors and Ergonomics Society Europe Chapter 2013 Annual Conference*; de Waard, D., Brookhuis, K., Wiczorek, R., di Nocera, F., Brouwer, R., Barham, P., Weikert, C., Kluge, A., Gerbino, W., Toffetti, A., Eds.; Human Factors and Ergonomics Society: Washington, DC, USA, 2014. Available online: http://hfes-europe.or (accessed on 25 March 2023).

24. Martínez-Buelvas, L.; Rakotonirainy, A.; Grant-Smith, D.; Oviedo-Trespalacios, O. A transport justice approach to integrating vulnerable road users with automated vehicles. *Transp. Res. Part D Transp. Environ.* **2022**, *113*, 103499. [CrossRef]

25. Corazza, M.V.; Di Mascio, P.; D'Alessandro, D.; Moretti, L. Methodology and evidence from a case study in Rome to increase pedestrian safety along home-to-school routes. *J. Traffic Transp. Eng.* **2020**, *7*, 715–727. [CrossRef]

26. Kirolos, H.; Alluri, P.; Gan, A. Analyzing pedestrian crash injury severity at signalized and non-signalized locations. *Accid. Anal. Prev.* **2015**, *81*, 14–23.

27. Sgarra, V.; Di Mascio, P.; Corazza, M.V.; Musso, A. An application of ITS devices for powered two-wheelers safety analysis: The Rome case study. *Adv. Transp. Stud.* **2014**, *B33*, 85–96. [CrossRef]

28. Di Mascio, P.; Fusco, G.; Grappasonni, G.; Moretti, L.; Ragnoli, A. Geometrical and functional criteria as a methodological approach to implement a new cycle path in an existing urban road network: A case study in Rome. *Sustainability* **2018**, *10*, 2951. [CrossRef]

29. Tranter, P.J. Speed Kills: The Complex Links Between Transport, Lack of Time and Urban Health. *J. Hered.* **2010**, *87*, 155–166. [CrossRef] [PubMed]

30. Cantisani, G.; Durastanti, C.; Moretti, L. Cyclists at roundabouts: Risk analysis and rational criteria for choosing safer layouts. *Infrastructures* **2021**, *6*, 34. [CrossRef]

31. Cantisani, G.; Moretti, L.; De Andrade Barbosa, Y. Risk analysis and safer layout design solutions for bicycles in four-leg urban intersections. *Safety* **2019**, *5*, 24. [CrossRef]

32. Hass-Klau, C. *An Illustrated Guide to Traffic Calming: The Future Way of Managing Traffic*; Friends of the Earth: London, UK, 1990.

33. Distefano, N.; Leonardi, S. Evaluation of the Benefits of Traffic Calming on Vehicle Speed Reduction. *Civ. Eng. Arch.* **2019**, *7*, 200–214. [CrossRef]

34. Damsere-Derry, J.; Lumor, R.; Bawa, S.; Tikoli, D. Effects of Traffic Calming Measures on Mobility, Road Safety and Pavement Conditions on Abuakwa-Bibiani Highway. *Front. Sustain. Cities* **2020**, *2*, 26. [CrossRef]

35. Loprencipe, G.; Moretti, L.; Pantuso, A.; Banfi, E. Raised pedestrian crossings: Analysis of their characteristics on a road network and geometric sizing proposal. *Appl. Sci.* **2019**, *9*, 2844. [CrossRef]

36. Paszkowski, J.; Herrmann, M.; Matthias, R.; Szarata, A. Modelling the Effects of Traffic-Calming Introduction to Volume–Delay Functions and Traffic Assignment. *Energies* **2021**, *14*, 3726. [CrossRef]

37. Distefano, N.; Leonardi, S. Evaluation of the Effectiveness of Traffic Calming Measures by SPEIR Methodology: Framework and Case Studies. *Sustainability* **2022**, *14*, 7325. [CrossRef]

38. Elvik, R. Area-wide urban traffic calming schemes: A meta-analysis of safety effects. *Accid. Anal. Prev.* **2001**, *33*, 327–336. [CrossRef] [PubMed]

39. Gonzalo-Orden, H.; Pérez-Acebo, H.; Linares Unamunzaga, A.; Rojo Arce, M. Effects of traffic calming measures in different urban areas. *Transp. Res. Procedia* **2018**, *33*, 83–90. [CrossRef]
40. Grava, S. *Urban Transportation Systems. Choices for Communities*; McGraw-Hill: New York, NY, USA, 2003; pp. 1–12.
41. Greze, F. *Villes Plus Sure, Quartiers Sans Accidents*; CERTU: Lyon, France, 1994.
42. Rosbach, O. *Miljøprioriterede Gennemfarter, Rapport n. 70*; Vejdirektorate: Copenhagen, Denmark, 1996.
43. Sanz Alduan, A. *Calmar el trafico*; Ministerio Obras Publicas, Transportes y Medio Ambiente: Madrid, Spain, 1996.
44. Corda, G.P. *Disegno di Strade e Controllo del Traffico*; Hoepli: Milano, Italy, 1994.
45. Huesler, W. *Langsamer und fluessiger fahren, Bericht 61 des NRP "Stadt und Verkehr"*; NRP Report; NRP: Zuerich, Switzerland, 1994.
46. Prinz, D. *Staedtbau: Staedbauliches Entwerfen*; Kohlhammer: Stuttgart, Germany, 1996.
47. Gehl, I.; Gemzøe, L. *Public spaces Public Life*; The Danish Architectural Press: Copenhagen, Denmark, 1996.
48. Engwicht, D. *Reclaiming Our Cities and Town*; New Society Publishers: Philadelphia, PA, USA, 1993.
49. Gunnarsson, O. Problems and needs of pedestrians. *IASST Res.* **1995**, *19*, 47–57.
50. Appleyard, D. *Liveable Streets*; University of California Press: Berkeley, CA, USA, 1982.
51. Lynch, K. *The Image of City*; MIT Press: Cambridge, MA, USA, 1960; ISBN 9780262120043.
52. Cullen, G. *Townscape*; Architectural Press: New York, NY, USA, 1961.
53. Alexander, C.; Ishikawa, S.; Silverstein, M. *A Pattern Language*; Oxford University Press: New York, NY, USA, 1977.
54. Hass-Klau, C. *The Pedestrian and City Traffic*; Bellhaven Press: London, UK, 1990.
55. Worpole, K. *Towns for People—Transforming Urban Life*; Open University Press: Buckingham, UK, 1992.
56. Huesler, W. *Flaechensparen im Strassenverkehr, Bericht 29 des Nationalen Forschungprogrammes "Boden"*; Liebefeld: Bern, Switzerland, 1989.
57. McEwan, S. *Environmental Policy and Soft Modes: An Affordable Choice*; Domping: London, UK, 1997.
58. Kruszyna, M.; Matczuk-Pisarek, M. The Effectiveness of Selected Devices to Reduce the Speed of Vehicles on Pedestrian Crossings. *Sustainability* **2021**, *13*, 9678. [CrossRef]
59. Riel, J. Urban Compatible Roads and Traffic Calming. In *Recent Advancements in Civil Engineering*; Laishram, B., Tawalare, A., Eds.; Lecture Notes in Civil Engineering; Springer: Singapore, 2022; Volume 172, pp. 753–763. [CrossRef]
60. Noland, R.B.; Gao, D.; Gonzales, E.J.; Brown, C. Costs and benefits of a road diet conversion. *Case Stud. Transp. Policy* **2015**, *3*, 449–458. [CrossRef]
61. Hui, N.; Saxe, S.; Roorda, M.; Hess, P.; Miller, E.J. Measuring the completeness of complete streets. *Transp. Rev.* **2018**, *38*, 73–95. [CrossRef]
62. Gearey, M. Nature-based solutions: Barcelona's "sustainable city". *Teach. Geogr.* **2019**, *44*, 7–10.
63. Ewing, R. Traffic calming in the United States: Are we following Europe's lead? *Urban Des. Int.* **2008**, *13*, 90–104. [CrossRef]
64. Jain, A. Air Quality and Pollution Control. In *Climate Resilient, Green and Low Carbon Built Environment*; Jain, A., Ed.; Green Energy and Technology; Springer: Singapore, 2023; pp. 31–43.
65. Masoud, A.R.; Idris, A.; Lovegrove, G. Quality of Life Outcomes of the Smarter Growth Neighborhood Design Principles: Case Study City of Kelowna. In *Proceedings of the Canadian Society of Civil Engineering Annual Conference 2021*; Wallbridge, S., Ed.; CSCE 2021; Lecture Notes in Civil Engineering; Springer: Singapore, 2022; Volume 250, pp. 457–470.
66. Rui, J.; Othengrafen, F. Examining the Role of Innovative Streets in Enhancing Urban Mobility and Livability for Sustainable Urban Transition: A Review. *Sustainability* **2023**, *15*, 5709. [CrossRef]
67. Staricco, L.; Vitale Brovarone, E. Livable neighborhoods for sustainable cities: Insights from Barcelona. *Transp. Res. Procedia* **2022**, *60*, 354–361. [CrossRef]
68. Appleyard, B. Livable streets for schoolchildren: A human-centred understanding of the cognitive benefits of Safe Routes to School. *J. Urban Des.* **2022**, *27*, 692–716. [CrossRef]
69. Gitelman, V.; Carmel, R.; Pesahov, F.; Chen, S. Changes in road-user behaviors following the installation of raised pedestrian crosswalks combined with preceding speed humps, on urban arterials. *Transp. Res. F Traffic Psychol. Behav.* **2017**, *46*, 356–372. [CrossRef]
70. Van Treese II, J.; Koeser, A.K.; Fitzpatrick, G.E.; Olexa, M.T.; Allen, E.J. Drivers' risk perception of roadside trees. *Arboric. J.* **2018**, *40*, 153–161. [CrossRef]
71. Berloco, N.; Coropulis, S.; Intini, P.; Ranieri, V. Effects of Berlin speed cushions in urban restricted speed zones: A case study in Bari, Italy. *Transp. Res. Procedia* **2022**, *60*, 180–187. [CrossRef]
72. Rahman, F.; Sakamoto, K.; Kubota, H. Decision making process of traffic calming devices: A comparative study. *IATSS Res.* **2007**, *31*, 94–106. [CrossRef]
73. Sołowczuk, A. Effect of Traffic Calming in a Downtown District of Szczecin, Poland. *Energies* **2021**, *14*, 5838. [CrossRef]
74. Sheykhfard, A.; Haghighi, F. Evaluation of Effectiveness of Horizontal Pavement Marking on Reduction of Vehicles' Speed. *J. Transp. Res.* **2018**, *15*, 111–122.
75. Jazcilevich, A.; Mares Vázquez, J.M.; Ramírez, P.L.; Pérez, I.R. Economic-environmental analysis of traffic-calming devices. *Transp. Res. Part D Transp. Environ.* **2015**, *36*, 86–95. [CrossRef]
76. Jasiūnienė, V.; Pociūtė, G.; Vaitkus, A.; Ratkevičiūtė, K.; Pakalnis, A. Analysis and evaluation of trapezoidal speed humps and their impact on the driver. *Balt. J.Road Bridge Eng.* **2018**, *13*, 104–109. [CrossRef]

77. Garcia-Castro, A.; Monzon, A.; Valdes, C.; Romana, M. Modelling different penetration rates of eco-driving in urban areas. Impacts on traffic flow and emissions. *Int. J. Sustain. Transp.* **2016**, *11*, 282–294. [CrossRef]
78. Desyllas, J.; Duxbury, E. Axial Maps and Visibility Graph Analysis. In Proceedings of the 3rd International Symposium on Space Syntax, Atlanta, GA, USA, 7–11 May 2001; Volume 27, pp. 27.1–27.13.
79. Pushkarev, B.S.; Zupan, J.M. *Urban Space for Pedestrians Urban Space for Pedestrians*; The MIT Press: Cambridge, MA, USA, 1975.

 *sustainability*

Article

Sustainable Complete Streets Design Criteria and Case Study in Naples, Italy

Alfonso Montella [1,*], **Salvatore Chiaradonna** [2], **Alessandro Claudi de Saint Mihiel** [3], **Gord Lovegrove** [4], **Pietro Nunziante** [3] and **Maria Rella Riccardi** [1]

1   Department of Civil, Architectural and Environmental Engineering, University of Naples Federico II, 80125 Naples, Italy
2   Municipality of Montella, 83048 Montella, Italy
3   Department of Architecture, University of Naples Federico II, 80134 Naples, Italy
4   School of Engineering, University of British Columbia, Okanagan, Vancouver, BC V6T 1Z4, Canada
*   Correspondence: almontel@unina.it or alfonso.montella@unina.it; Tel.: +39-3389996765

Abstract: Background: A growing number of communities are re-discovering the value of their streets as important public spaces for many aspects of daily life, creating the need for a transformation in the quality of those streets. An emerging concept of 'complete streets' is to accommodate all users of the transportation system. Methods: In this paper, we present sustainable complete streets design criteria that integrate complete streets by adding socio-environmental design criteria related to the aesthetics, environment, liveability, and safety. To help set priorities, identify the street design features, and create intuitive multimodal networks throughout the city, we have defined a list of the general and specific criteria to be addressed for sustainable complete streets. Results: The proposed design criteria provide a street network with improvements in its aesthetics, to recover the historical urban character and realize historical area planning goals; the environment, to increase the permeable surfaces, reduce the heat island effect, and to absorb traffic-related air pollution; the liveability, to create a public space destination in the urban landscape; and safety, to improve the safety of all road users. The design scenarios proposed in the study were conceived to help practitioners to consider these context-based uses and design accordingly by gaining knowledge from past experiences to benefit future projects. Conclusions: The case study of the urban rehabilitation of the "Mostra d'Oltremare" area and its cultural and architectural assets in Naples, Italy, highlights the practical application of the proposed criteria and the possibility of using these criteria in other urban contexts.

Keywords: urban streets; aesthetics; environment; liveability; safety; vulnerable road users

Citation: Montella, A.; Chiaradonna, S.; Mihiel, A.C.d.S.; Lovegrove, G.; Nunziante, P.; Rella Riccardi, M. Sustainable Complete Streets Design Criteria and Case Study in Naples, Italy. *Sustainability* **2022**, *14*, 13142. https://doi.org/10.3390/su142013142

Academic Editors: Salvatore Leonardi and Natalia Distefano

Received: 8 August 2022
Accepted: 1 October 2022
Published: 13 October 2022

Publisher's Note: MDPI stays neutral with regard to jurisdictional claims in published maps and institutional affiliations.

1. Introduction

A growing number of communities are re-discovering the value of their streets as important public spaces for many aspects of daily life, thus creating the need for a transformation in the quality of streets. People need streets that are safe to cross or walk along, that offer places to meet, and have a vibrant mix of retail. Additionally, more people must be able to walk and ride bicycles in their neighbourhoods, as communities pursue more sustainable development and lifestyle patterns. As a result, an increasing number of cities are looking at modifying the way they design their streets. Furthermore, there is a need for flexibility in applying the current design guidelines and the use of creative design in addressing site-specific project needs [1]. In North America, an emerging concept is to accommodate all users of the transportation system, be they pedestrians, cyclists, public transit users, or motor vehicle drivers and passengers, a concept that has been variously labelled 'complete streets' and 'walkable thoroughfares' [2–9]. The European Commission has been actively supporting sustainable urban mobility planning (SUMP) over the last years. Its core goal is to improve the accessibility and quality of life by achieving a shift towards sustainable mobility [10]. Recently, several cities have prepared guidelines as well as sets of best practices to support and encourage the tailoring of designs towards the creation of a community that promotes safety, connectivity, and attractiveness through a transportation network that accommodates all modes, all ages, and all abilities (i.e., Alexandria, [11]; Denver, [12]; Edmonton, [13]; London, [14]; Montgomery, [15]). Despite the rapid increase in the interest in the complete streets design criteria and the European Commission's support in sustainable mobility planning, only a few European cities have provided transportation engineers, city planners, urban designers, and architects with policies incorporating such an approach to address the change from the past vehicle-focused roadway design philosophy to user-focused street planning.

At the same time as communities re-discover their streets as public spaces and complete streets design principles, they are also realizing that street design significantly impacts the transport and land use system and overall community sustainability. It has long been understood that how we design our communities impacts the demand for travel and the choice of the travel mode [16,17]. In turn, the resultant travel demand impacts non-renewable energy consumption, pollutant and greenhouse gas emissions, obesity, congestion, collisions, and other factors that impact community [18]. Previous research found an association between promoting active and sustainable mobility (i.e., walking and cycling) with physical and mental well-being [19]. In a "complete street" intervention in Salt Lake City (Utah), a street was renovated to be more supportive of active travel by pedestrians, cyclists, and transit riders. This renovation included five new rail stops, improved sidewalks, bike lanes, and landscaping. After the intervention, residents who started using the complete street were found to have increased their physical activity and reduced their body mass indexes [20]. Another research carried out in Madrid found a 15% reduction of the nitrogen dioxide pollution in just three months after establishing low emission zones [10].

Road safety also benefits from encouraging active modes of transport. The European Traffic Safety Council [21] declared that 9500 people are killed on average on urban roads in the EU. Only in 2017, the people killed in the urban area accounted for 38% of all road deaths. Roughly 70% of those killed on urban roads were vulnerable road users: 39% were pedestrians, 12% were cyclists and 19% were powered-two-wheeler (PTW) riders, whereas car occupants accounted for 25% of all road deaths. Road regenerations and sustainable mobility measures can further contribute to tackling a city's road safety problems and help to reach the EU target of halving the road deaths and serious injuries by 2030 [22].

Therefore, it is of critical importance that all street designs and street reconstruction projects follow complete streets design principles. The need to promote the healthy and safer development of cities has become even more urgent with the outbreak of the COVID-19 pandemic, stressing the importance for policymakers and urban planners to shift their focus to more user-oriented planning and urban renewal [23].

To address these emerging issues, the purpose of this paper is two-fold: (1) to present sustainable complete streets design criteria that integrate the concept of complete streets with the principles of socio-environmental design criteria related to the aesthetics, environment, liveability, and safety; and (2) to present an Italian case study to highlight the practical application of these sets of criteria, which taken together we will label as 'sustainable complete streets' design criteria. The design scenarios that will be provided in this paper have the purpose of showing how the sustainable design criteria may vary by street type and represent possible alternatives to challenge the primary accommodation of automobiles, offering equally safe and comfortable mobility among all users.

Providing practical applications in a series of problematic locations, the paper also aims to seek out opportunities to design and retrofit streets in one of the most complex urban contexts in Naples. The sustainable complete streets design criteria can be employed in retrofits of urban areas and help to enhance the overall community sustainability.

2. Literature Review

The sustainable complete streets are streets that aim to reflect and accommodate all users' needs while incorporating sustainable elements including street trees, landscaping, and environmental and social perspectives, so that people walking, cycling, taking transit, and scooting have the same access to safe and comfortable streets as those driving a motor vehicle. Furthermore, the term sustainable also refers to the mobility provision of active transportation infrastructures ensuring a connective network of transit options, reducing car trips, and promoting healthier active transport choices.

Aesthetics, environment, liveability, and safety are all part of the very complex land use and transport system. The quality of urban life—the space designed for pedestrians, cyclists, and bus traffic together with that of private and commercial vehicles—is one of the major urban street design issues of our time. Collectively, we refer to it as aesthetics or functional ambience [24].

Historically, streets and public squares were designed for symbolic representations of government and/or monarchy power: see for example Washington, D.C.; Paris, France; and Naples, Italy. Streets were defined by building facades and facing activities. Architectural design defined the quality of a building and an urban landscape surrounding it. With most cities now built, the emerging practise has been to search for open spaces to retrofit, thereby helping to improve the quality and ambience of city life with an enhanced sense of place, benefiting local commuters, businesses, and property owners [14].

Today, an increased population and traffic density require streets as a means of connection between city destinations. This evolution in a street's functional ambience, from a local support space to a connector of disparate non-local functions, has frustrated local planning and built-environment design processes. Moreover, the volume and speed of traffic has increased in areas historically designed for much lower traffic demands, eroding the intrinsic benefits of restorative, resting and/or green spaces on streets.

Streets have also needed to deal with the context within which they are located. Since 1986, the United Nations have introduced the principle of incorporating environmental and sustainable practices [25] to space planning. Identified by the term eco-design [26], this approach assumes that any size and shape of the project also impacts the landscape and environment, other than just the society. Thus, it places attention on an economic, optimized, proportionate, and essential design aimed at a particular utility, use or purpose. Moreover, when talking about the design of more sustainable communities, public space considerations often dominate, and lead to urban liveability design discussions on population density, mobility, infrastructure, and green space [27]. Ensuring healthy lives and making cities and human settlements inclusive, safe, resilient, and sustainable are targets that also include road safety (e.g., targets 3.6 and 11.2 [28]). Lowering the number of road user casualties is the key to improve the overall performance of the transport system and to meet the citizens' and companies' needs and expectations. Furthermore, more than half of fatalities are among vulnerable road users (VRUs) such as the riders of mopeds, pedestrians, cyclists, and motorcyclists, who represent an important challenge for road safety [29]. Undesirable road features may represent potential hazards for road users, contributing to future crashes. According to a preventive and systemic approach to road safety, road inspections are essential to identify treatments to improve the safety of the existing network before crashes happen [30–33]. However, safety should be considered in all design stages, and road safety audits should form an integral part of the design process during the draft design, detailed design, and pre-opening and early operation stages [30]. Among the factors contributing to road crashes, speeding or travelling at inappropriate speeds for the road environment, the prevailing weather, light, traffic and road conditions, are considered the most critical [34–39]. Higher speed reduces the available reaction time for drivers and, therefore, creates a greater crash risk. Moreover, a high collision speed is an aggravating factor in all crashes. An increased collision speed is associated with more severe consequences in terms of injury and the material damage. To reduce the volume of auto trips and the exposure to a crash risk, the SMARTer Growth Neighbourhood design [27,40,41] has been used in the urban area as a system that facilitates lower auto volumes, lower vehicle speeds, and a lower conflict frequency between a pedestrian/cyclist/vehicle.

The risk of pedestrian crashes also increases when the crosswalk locations are not in line with the desired pedestrian paths [32] or are not appropriate for the road alignment [42,43]. Parking, cycle facilities, and bus stops may create a significant increase in the pedestrian risk. In some circumstances, parking is allowed very close to the crosswalk, thus creating conflicts between pedestrians and vehicles. Pedestrian accessibility is another important safety issue, including for mothers with baby strollers, seniors with walking aids, wheelchair users, visually impaired users, and tourists with luggage. Crosswalks that are not designed for accessibility have been associated with an increased number in people crossing away from a crosswalk, and not using the adjacent sidewalks at all, further degrading safety [33]. Recently, to encourage a healthy and environmentally sustainable lifestyle, the promotion of bicycle use has been embraced by urban and transport planners; however, there is evidence of an increasing crash risk for cyclists rather than for motorized vehicles [43–45], mainly due to infrastructure characteristics leading to a higher number of conflicts between cyclists and other road users. Cyclists are legitimate users of the roadway and an integral part of the transportation system so that the facilities for cyclists, preferably separated and continuous cycle tracks on both sides of the street, should be considered an integral part of any proper road planning process and network design [46].

An important safety issue is the sight distance at urban intersections, cycle crossings, crosswalks, and even along sidewalks, as there are often obstacles that obstruct the views and distract road users. To warn motorists of hazards, and to regulate and guide traffic (e.g., performing such functions as assigning the right-of-way), traffic control devices are extensively used. Their effectiveness is proportional to their clarity and visibility at any time of the day, in all seasons, and through each weather condition, whereas their deficiencies have been found to be influencing drivers' unsafe behaviours [47,48], contributing to a considerable proportion of crashes. Adequate lighting can improve visibility during the night-time [42,49–51] and this is particularly important to improve the pedestrian/cyclist visibility on or beside roads, as well as on off-road paths, and the visibility of other road users for pedestrians and cyclists. Previous research results encourage the use of light-emitting diode (LED) sources during the night-time as they are effective for increasing drivers' attention [52,53].

3. Methodology

3.1. General Issues

Efforts to transform streets into sustainable complete streets (or from traffic-based to accessibility-based designs) face resistance at times, from both professional communities of traffic engineers and planners and from policy makers and the general public who feel new designs do nothing more than reduce automobile access [9]. However, re-thinking and re-shaping streets may have a significant impact on street functionality providing evidence that streets can do more than just move cars. The proposed sustainable complete streets design criteria aim to provide a transport system in line with the contemporary social, economic, and environmental priorities of urban mobility and access, including care in providing quality and comfortable (i.e., liveable) land uses. The aesthetics, environment, liveability, and safety are all invoked to re-think the streets and to re-design them to meet changing preferences and future needs. The implementation of such design principles may help in [8–15,25–27]:

- Reducing traffic crashes and injuries, of both motorized and vulnerable road users;
- Reducing motorized traffic volume and speed;
- Increasing pedestrian, cyclist, and public transport traffic volume;
- Improving the quality of the urban landscape space, for example, its aesthetics and functional ambience;
- Recovering the urban character lost in a historical evolution due to a lack of conscious, coordinated urban planning;

- Creating a homogeneous road environment with regards to the choice of materials, colours, lighting, and treatment of green spaces;
- Increasing 'green space' opportunities for pedestrians and cyclists to stop, shop, and socialize;
- Reducing non-renewable energy use and its associated air pollution emissions;
- Reducing nuisances due to traffic noise and urban heat island effects.

Nevertheless, the selection of the required actions to shift ordinary streets into streets that serve multiple purposes and multiple modes of transportation is not easy. Based on a literature review and on the most recent guidelines [11–15] that support and encourage tailoring designs towards sustainability and complete streets principles, the general and specific criteria to be addressed are proposed, identified, and briefly explained in Table 1.

Table 1. Synthesis of the sustainable complete streets design criteria.

General Criteria (Section #)	Specific Criteria	Integration with Complete Streets Criteria
Aesthetics (3.2)	A1 Enhancement of the urban environment to recover historical urban character and realize historical area planning goals	Full integration between street design and the planning for redevelopment of the public space
	A2 Integration of materials, colours, lighting, and treatment of green spaces	Sustainable approach to integrate the street design and the built environment
	A3 Durable materials	Requirements for the use of durable pavements, markings, signs, lightings, and street furniture to reduce the unpleasantness caused by deterioration during its life cycle
Environment (3.3)	E1 Eco-design	Life Cycle Design and Life Cycle Assessment of building products
	E2 Planting and trees	Use of green space to form a natural filter and buffer from pollutant gases, dust, and noise as well as to reduce urban heat islands
	E3 Reduction in traffic-related air pollution and noise	Strengthening of the community system design approach to reduce auto trips and promote healthier active transport choices
Liveability (3.4)	L1 Usability and comfort of the public outdoor space system	Attention to the functional and environmental design aspects of public space to ensure that end user and lifecycle impacts are addressed with a view toward sustainability
	L2 Creation of a public space destination in the urban landscape which affects activities and social relations	Coordination between urban planners, landscape architects, and transportation engineers in developing the design of the street as a public space destination in the urban landscape
Safety (3.5)	S1 Road safety inspections	Introduction of formal road safety inspections of the existing network
	S2 Road safety audits	Introduction of formal road safety audit of all plans and designs prior to construction
	S3 SMARTer Growth Neighbourhood design	Safe system design of community land use and transport to reduce auto use and speeding
	S4 Safe pedestrian routes and crosswalks	Specific criteria to provide a comprehensive, safe pedestrian route network
	S5 Safe cycle routes and crossings	Specific criteria to provide a comprehensive, safe, and continuous cycle route network
	S6 Sight distance	Explicit consideration of visibility from the perspective of all roadway users
	S7 Transit and pedestrian accessibility	Link between accessibility and safety
	S8 Traffic control devices	High-performance traffic control devices as explicit safety criterion (e.g., roundabouts)
	S9 Lighting	Use of high-performance lighting as explicit safety criterion (e.g., solar power LEDs)

Practical applications of the sustainable complete streets design criteria are presented with reference to the Mostra project in Italy (Section 4). A survey was made among engineers with relevant expertise in the field of road design and transportation planning and architects with similar experience to understand how they would evaluate the practical application of the sustainable complete streets design criteria in the four specific redesign locations presented in the case study.

In Figure 1 we identify the main steps towards the implementation of sustainable complete streets design criteria.

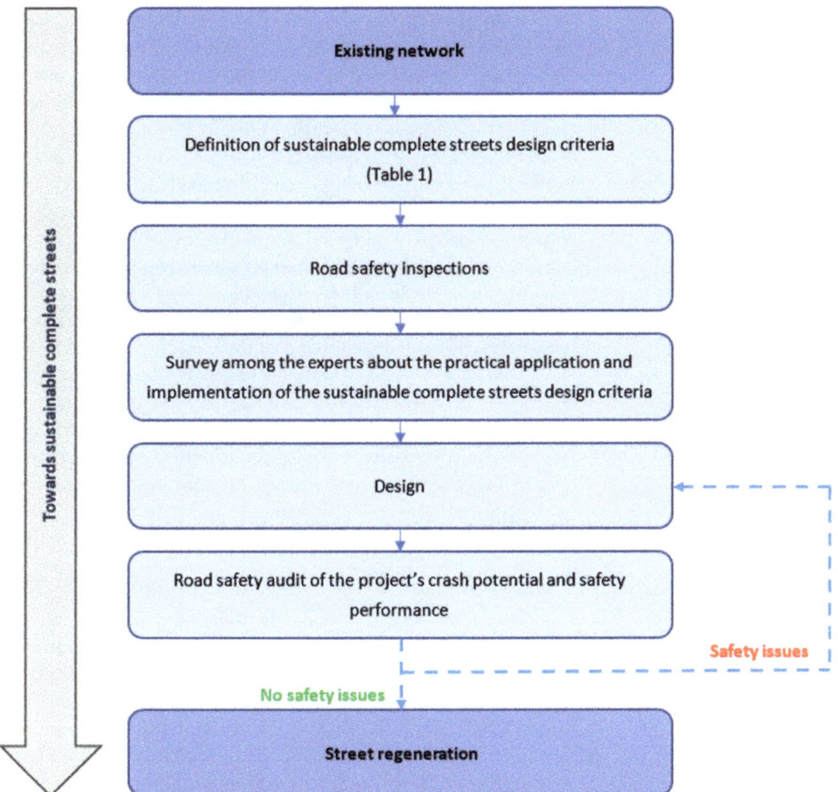

Figure 1. Methodology workflow.

Compared to the existing design criteria discussed in Section 2, the major novelties of the proposed criteria are the integration of sustainability and the complete streets criteria as well as the definition of specific criteria to provide a street network adapted to contemporary requirements for community sustainability, quality of life, and vulnerable road user safety, while providing for traditional urban mobility, the historic context, and efficiency.

3.2. Aesthetics

Planning the redevelopment of public space must begin with defining the principles that will relate to the quality of the open and public areas. Indeed, the aesthetics reflect the quality of urban life and includes the space designed for pedestrians, cyclists, and bus traffic together with that of private and commercial vehicles in a harmonious way [13,14]. The rehabilitation of the public space, including complete street re-design, provides an opportunity for a sustainability-oriented, eco-design approach to project design. For example, it might include the materials selection to maximize the accessibility, sustainability, durability, drainage, and aesthetic appropriateness to be integrated with colours, lighting, green space and ecology in the project [12,13]. However, steps toward realizing this opportunity must deal with the problems that the unruly urban phenomenon of traffic proliferation has generated. No literature demonstrating this eco-design retrofit approach has been found.

3.3. Environment

Environmental criteria describe the relative importance of maximising the positive environmental impacts and minimising the negative environmental impacts in the design of a street [14,25]. The eco-design aims to control the design and assessment of impacts over the entire life cycle of the design 'product'. This eco-design concept of sustainability-based design principles leads to an integration of policy, inputs, and production to ensure green procurement, the durability of materials, green construction, the choice of urban furniture, and high-performance lighting [26].

Green space can be used to form a natural barrier to pollutant gases, dust, and noise as well as to reduce urban heat islands. Additionally, through the choice of appropriate materials it is possible to reduce urban heat effects, such as choosing no-fines paving, and rough or smooth functional strips. As another example, through the road surface material treatment, results can be achieved toward speed reduction for safety, energy and resource savings, in both the up-front and long-term benefits [12,13,26]. Moreover, the reduction in maintenance costs (typically over half of a project's life-cycle costs), can be achieved by careful resource choices made at the design phase of the project [54].

The factors of environmental and architectural integrity of use provide an urban space with different degrees of aesthetic quality. A perception of environmental comfort can be attributed to the perceived factors and can shape positive behaviour. In this sense, the aesthetic quality of the space must be seen increasingly as an integration between perceptual factors, signage, graphic communication and environmental comfort combined with vegetation. Aesthetic integrity represents how well a design solution's appearance and use behaviour can integrate with the multiple functions of an urban space.

3.4. Liveability

Impacting the quality of life, liveability refers to the significance of a street to its users as a social or recreational destination [9,14]. The characteristics of the broader reference context is a common concern of the community's architectural image at both the city-wide and local-area scales, which intersect via local sustainability-oriented designs. For example, consider in broader contexts the relationship between buildings and their environment characterized by the presence of historically established features and activities. Urban planners and landscape architects must work with transportation engineers in developing the design of a street as a public space destination in the urban landscape, not just as a thoroughfare, including how its physical elements affect the activities and social relations in those two (local and city-wide) contexts.

These two non-engineering disciplines rightfully pay attention to public space in terms of [24]:

- Ambience, i.e., the conditions of usability, comfort and safety of the public outdoor space system, including its quantity and quality of green amenities; its environmental and ecological function; and its coherent configuration;
- Function, highlighting the added value of high environmental performance in terms of efficiency, functionality, maintenance, durability and recyclability, including a reduction in energy consumption at the urban and building scale; reductions in air, water, noise and visual pollution; and the mitigation of impacts generated by infrastructure systems.

The attention to these functional and environmental design aspects of public space helps to ensure that user and lifecycle impacts are addressed with a view toward sustainability—pedestrians and cyclists; transit; parking; innovative technologies; renewable energy sources; and the integration of explicit (e.g., signs) and implicit (e.g., surface colouration) information and communications technologies.

Beyond public space as a destination, road (re-)designs must also consider other fundamental functions, including that of moving people. Road network rehabilitation is an opportunity for a wider urban and environmental regeneration, but only where it is possible to have an integrated, team approach within the fields of environment, engineering, and planning [9,13,15]. It means an integrated vision of design measures aimed at improving the quality of urban life, developing approaches that encourage public–private partnerships and involving a wider variety of stakeholders in the process of urban transformation than being only historically engaged [55]. A diverse team for these projects in which complexity reigns is a strategic and success-critical resource to be cherished and supported. In this regard, it is strategically important to identify the most appropriate regulatory and administrative frameworks to implement the configured actions, by not only checking their compatibility, but also proposing innovative and advanced solutions that best meet well-defined community values and visions.

3.5. Safety

To make cities inclusive, safe, resilient and sustainable, safe transportation facilities for all users and mobility provisions have a pivotal role. Safety means lowering the number of road user casualties, improving the overall performance of the transport system and meeting citizens' and companies' needs and expectations. A sustainable complete street intervention should also include a consideration of the pedestrian priority, bicycle priority, and transit priority, when designing the new shape of a street or designing a future street.

3.5.1. Road Safety Inspection

To improve the safety of the existing network, the sustainable complete streets design criteria require routine safety inspections of the road network to identify potential hazards for all road users. These are assessed by measuring the risk in relation to road features that may lead to future crashes, so that the remedial treatments may be implemented before crashes happen according to a preventive and systemic approach to road safety [30–34]. Traditionally, safety improvements have been realised only at hotspots after a high number of crashes have occurred [49,56–59]. If only the hotspots are treated, many locations that can be expected to experience crashes in the future will remain untreated. Waiting for crashes to occur to warrant action carries a high price, as vulnerable road user crashes tend to be severe. Furthermore, vulnerable road user crashes are generally widespread in the road network and under-reported, thus creating difficulties in hotspot identification based only on crash statistics.

3.5.2. Road Safety Audits

The designers should investigate how the road environment is perceived, and ultimately utilized, by each user group—pedestrians, cyclists, and motorists—and within each user group, namely, seniors, children, persons with disabilities, and the drivers of powered two-wheelers, trucks, and cars. Several issues must be considered to ensure a safe design and it is not possible to define specific rules which guarantee a safe design. Standards are an important starting point with any road design and a designer should be familiar with the relevant standards, attempt to comply with them and be aware of where any standard cannot be achieved.

However, road designers will agree that standards do not guarantee safety, for the following reasons:

- Standards are often a minimum requirement and combining a series of minimums can leave no room for error, either on the part of the designer, the builder or the final users;
- Some situations require specific expert judgements, which might not be covered by the standards;
- Individual road elements, designed to standard, may be quite safe in isolation but may be unsafe when combined with other standard elements;
- Driver errors, which contribute to 93% of all crashes [49], are bound to happen.

A road safety audit is a formal examination of a future road or traffic project in which an independent, qualified team reports on the project's crash potential and safety performance [56]. Especially in urban areas, where there are often several competing objectives, a road safety audit ensures that safety is not overlooked in satisfying the many competing and complex objectives.

For urban area design processes, recommended principles to promote safer road use are provided below.

3.5.3. Smarter Growth Neighbourhood Design

Key issues of the SMARTer Growth Neighbourhood design are:

- Roundabouts to reduce conflicts and crashes [60–62];
- Traffic calming to lower speeds and short-cutting on local streets, and improve ped/bike safety;
- Continuous off-road ped/bike paths, such that biking across a neighbourhood is faster than driving;
- Fewer road lanes of narrower width, with lower speed limits to address the crossing distance and safety;
- Comprehensive bicycle and pedestrian route networks in accord with active travel demand and desire lines, integrated with compact, connected, and coordinated mixed land uses;
- Keeping major mobility roads to perimeter one-way couplets containing service commercial blocks and controlled by roundabouts to reduce the crossing risk and speeds;
- Interspersed public, open, green, restorative spaces throughout, such that no dwelling is longer than a one-minute walk away from public open space, with a major central, car-free piazza that promotes both local and community-wide social interaction;
- Convenient, affordable, and accessible, high-capacity public transit connecting this neighbourhood with major civic destinations.

3.5.4. Pedestrian Crosswalks

Crosswalk locations must be evaluated in relation to the road alignment and the desired pedestrian paths and a special emphasis should be given to the interaction with parking, bicycle facilities, and bus stops along major roads. The use of bulb-outs to reduce the interaction between pedestrians, bicycles, bus stops, and parked vehicles is strongly encouraged, such as that in the Dutch bicycle design guidelines [46]. Mid-block crosswalks located after bus stops may induce pedestrians to cross in front of the stopped bus, with a detrimental safety effect due to the visibility of pedestrians obscured by the bus.

Vehicles turning into and out of driveways may conflict with pedestrians walking along roadways. These conflicts can be reduced by a consideration of pedestrians during the planning stages of a project, and by consolidating existing driveways. Removed driveways, or the consolidation of driveways, can create space for other uses such as seating opportunities that might support adjacent businesses, or planting areas that provide buffers between vehicles and pedestrians.

In the case of wider roads, the presence of an adequate median 'block' and/or refuge should be carefully considered. Painted crosswalks aligned with pedestrian desire lines encourage pedestrians to cross within the crosswalk, where drivers are more likely to expect them.

3.5.5. Pedestrian Accessibility

Designers should consider the needs of all road users looking in detail at the accessibility needs in each local context; therefore, the important accessibility issues to consider include [55,63]:

- The presence and slope of curb ramps (e.g., crosswalks without ramps or with ramps with an excessive grade are inaccessible for wheelchairs and baby carriages);
- The height of the curbs separating sidewalks from roads (e.g., an excessive height poses significant problems of accessibility—falls and trips—to older people moving between parking spaces and the sidewalk);
- Rumble-strips and tactile-strips for visually-impaired pedestrians (e.g., ramps without preceding 'warning' strips are a significant hazard for blind people);
- The accessibility of median breaks (e.g., inaccessible breaks make a refuge ineffective).

3.5.6. Cycle Routes

The context of the road for a bicycle facility is a key element that should be considered in a design. Cyclists should be provided with a complete and connected bicycle network, without any horizontal or vertical obstructions (either temporary or permanent) along the facilities, which offers safe routes to destinations. The riding surface should be smooth, stable, free of debris, and with adequate drainage. In many cities, retrofitting to meet complete street principles for bicycles is a difficult problem. A variety of innovative and effective design solutions is emerging in complete streets design guides, including road diets, parallel 'bicycle-friendly' streets, and protected, buffered bicycle lanes [46,64–67].

3.5.7. Sight Distance

An insufficient sight distance increases the crash risk by reducing reaction times and stopping distances. An adequate sight distance provides drivers with sufficient time to identify and appropriately react to all elements of the road environment, including other road users and hazards [68]. Designers should evaluate the visibility from the perspective of all roadway users, especially children and persons in wheelchairs, who may be lower to the ground. The sight lines between all users should be free from obstructions. Typical obstructions the designer must deal with carefully include lighting and traffic signal posts, traffic control and bus stop posts, traffic signal controller kiosks, utility kiosks, newspaper kiosks, litter and recycling kiosks, and trees.

3.5.8. Traffic Control Devices

High-quality road signs and road markings are crucial to support drivers [55]. As regards the road signs, the use of fluoro-reflective, micro-prismatic sheets is recommended. Fluoro-reflective, micro-prismatic sheeting allows for [69,70]:

- Greater daytime visibility;
- Greater night-time visibility;
- Greater visibility in the presence of light pollution (e.g., fog);
- Optimum angularity;
- Excellent visibility at wide entrance and observation angles;
- Greater visibility in critical conditions, such as in rain, snow, cloudy weather, at dawn, and at dusk.

As regards the markings, it is recommended to use high performance markings such as cold-hardened materials, which provide for enhanced safety through effective daylight and night-time visibility, as measured by their luminance coefficient under diffuse illumination, and by retro-reflectivity (i.e., their ability to reflect light from a vehicle's headlights back to a driver's eye) [71].

3.5.9. Lighting

Optimal lighting allows for improved perception and visibility of pedestrians, the mutual sighting of vehicles, the right perception of the road environment, and the visibility of potential hazards.

To obtain the best safety and energy performances, the use of LEDs is recommended. They are solid-state devices that produce blue light in combination with phosphors that convert some of the blue light to yellow light, with the resulting mixture appearing white. LEDs allow for very long-rated lives and a good lumen maintenance. Some of the main advantages of LEDs are a more uniform light distribution; lack of warm-up time; energy savings; reduction in the frequency of maintenance; directional light; reduced light pollution; environment-friendly characteristics; and breakage and vibration resistance [72].

An important safety requirement is to provide an adequate pedestrian visibility distance at crosswalks, defined as the distance at which a driver can see a pedestrian well enough to be able to respond appropriately to the pedestrian's presence. The greater the visibility distance, the more time a driver will have to react to the pedestrian before a conflict occurs [52]. In some instances, energy-saving, automated crosswalk lighting can cause hazards if triggered too late after a pedestrian is present. Overly delayed crosswalk lighting activation can restrict the visibility of crossing pedestrians; therefore, the timing of lighting activation is a safety-critical design issue.

4. Case Study

4.1. Overview

The Campania Region in Italy was chosen as part of the EU's Regional Operational Programme framework to identify major projects in the fields of sustainable transport, environment, infrastructure, and tourism. On 28 March 2011, Resolution 122 of the Regional Council established several major projects in the city of Naples, including one project related to the urban rehabilitation of the "Mostra d'Oltremare" area and its cultural and architectural assets. The design of this urban rehabilitation has employed the principles and criteria of sustainably complete streets design, as summarized below.

The study area is characterised by the presence of important attractors: buildings of significant architectural value in the Overseas Exhibition, the university, and in sports facilities. Taken together with the major regional road and railway networks, this area is one of Naples main activity centres. Its urban development occurred mainly in the late 1920s and 1930s, and in 1936 Mussolini awarded Naples the Triennial Overseas Exhibition, in Italian referred to as the "Mostra d'Oltremare", or locally known as Mostra. Typical of Mussolini-period architecture, Mostra contains a monumental centre, and was intended to be representative of the historical centre of a modern-era city. Mostra is characterized by wide open spatial and environmental qualities, thanks to the purposeful design of green and open spaces, which were intended to connect the two historical centres (i.e., Mostra and historical Naples). It remains of nostalgic importance for city life as the main centre for trade shows and sports events, as well as for the presence of numerous national research institutes, university departments, and government offices.

The urban context within which Mostra is situated (Figure 2) is complex, where over the past 70 years Naples has been impacted by a series of civic policy decisions that, in recent decades, have created major road traffic congestion problems. Only in recent years has Naples started to consider ways to close gaps in its pedestrian and bicycle networks. Moreover, the contemporary design and civic policies have evolved to embrace the dichotomies of everyday life constantly experienced in Naples, including private auto use versus public transit, accessibility versus mobility, private versus public spaces, and day versus night use. Design criteria were required to rehabilitate this pluralistic area, with its modern economy based heavily on international tourism that 'never sleeps'.

Figure 2. Study area.

The Mostra environmental and architectural rehabilitation was aimed at recovering its urban character in the context of its historical significance, while addressing its growing circulation, access, and aesthetics needs, including changes to the streetscape and networks. Specific problems that the project had to address related to:

- Urban streets with the features of rural roads (D1 in Figures 2, 3a and 4a);
- The layouts of urban spaces with poor consideration of the environment, aesthetics, and liveability (D2 in Figures 2, 5a and 6a);
- Inadequate and discontinuous pedestrian and bicycle paths (D3 in Figures 2, 7a and 8a);
- Pedestrian and bicycle routes in the layout of roads for motorized vehicles (D4 in Figures 2, 9a and 10a).

(**a**) Existing scenario.

(**b**) Design scenario.

Figure 3. Urban street with the features of a rural road: example of design improvement.

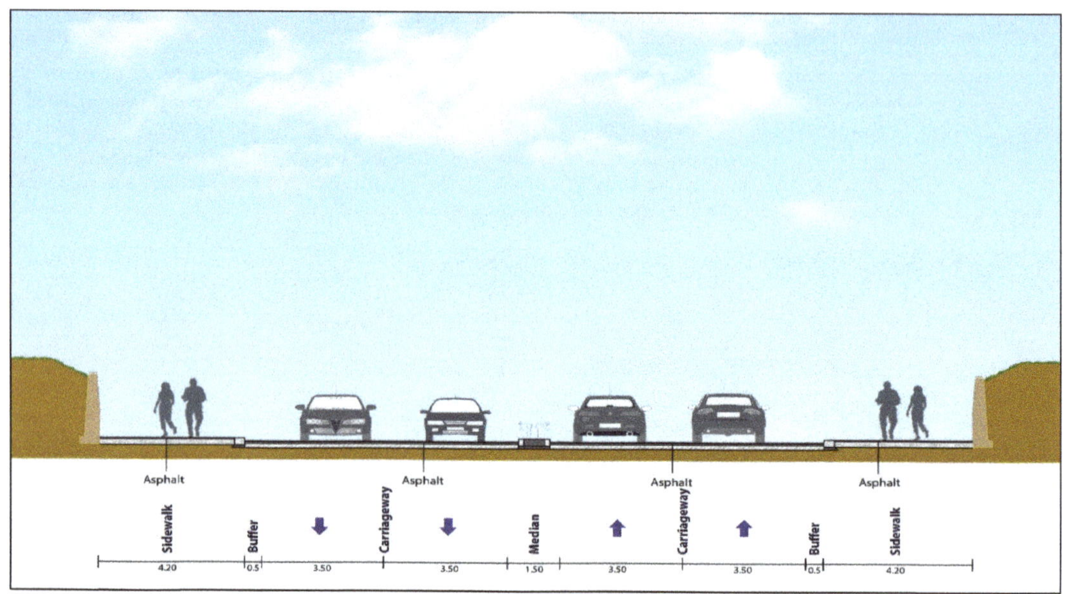

(**a**) Existing scenario.

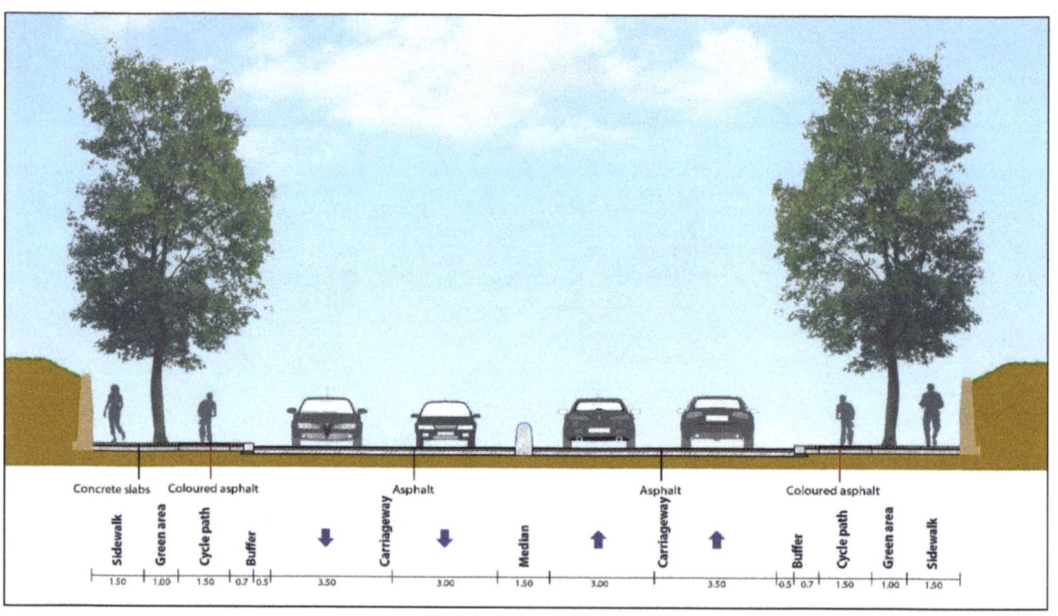

(**b**) Design scenario.

Figure 4. Urban street with features of a rural road: existing (**a**) and design (**b**) cross section.

(**a**) Existing scenario.

(**b**) Design scenario.

Figure 5. Layout of urban space with poor consideration of environment, aesthetics, and liveability: example of design improvement.

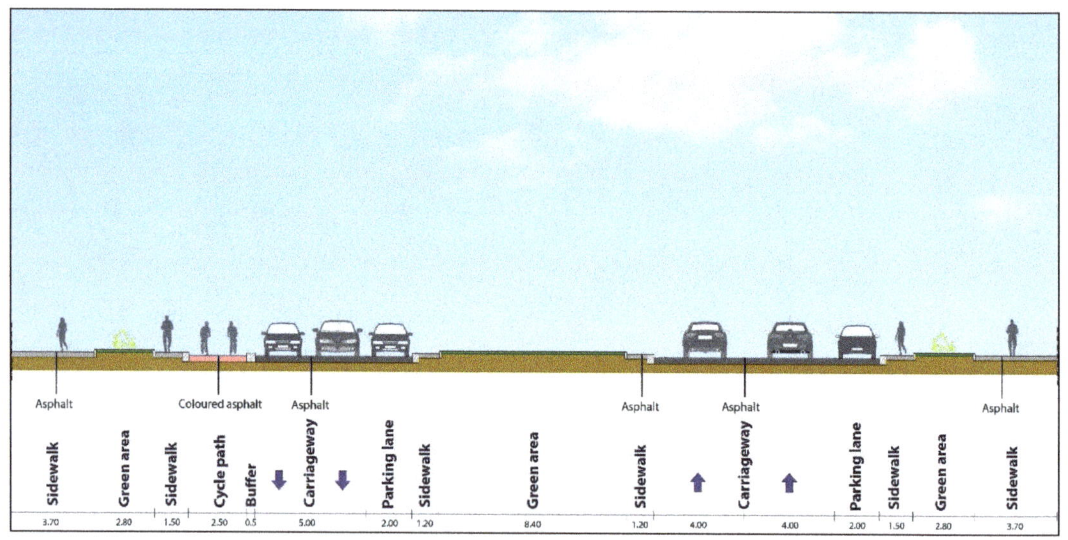

(**a**) Existing scenario.

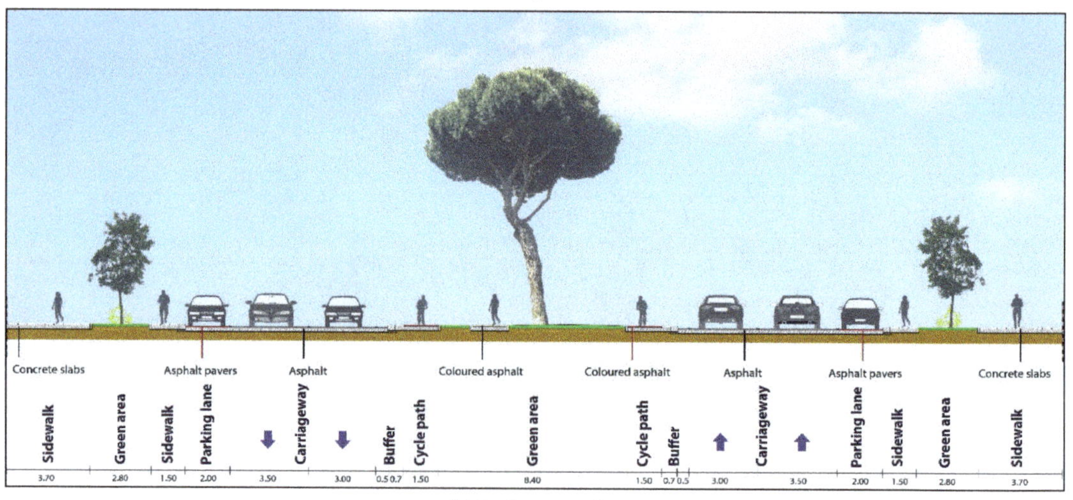

(**b**) Design scenario.

Figure 6. Layout of urban space with poor consideration of environment, aesthetics, and liveability: existing (**a**) and design (**b**) cross section.

(**a**) Existing scenario.

(**b**) Design scenario.

Figure 7. Inadequate and discontinuous pedestrian and bicycle paths: example of design improvement.

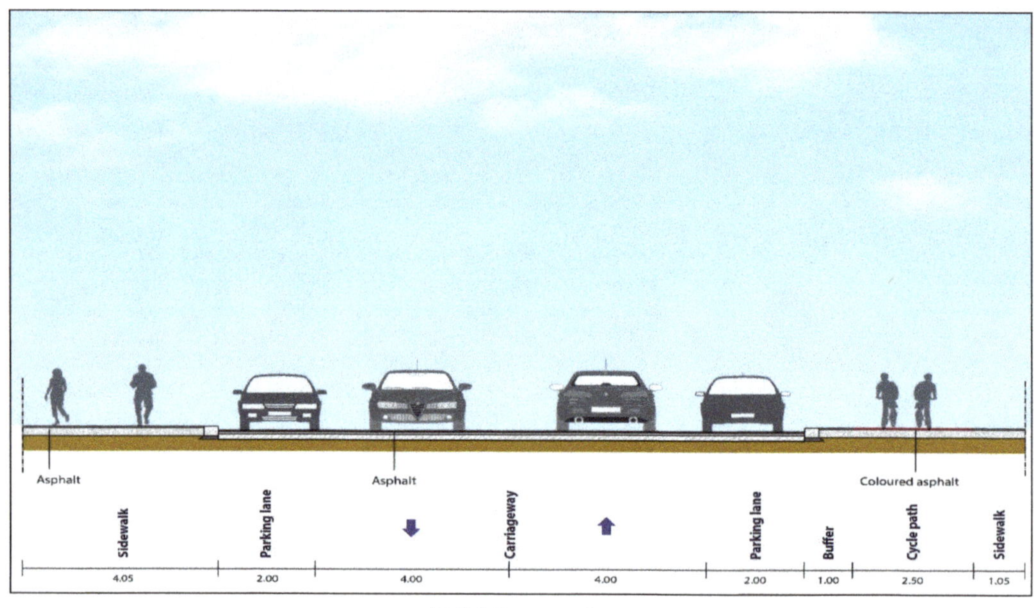

(**a**) Existing scenario.

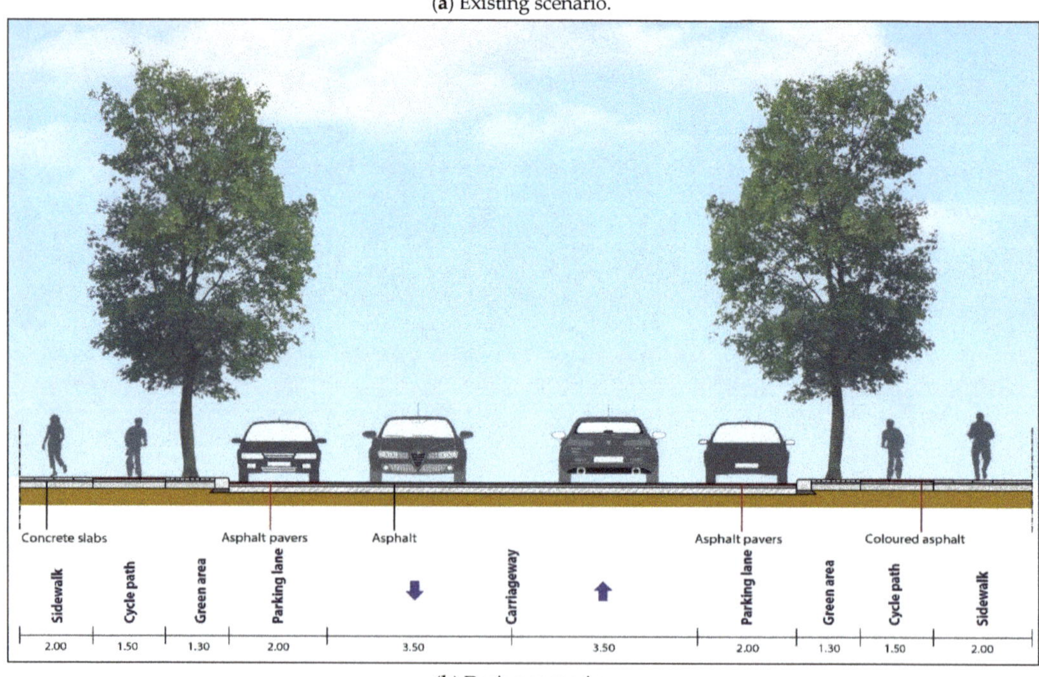

(**b**) Design scenario.

Figure 8. Inadequate and discontinuous pedestrian and bicycle paths: existing (**a**) and design (**b**) cross section.

(**a**) Existing scenario.

(**b**) Design scenario.

Figure 9. Pedestrian and bicycle routes in the layout of roads for motorized vehicles: example of design improvement.

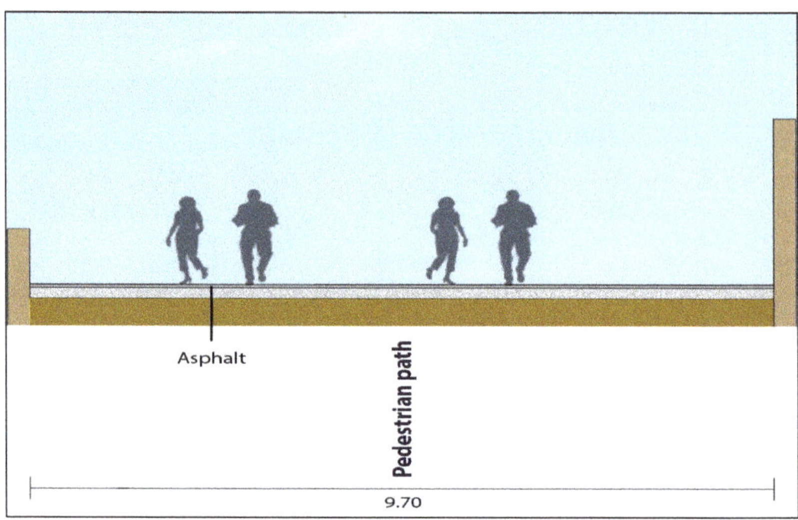

(**a**) Existing scenario.

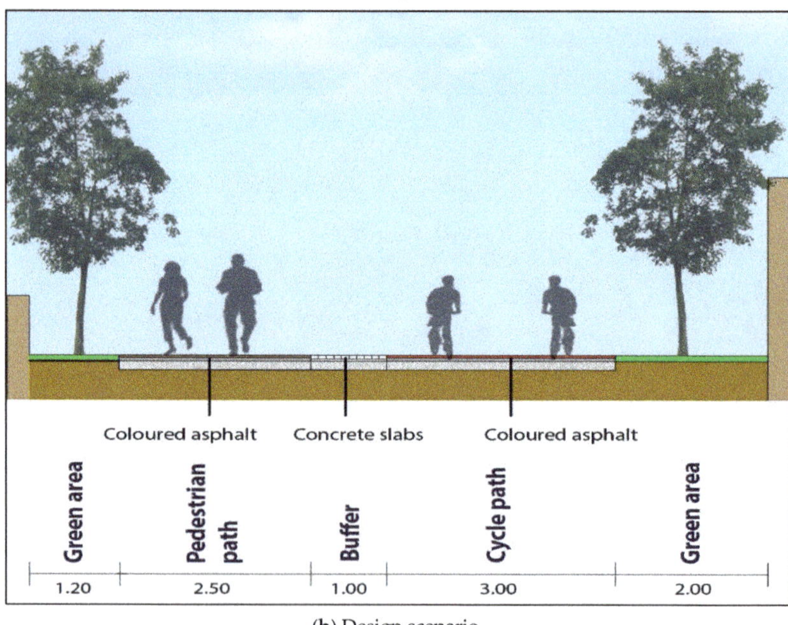

(**b**) Design scenario.

Figure 10. Pedestrian and bicycle routes in the layout of roads for motorized vehicles: existing (**a**) and design (**b**) cross section.

Practical application of the sustainable complete streets design criteria is presented with reference to four specific designs of the Mostra project which are aimed at solving the issues presented above. To understand how independent experts would evaluate in each case study the application of the sustainable complete streets design criteria reported in Table 1, a survey was made among engineers with relevant expertise in the field of road design and transportation planning and architects with similar experience. The main experts' opinions were considered to re-shape the four locations. In Table 2 we reported a synthesis of the specific criteria, whose implementation was suggested by the experts, and that we were able to introduce in the designs of the four case studies depending on the available section space.

Table 2. Sustainable complete streets design criteria.

General Criteria	Specific Criteria	Design 1	Design 2	Design 3	Design 4
Aesthetics	A1 Enhancement of the urban environment to recover historical urban character and realize historical area planning goals	√	√		√
	A2 Integration of materials, colours, lighting, and treatment of green spaces	√	√	√	√
	A3 Durable materials	√	√	√	
Environment	E1 Eco-design	√	√	√	√
	E2 Planting and trees	√	√	√	√
	E3 Reduction in traffic-related air pollution and noise	√	√	√	√
Liveability	L1 Usability, comfort and safety of the public outdoor space system		√	√	√
	L2 Creation of a public space destination in the urban landscape which affects activities and social relations		√		√
Safety	S1 Road safety inspections	√	√	√	√
	S2 Road safety audits	√	√	√	√
	S3 Fused grid neighbourhood pattern	√	√	√	√
	S4 Safe pedestrian routes and crosswalks	√	√	√	√
	S5 Safe cycle routes and crossings	√	√	√	√
	S6 Sight distance		√	√	
	S7 Transit and pedestrian accessibility		√	√	
	S8 Traffic control devices	√	√		
	S9 Lighting	√	√	√	√

4.2. Design 1—Urban Streets with the Features of Rural Roads

The rural character (Figures 3a and 4a) of some streets in Mostra have a significant impact on its liveability. They encourage high speeds, discourage pedestrian and bicycle traffic, and generally dampen human-scale activities. To recover the urban character and improve access and visibility to the entire Mostra area, the design principles were built on a homogeneous choice of materials, colours, lighting and treatment of green spaces, including (Figures 3b and 4b):

- The steel median safety barrier was replaced with a blue colour concrete median barrier with a rounded shape (criterion A1, see Table 1);
- Pedestrian crossings with white zebra stripes were embedded in a red colour surface with printed bituminous concrete appearing as brick pavers (criteria S3, S4 and S8);
- The uneven and potholed pavement was replaced with an even, antiskid surface (criteria A1, A3, and E1);
- The carriageways without markings were marked with high-performance cold-hardened road markings delineating 3.50 m wide traffic lanes (criteria A3, S3, and S8);
- Both sides of the road had one-way separated bicycle lanes installed (criteria A2, E1, E2, E3, S3, S5, and S9), with these features:
  - Raised from the street and flush with the sidewalk,
  - Width equal to 1.50 m,
  - Red colour bituminous concrete surface,
  - A 0.70 m buffer from the street via a basalt curb and concrete square slabs,
  - A 1.00 m buffer from the sidewalks made by concrete square slabs, trees with regular spacing, and lighting poles with LED sources;

- Continuous and accessible sidewalks 1.50 m wide were installed (criteria S3 and S5);
- The existing, obstructive advertising signs were removed from sidewalks (criteria A1 and A2).

4.3. Design 2—Inadequate Layout

Viale di Augusto, the most important street in the Mostra area, is characterised by a layout with poor consideration of the environment, aesthetics, and liveability (Figures 5a and 6a). Its wide median is not used by pedestrians or cyclists, and its aesthetic quality is very poor. Moreover, the historical palms have died, and car parking is allowed in the median, creating dangerous conflicts between the through-traffic and people crossing the street to reach their parked vehicles. At some point in the past, a separate two-way bicycle lane had been installed in one carriageway, creating an abnormal-width carriageway of two 2.25 m lanes (plus a 0.50 m shoulder). Most importantly, the bicycle lane had been used more by pedestrians and parents with strollers due to inadequate alternatives and obstructions in historical sidewalk areas (e.g., parked cars, scooters, and street furniture). To solve these issues, a combination of measures was designed (Figures 5b and 6b), including:

- The median was completely redesigned with new trees (indigenous and hardy species) and pedestrian and bicycle paths (criteria A1, A2, E2, E3, S3, S4, and S5);
- In the median, a continuous pedestrian route was created, with central plazas and benches at regular intervals (criteria L1 and L2);
- In the median, one-way separated bicycle lanes were introduced, flush with the sidewalk, 1.50 m wide, with a red colour bituminous concrete surface, and a 0.70 m buffer from the street (criterion S5);
- Parking in the median was removed (criterion S3);
- The former counter-flow bicycle lane was removed (criterion L1);
- Parallel parking was introduced in the nearside with a different surface, i.e., a red colour surface with printed bituminous concrete appearing as brick pavers (criteria A3 and E1);
- Bulb-outs were introduced in all the pedestrian crossings (criteria S4 and S6);
- Lighting poles with LED sources were employed (criteria L2 and S9);
- Antiskid surfaces were added (criteria A3 and E1);
- High-performance cold-hardened road markings were specified, delineating 3.50 m wide outer lanes (where buses are expected) and 3.00 m wide inner lanes (criteria A3, S3, and S8);
- On each side of the Viale, new continuous and accessible sidewalks with concrete square slabs were installed (criteria L1 and S7).

4.4. Design 3—Inadequate Pedestrian and Bicycle Paths

On another part of the road network, the Viale Kennedy, a two-way bicycle lane had been installed over the existing sidewalks, which imperilled pedestrians, and many parts of the Viale had no street trees (Figures 7a and 8a). To solve these issues, the design included (Figures 7b and 8b):

- The replacement of the existing two-way bicycle lane with one-way separated bicycle lanes on both sides (criteria L1, S3, and S5) that were flush with the sidewalk, including a 1.50 m wide, red coloured, bituminous concrete surface, separated by a 1.30 m basalt curb buffer, street trees, and LED street lights (criteria A2, E1, E2, E3, S3, S5, and S9);
- Continuous and accessible pedestrian paths 2.00 m wide (criteria S4 and S7);
- Bulb-outs in all the pedestrian crossings (criteria S4 and S6);
- Lane widths that were reduced from 4.00 m to 3.50 m (criterion S3);
- Antiskid surfaces (criteria A3 and E1);
- High-performance cold-hardened road markings (criteria A3, S3, and S8).

4.5. Design 4—Conflicts between Pedestrians, Cyclists, and Cars

In Via Claudio (Figures 9a and 10a), the design needed to address high pedestrian and bicycle flows, as the street connects the north and the south campuses of the School of Engineering at the University of Naples. It also connects the north campus with a bus stop and the railway station. Because of its high use by vulnerable users, this street had been designated as a shared pedestrian and bicycle route closed to car traffic. However, a lack of police and physical enforcement (i.e., no self-enforcing physical barriers) had allowed the street to be historically used as a shortcutting route for auto traffic, despite having no separated sidewalks nor bicycle lanes, which has caused continuing VRU/auto conflicts (Figure 8a). A combination of countermeasures was designed, including (Figures 9b and 10b):

- The removal of the concrete pavement (criteria A1 and A2);
- The installation of green strips on both sides with grass, trees, benches, and lighting poles with LED sources (criteria E1, E2, E3, L1, L2, and S9);
- Installation of a two-way bicycle lane, flush with the sidewalk, 3.00 m wide, with a red colour bituminous concrete surface (criteria L1, L2, S3, and S5);
- Installation of a 1.00 m wide buffer between the bicycle lane and sidewalk, using concrete square slabs (criteria S3, S4, and S5);
- Installation of a 2.00 m wide coloured sidewalk (criteria S3 and S4);
- Installation of 'gateway' architectural features at its ends (criteria A1, L1, and L2).

5. Results and Discussion

The sustainable complete streets criteria were applied to intervene in a series of problematic locations in the city of Naples, Italy, to seek out opportunities to design and retrofit streets in one of the most complex urban contexts of the city. The case study presented a proposal for street regenerations towards multifunctional, smart, safe, walkable, and liveable streets. Given the legacy of planning, engineering, and infrastructure in facilitating automobile movement, the shift toward the accommodation of other transport modes is not a simple change. The scenarios reported in this paper in fact challenge the primary accommodation of automobiles. Each one was reported to provide example practices to inform how existing streets may be transformed into complete streets.

To help set priorities, identify street design features, and create intuitive multimodal networks throughout the city, a checklist was created identifying all the specific criteria. Then, a survey was made among engineers and architects with relevant expertise in the field of road safety and transport systems to understand how those experts would address the application of the criteria in the case study.

The design elements were built based on a homogeneous choice of materials, colours, lighting, and treatment of green spaces, and have a potential significant impact on the liveability of the area and its aesthetics. Indeed, previous studies have demonstrated how an improvement in the quality and ambience of city life is strongly associated with its landscaping [5,20]. Furthermore, different users' needs are considered to ensure that mobility is equally safe and comfortable for all users. The presence of parts of the streets dedicated to pedestrians and cyclists improves their access and visibility, discouraging high vehicle speeds and promoting human-scale activities.

The benefits of adopting sustainable principles in complete streets design criteria are substantial. The design of community land use and transport encourages active transportation with its health benefits for individual users and for the community. Furthermore, active transportation has environmental benefits from producing no air pollution [66]. Complete streets also have economic benefits. Safe and convenient pedestrian amenities boost foot traffic, which can increase retail sales benefiting local retailers [73].

The sustainable complete streets design criteria can be employed in the retrofits of urban areas and help to enhance overall community sustainability. The design scenarios proposed in this study were conceived to help practitioners to consider these context-based uses and design accordingly, gaining knowledge from past experiences to benefit future projects; however, not every street is intended or is suitable for the accommodation of every user mode. The complete streets design criteria have large sets of potential competing priorities, where the importance of each priority will vary depending on the street context and its role in the network [74]. Hence, the identification of the different priorities by street type is necessary to begin to understand the trade-offs among a street's required function, its space organization, and traffic volume.

Nevertheless, a reduction in the vehicle-destined spaces is not easy to understand and digest for habitual road users. Hence, national, provincial, and municipal policies should work on public acceptance and emphasize the City's interest and investment in developing safe and accessible streets that are accommodating to multiple modes and that provide an attractive public realm that allows for safe movements. Another impediment to implement sustainable complete streets relies in the resources available that should be allocated for sidewalks, bike lanes, and other complete street elements. A deep belief is that the construction costs of sustainable complete streets projects exceed the construction costs of traditional road projects. Considering the small percentage of project budgets required to include complete street elements and the significant market fluctuation in historical project construction costs, a study carried out by the Charlotte Department of Transportation [75] concluded that the construction costs of complete streets do not necessarily represent an incremental cost compared to traditional designs.

6. Conclusions

Streets are important public spaces for many aspects of daily life and have significant effects on the liveability of a city, but road designs and functions have evolved from a local-context focus to a broader connecting focus such that they no longer consider these local contextual needs. However, the need to promote the healthy, safer, and sustainable development of cities has become even more urgent with the outbreak of the COVID-19 pandemic, stressing the importance for policymakers and urban planners to shift their focus to more user-oriented planning and urban renewal.

This paper has proposed and discussed sustainability-oriented, eco-design urban street design criteria that are meant to complement and build on the existing complete streets design criteria. Specifically, the term 'sustainably complete streets' design criteria has been coined to denote the addition of socio-environmental design criteria related to the aesthetics, environment, liveability, and safety. Compared to the existing design criteria, the major novelties of the proposed criteria are the integration of sustainability and complete streets criteria as well as the definition of specific criteria to provide a street network adapted to contemporary requirements for community sustainability, quality of life, and vulnerable road user safety, while providing for traditional urban mobility, a historical context, and efficiency. The scenarios reported in this paper provide example practices on how existing streets may be transformed into complete streets. The sustainable complete streets criteria should always be included during the planning, design, and operation of roadways by government agencies, consultants, and practitioners, when designing for future streets or regenerated streets in an area experiencing land development, when implementing a capital improvement project (i.e., the construction or reconstruction of a street, intersection, or bridge), and during maintenance treatments (i.e., resurfacing a street or conducting major work in the street). This may create an opportunity to reconsider some aspects of a street's design. The importance of considering these criteria comes early in the roadway planning and design process to identify the desired characteristics of a roadway. Street design is complex and must respond to a series of local conditions and site constraints, and to set the users' priorities, an expert judgement and coordination among the architects and engineers is paramount for making decisions for specific streets and for establishing community priorities and strategies that will harmonize the transportation system, land use, and redevelopment.

If a systematic approach is taken as advocated, the benefits will include reduced motorized traffic, a reduced energy consumption, reduced traffic noise, reduced air pollution, and increased pedestrian, bicycle, and public transport traffic and safety. At the same time, criteria are included to improve the aesthetic and liveability qualities of public and/or green spaces. This is obtained through urban design solutions for structured and consistent thematic areas aimed at a unified picture of homogeneous urban zones.

The proposed criteria look at the environmental and architectural restoration to recover urban character in respect of the historical evolution and urban planning of an area. Using a systematic and team-based approach, the design solution can create a road environment that serves both functional and local needs. The local context and architectural design principles support a homogeneous, unified approach as regards the choice of materials, colours, lighting, and treatment of the green spaces. An eco-design approach has been proposed, and special emphasis is given to planting tree species and to the implementation and increase in permeable surfaces in order to reduce the heat island effect and absorb the air pollution caused by vehicular traffic.

The case study of the urban rehabilitation of the "Mostra d'Oltremare" area and its cultural and architectural assets in Naples, Italy, highlights the practical application of these proposed criteria and the possibility of using these criteria in other urban contexts. The case study shows both how each specific criteria were applied and how the concept of sustainable complete streets can contribute to urban aesthetics, the environment, liveability, and safety.

Author Contributions: Conceptualization, S.C, A.C.d.S.M., G.L., A.M. and P.N.; methodology, S.C., A.C.d.S.M., G.L., A.M., P.N. and M.R.R.; validation, S.C. and A.M.; investigation, S.C., A.C.d.S.M., A.M. and P.N.; writing—original draft preparation, S.C., A.C.d.S.M., A.M. and P.N.; writing—review and editing, S.C., A.C.d.S.M., A.M., G.L., P.N. and M.R.R.; supervision, A.M. All authors have read and agreed to the published version of the manuscript.

Funding: This research received no external funding.

Institutional Review Board Statement: Not applicable.

Informed Consent Statement: Not applicable.

Data Availability Statement: Not applicable.

Acknowledgments: The authors acknowledge Mario Calabrese, councillor for public works of the City of Naples, for his technical support.

Conflicts of Interest: The authors declare no conflict of interest.

References

1. Transportation Research Board. International Perspectives of Urban Street Design. Transportation Research Circular E-C097. 2006. Available online: http://onlinepubs.trb.org/onlinepubs/circulars/ec097.pdf (accessed on 1 September 2022).
2. Gregg, K.; Hess, P. Complete streets at the municipal level: A review of American municipal complete street policy. *Int. J. Sustain. Transp.* **2019**, *13*, 407–418. [CrossRef]
3. Institute of Transportation Engineering. *Designing Walkable Urban Thoroughfares: A Context Sensitive Approach*; Publication No. RP-036A; Institute of Transportation Engineering: Washington, DC, USA, 2010; Available online: http://library.ite.org/pub/e1cff43c-2354-d714-51d9-d82b39d4dbad (accessed on 1 September 2022).
4. Institute of Transportation Engineering. Integration of Safety in the Project Development Process and Beyond: A Context Sensitive Approach. 2015. Available online: http://library.ite.org/pub/e4edb88b-bafd-b6c9-6a19-22e98fedc8a9 (accessed on 1 September 2022).
5. McCann, B. Completing Our Streets: The Transition to Safe and Inclusive Transportation Networks. Island Press: Washington, DC, USA, 2013.
6. National Association of City Transportation Officials. *Urban Street Design Guide*; National Association of City Transportation Officials: New York, NY, USA, 2013; Available online: https://nacto.org/publication/urban-street-design-guide/ (accessed on 1 September 2022).
7. National Complete Streets Coalition. Introduction to Complete Streets. 2015. Available online: https://smartgrowthamerica.org/resources/introduction-to-complete-streets/ (accessed on 1 September 2022).
8. New York City Department of Transportation. Street Design Manual, Updated Second Edition. 2015. Available online: https://nacto.org/wp-content/uploads/2016/05/2-6_NYCDOT-Street-Design-Manual-2nd-ed-ch-2_2015.pdf. (accessed on 1 September 2022).
9. Schlossberg, M.; Rowell, J.; Amos, D.; Sanford, K. Rethinking Streets: An Evidence-Based Guide to 25 Complete Street Transformations. Sustainable Cities Initiative, University of Oregon: Eugene, OR, USA, 2013.
10. Rupprecht, S.; Brand, L.; Böhler-Baedeker, S.; Brunner, L.M. *Guidelines for Developing and Implementing Sustainable Urban Mobility Plan*, 2nd ed.; Rupprecht Consult Editor: Cologne, Germany, 2019.

11. Alexandria Department of Transportation and Environmental Services. Alexandria Complete Streets Design Guidelines. 2022. Available online: https://www.alexandriava.gov/transportation-planning/complete-streets-design-guidelines (accessed on 1 September 2022).

12. Pulsipher, D.; Lamie, R.; Johnson, K.; Price, B. Denver Complete Streets Design Guidelines. 2020. Available online: https://www.denvergov.org/files/assets/public/doti/documents/standards/doties-017.0_complete_streets_guidelines.pdf (accessed on 1 September 2022).

13. Stantec Consulting Ltd and the City of Edmonton. Complete Streets Design and Construction Standards, City of Edmonton. 2018. Available online: https://bicycleinfrastructuremanuals.com/manuals4/CompleteStreets_DesignStandards_Sept2018.pdf (accessed on 1 September 2022).

14. London Canada. London Complete Streets Design Manual. 2018. Available online: https://london.ca/sites/default/files/2020-09/Complete%20Streets%20Design%20Manual.pdf (accessed on 1 September 2022).

15. Montgomery County Department of Transportation. Montgomery County Complete Streets. 2021. Available online: https://montgomeryplanning.org/wp-content/uploads/2022/03/Montgomery-County-CSDG_Approved-2021.pdf (accessed on 1 September 2022).

16. Department for Transport. *Manual for Streets*; Thomas Telford Publishing: London, UK, 2007. Available online: www.gov.uk/government/uploads/system/uploads/attachment_data/file/341513/pdfmanforstreets.pdf (accessed on 1 September 2022).

17. Transport for London. *Streetscape Guidance*, 4th ed; Transport for London: London, UK, 2019. Available online: https://content.tfl.gov.uk/streetscape-guidance-.pdf. (accessed on 1 September 2022).

18. Wegman, F.; Aarts, L. Advancing Sustainable Safety: National Road Safety Outlook for 2005–2020. SWOV Institute for Road Safety Research, Leidschendam, The Netherlands. 2006. Available online: www.swov.nl/sites/default/files/publicaties/rapport/dmdv/advancing_sustainable_safety.pdf (accessed on 1 September 2022).

19. Sandt, L.; Brookshire, K.; Heiny, S.; Blank, K.; Harmon, K. Toward a Shared Understanding of Pedestrian Safety: An Exploration of Context, Patterns, and Impacts. Pedestrian and Bicycle Information Center. 2020. Available online: https://www.pedbikeinfo.org/cms/downloads/PBIC_Pedestrian%20Safety%20Background%20Piece_7-2.pdf (accessed on 1 September 2022).

20. Brown, B.B.; Werner, C.M.; Tribby, C.P.; Miller, H.J.; Smith, K.R. Transit use, physical activity, and body mass index changes: Objective measures associated with complete street light-rail construction. *Am. J. Public Health* **2015**, *105*, 1468–1474. [CrossRef]

21. ETSC. PIN FLASH Report 37: Safer Roads, Safer Cities: How to Improve Urban Road Safety in the EU. 2019. Available online: https://etsc.eu/wp-content/uploads/PIN-FLASH-37-FINAL.pdf (accessed on 1 September 2022).

22. European Commission. EU Road Safety Policy Framework 2021–2030, Next Steps towards "Vision Zero". 2020. Available online: https://op.europa.eu/en/publication-detail/-/publication/d7ee4b58-4bc5-11ea-8aa5-01aa75ed71a1 (accessed on 1 September 2022).

23. Song, X.; Cao, M.; Zhai, K.; Gao, X.; Wu, M.; Yang, T. The effects of spatial planning, well-being, and behavioural changes during and after the COVID-19 pandemic. *Front. Sustain. Cities* **2021**, *3*, 686706. [CrossRef]

24. Verheijen, M. Functional Ambiance. De Urbanisten. 2016. Available online: www.urbanisten.nl/pdf/Functional%20Ambiance%20-%20artikel%20Huig.pdf (accessed on 1 September 2022).

25. United Nations. Report of the World Commission on Environment and Development: Our Common Future. 1987. Available online: http://www.un-documents.net/our-common-future.pdf (accessed on 1 September 2022).

26. Vezzoli, C.; Manzini, E. *Design for Environmental Sustainability*; Springer: London, UK, 2008.

27. Grammenos, F.; Lovegrove, G. Remaking the City Street Grid—A Model for Urban and Suburban Development. McFarland Publishers: Jefferson, NC, USA, 2015.

28. United Nations. Resolution 74/299 Improving Global Road Safety. 2020. Available online: https://documents-dds-ny.un.org/doc/UNDOC/GEN/N20/226/30/PDF/N2022630.pdf?OpenElement. (accessed on 1 September 2022).

29. World Health Organization. *Global Status Report on Road Safety 2018*; World Health Organization: Geneva, Switzerland, 2018.

30. Austroads. Guide to Road Safety Part 6: Managing Road Safety Audits. Austroads Publication AGRS06/19: Sydney, NSW, Australia, 2019.

31. FHWA. A Systemic Approach to Safety: Using Risk to Drive Action. Report FHWA SA-15-054. 2016. Available online: https://safety.fhwa.dot.gov/systemic/fhwasa15054/apprch.pdf. (accessed on 1 September 2022).

32. Montella, A.; Guida, C.; Mosca, J.; Lee, J.; Abdel-Aty, M. Systemic approach to improve safety of urban unsignalized intersections: Development and validation of a Safety Index. *Acc. Anal. Prev.* **2020**, *141*, 105523. [CrossRef]

33. Thomas, L.; Sandt, L.; Zegeer, C.; Kumfer, W.; Lang, K.; Lan, B.; Horowitz, Z.; Butsick, A.; Toole, J.; Schneider, R. Systemic Pedestrian Safety Analysis. NCHRP Research Report 893. 2018. Available online: https://www.nap.edu/download/25255 (accessed on 1 September 2022).

34. Aarts, L.; van Schagen, I. Driving speed and the risk of road crashes: A review. *Accid. Anal. Prev.* **2006**, *38*, 215–224. [CrossRef]

35. Hauer, E. Speed and safety. *Transp. Res. Rec.* **2009**, *2103*, 10–17. [CrossRef]

36. Montella, A.; Imbriani, L.L.; Marzano, V.; Mauriello, F. Effects on speed and safety of point-to-point speed enforcement systems: Evaluation on the urban motorway A56 Tangenziale di Napoli. *Acc. Anal. Prev.* **2015**, *75*, 164–178. [CrossRef]

37. Montella, A.; Mauriello, F.; Permetti, M.; Rella Riccardi, M. Rule discovery to identify patterns contributing to overrepresentation and severity of run-off-the-road crashes. *Acc. Anal. Prev.* **2021**, *155*, 106119. [CrossRef]

38. Neuman, T.R.; Slack, K.L.; Hardy, K.K.; Bond, V.L.; Potts, I.; Alberson, B.; Lerner, N. NCHRP Report 500: Guidance for Implementation of the AASHTO Strategic Highway Safety Plan: A Guide for Reducing Speeding-Related Crashes. Transportation Research Board: Washington, DC, USA, 2009; Volume 23.
39. National Highway Traffic Safety Administration. *MMUCC Guideline: Model Minimum Uniform Crash Criteria*, 5th ed.; Report DOT HS 812 433; U.S. Department of Transportation: Washington, DC, USA, 2017.
40. Masoud, A.; Lee, A.; Faghihi, F.; Lovegrove, G. Building sustainably safe and healthy communities with the fused grid development layout. *Can. J. Civ. Eng.* **2015**, *42*, 1063–1072. [CrossRef]
41. Masoud, A.R.; Idris, A.O.; Lovegrove, G. Modelling the impact of fused grid network design on mode choice behavior. *J. Transp. Health* **2019**, *15*, 100627. [CrossRef]
42. Rella Riccardi, M.; Mauriello, F.; Sarkar, S.; Galante, F.; Scarano, A.; Montella, A. Parametric and non-parametric analyses for pedestrian crash severity prediction in Great Britain. *Sustainability* **2022**, *14*, 3188. [CrossRef]
43. Prato, C.G.; Kaplan, S.; Rasmussen, T.K.; Hels, T. Infrastructure and spatial effects on the frequency of cyclist-motorist collisions in the Copenhagen Region. *J. Transp. Saf. Secur.* **2016**, *8*, 346–360. [CrossRef]
44. Goerke, D.; Zolfaghari, E.; Marek, A.P.; Endorf, F.W.; Nygaard, R.M. Incidence and profile of severe cycling injuries after bikeway infrastructure changes. *J. Community Health* **2020**, *45*, 542–549. [CrossRef] [PubMed]
45. Brijs, T.; Mauriello, F.; Montella, A.; Galante, F.; Brijs, K.; Ross, V. Studying the effects of an advanced driver-assistance system to improve safety of cyclists overtaking. *Acc. Anal. Prev.* **2022**, *174*, 106763. [CrossRef] [PubMed]
46. CROW. *Design Manual for Bicycle Traffic*; CROW: Ede, The Netherlands, 2017.
47. Pulvirenti, G.; Distefano, N.; Leonardi, S.; Tollazzi, T. Are double-lane roundabouts safe enough? A chaid analysis of unsafe driving behaviors. *Safety* **2021**, *7*, 20. [CrossRef]
48. López, G.; de Oña, J.; Garach, L.; Baena, L. Influence of deficiencies in traffic control devices in crashes on two-lane rural roads. *Acc. Anal. Prev.* **2016**, *96*, 130–139. [CrossRef]
49. American Association of State Highway and Transportation Officials. *Highway Safety Manual*, 1st ed.; American Association of State Highway and Transportation Officials: Washington, DC, USA, 2010.
50. Montella, A.; de Oña, R.; Mauriello, F.; Rella Riccardi, M.; Silvestro, G. A data mining approach to investigate patterns of powered two-wheeler crashes in Spain. *Acc. Anal. Prev.* **2020**, *134*, 105251. [CrossRef]
51. Moral Garcia, S.; Castellano, J.G.; Mantas, C.J.; Montella, A.; Abellán, J. Decision tree ensemble method for analyzing traffic accidents of novice drivers in urban areas. *Entropy* **2019**, *21*, 360. [CrossRef]
52. Lantieri, C.; Costa, M.; Vignali, V.; Acerra, E.M.; Marchetti, P.; Simone, A. Flashing in-curb LEDs and beacons at unsignalised crosswalks and driver's visual attention to pedestrians during nighttime. *Ergonomics* **2022**, *64*, 330–341. [CrossRef]
53. Rella Riccardi, M.; Mauriello, F.; Scarano, A.; Montella, A. Analysis of contributory factors of fatal pedestrian crashes by mixed logit model and association rules. *Int. J. Inj. Control. Saf. Promot.* 2022; *Epub ahead of print*. [CrossRef]
54. Langdon, D.; Management Consulting. Life Cycle Costing (LCC) as a Contribution to Sustainable Construction. Guidance on the Use of the LCC Methodology and Its Application in Public Procurement. Final Guidance. 2007. Available online: https://ec.europa.eu/docsroom/documents/5060/attachments/1/translations/en/renditions/pdf (accessed on 4 October 2022).
55. Transport for London. Walking Action Plan. Making London the World's Most Walkable City. 2018. Available online: https://content.tfl.gov.uk/mts-walking-action-plan.pdf (accessed on 23 September 2022).
56. European Parliament. *Directive 2008/96/EC on Road Infrastructure Safety Management*; European Parliament: Brussels, Belgium, 2008.
57. European Parliament. *Directive (EU) 2019/1936 of the European Parliament and of the Council of 23 October 2019 Amending Directive 2008/96/EC on Road Infrastructure Safety Management*; European Parliament: Brussels, Belgium, 2019.
58. Montella, A. A comparative analysis of hotspot identification methods. *Acc. Anal. Prev.* **2010**, *42*, 571–581. [CrossRef] [PubMed]
59. World Road Association. Road Safety Manual: A Guide for Practicioners. 2019. Available online: http://roadsafety.piarc.org/en. (accessed on 4 October 2022).
60. Montella, A. Roundabouts. In *Safe Mobility: Challenges, Methodology and Solutions*; Lord, D., Washington, S., Eds.; Transport and Sustainability; Emerald Publishing Limited: Bingley, UK, 2018; Volume 11, pp. 147–174. [CrossRef]
61. Montella, A.; Turner, S.; Chiaradonna, S.; Aldridge, D. International overview of roundabout design practices and insights for improvement of the Italian standard. *Can. J. Civ. Eng.* **2013**, *40*, 1215–1226. [CrossRef]
62. Rodegerdts, L.; Bansen, J.; Tiesler, C.; Knudsen, J.; Myers, E.; Johnsonm, M.; Moule, M.; Persaud, B.; Lyon, C.; Hallmark, S.; et al. *Roundabouts: An Informational Guide*, 2nd ed.; NCHRP Report 672; Transportation Research Board: Washington, DC, USA, 2010.
63. Federal Highway Administration. *Road Safety Audits Guidelines*; Report No. FHWA-SA-06-06; U.S. Department of Transportation: Washington, DC, USA, 2006.
64. Federal Highway Administration. *Separated Bike Lane: Planning and Design Guide*; Report No. FHWA-HEP-15-025; U.S. Department of Transportation: Washington, DC, USA, 2015.
65. Massachusetts Department of Transportation. *Separated Bike Lane Planning & Design Guide*; Massachusetts Department of Transportation: Boston, MA, USA, 2015.
66. National Association of City Transportation Officials. *Urban Bikeway Design Guide*, 2nd ed.; National Association of City Transportation Officials: New York, NY, USA, 2014.

67. IBI GROUP. Background Report D: Final Report June 2020. Thorold Transportation Master Plan Complete Streets Strategy. 2020. Available online: https://www.thorold.ca/en/city-hall/resources/D-CompleteStreetsStrategy_FINAL_06-2020.pdf (accessed on 9 September 2022).
68. IRAP. Road Safety Toolkit. Available online: https://toolkit.irap.org/safer-road-treatments/sight-distance-obstruction-removal/ (accessed on 20 September 2022).
69. United Nations. Vienna Convention on Road Signs and Signals E/CONF.56/17/Rev.1. 1968. Available online: https://unece.org/DAM/trans/conventn/Conv_road_signs_2006v_EN.pdf. (accessed on 20 September 2022).
70. Waka Kotahi NZ Transportation Agency. Traffic Control Devices Manual Part 1. 2010. Available online: https://www.nzta.govt.nz/assets/resources/traffic-control-devices-manual/docs/part-1-general-requirements.pdf (accessed on 20 September 2022).
71. Waka Kotahi NZ Transportation Agency. Specification for High Performance Roadmarking, NZTA P30. 2009. Available online: https://www.nzta.govt.nz/assets/resources/high-performance-roadmarking/docs/high-performance-roadmarking.pdf (accessed on 20 September 2022).
72. Zalesinska, M.; Wandachowicz, K. On the quality of street lighting in pedestrian crossings. *Energies* **2021**, *14*, 7349. [CrossRef]
73. Transport Canada. Complete Streets: Making Canada's Roads Safer for All. 2009. Available online: https://publications.gc.ca/collections/collection_2012/tc/T41-1-72-eng.pdf (accessed on 9 September 2022).
74. Sousa, L.R.; Rosales, J. Contextually complete streets. In Proceedings of the Green Streets and Highways Conference 2010, Denver, CO, USA, 14–17 November 2010. [CrossRef]
75. Shapard, J.; Cole, M. Do complete streets cost more than in complete streets? *Transp. Res. Rec.* **2013**, *2393*, 134–138. [CrossRef]

 *sustainability*

Article

Pedestrian Crossings as a Means of Reducing Conflicts between Cyclists and Pedestrians in Shared Spaces

Chrysanthi Mastora [1], Evangelos Paschalidis [2], Andreas Nikiforiadis [1] and Socrates Basbas [1,*]

1   School of Rural & Surveying Engineering, Faculty of Engineering, Aristotle University of Thessaloniki, 54124 Thessaloniki, Greece; chrysamast@gmail.com (C.M.); anikiforiadis@topo.auth.gr (A.N.)
2   Transport and Mobility Laboratory, School of Architecture, Civil and Environmental Engineering (ENAC), École Polytechnique Fédérale de Lausanne (EPFL), CH-1015 Lausanne, Switzerland; evangelos.paschalidis@epfl.ch
*   Correspondence: smpasmpa@auth.gr

Abstract: One significant and simultaneously interesting problem in urban mobility has to do with the study of shared spaces where various categories of users coexist and act together. This paper aims to examine the behavior and preferences of pedestrians and cyclists, who both coexist in a shared space infrastructure along the seafront (which has a length of around 4.0 km) of the city of Thessaloniki, Greece. Furthermore, the problems caused by the coexistence, such as at the locations where there are pedestrian crossings on the bicycle lane, are recorded and evaluated. Traffic calming measures aimed at improving the existing situation in terms of safety and comfort for both pedestrians and cyclists are also explored. Data were collected through a web-based questionnaire survey, which was distributed via email to students and employees of Aristotle University of Thessaloniki. A total of 1194 questionnaires were collected in the framework of the survey during the year 2021, including responses from both pedestrians and cyclists. The questionnaires were analyzed through the use of descriptive and inferential statistics; the latter method suggested several significant differences in how each group of users (pedestrians or cyclists) perceived their behavior compared with the other. Latent variable and path models were estimated to investigate the behavior and attitude of users towards the crossings, examined as a function of their perception towards the other group; perception about the benefits of the infrastructure; preference for additional interventions; and overall opinion about the quality of the shared space area. The results suggest that forms of aggressive behavior, preference towards using the crossings, and the perceived safety are affected by the aforementioned factors. The results of this study can inform decision takers and decision makers in the area of land use regarding policy recommendations for facilitating interactions between pedestrians and cyclists in shared spaces.

Keywords: vulnerable road users; cyclists; pedestrians; shared space; urban road safety; pedestrian crossings

check for
updates

Citation: Mastora, C.; Paschalidis, E.; Nikiforiadis, A.; Basbas, S. Pedestrian Crossings as a Means of Reducing Conflicts between Cyclists and Pedestrians in Shared Spaces. *Sustainability* **2023**, *15*, 9377. https://doi.org/10.3390/su15129377

Academic Editors: Salvatore Leonardi and Natalia Distefano

Received: 4 May 2023
Revised: 31 May 2023
Accepted: 8 June 2023
Published: 10 June 2023

1. Introduction

The concept of shared streets and spaces, where different categories of users share a single infrastructure, is not a recent idea. In fact, streets have historically been a location for people to interact, a place where social, cultural, and economic activities of cities have taken place [1]. The advent of motor vehicles and subsequent growth of the automotive industry in the past century have presented novel challenges for transportation and urban planning. The primary objective has been to cater to the increasing volumes of motorized traffic and enable the faster movement of vehicles. These objectives and the respective policies have led to the formation of separate lanes for each road user category and to the allocation of more space to motorized vehicles rather than to active modes [2].

Transportation planners and policy makers have come to realize in recent times that creating separate lanes for each type of road user is a daunting task due to the scarcity of public space [3]. Furthermore, they have come to acknowledge the importance of enhancing the street's role in facilitating social interactions [1]. As of now, there is a trend towards reallocating urban space in a manner that prioritizes pedestrians and other forms of sustainable transportation while also transforming streets into hubs of social activity and human interaction. This is evidenced by the most recent edition of the Sustainable Urban Mobility Plans guidelines [4]. To this effect, cities are increasingly adopting shared infrastructure for pedestrians and bicycles.

Numerous studies have been conducted to determine the safety and desirability of pedestrian–cyclist coexistence and shared infrastructure compared with situations where cyclists share the road with motorized vehicles. Aultman-Hall and LaMondia conducted one of the initial research studies that examined the safety of shared infrastructures for pedestrians and cyclists [5]. The authors utilized a questionnaire survey to gather data on accidents and exposure, which allowed them to calculate indicators for three infrastructure types located in the United States. The findings indicate that falls are the most common type of accident in the shared pedestrian–cyclist infrastructures, whereas collisions between pedestrians and cyclists are infrequent [5]. Chong et al. conducted a study using the mortality data for New South Wales from the Australian Bureau of Statistics, as well as injury data from all public and private hospitals in the state [6]. Their aim was to compare the severity of collisions between bicycles and motor vehicles with those between bicycles and pedestrians. After conducting a statistical analysis, Chong et al. concluded that the risk of injury is greater for cyclists who are involved in collisions with motor vehicles, but collisions between bicycles and pedestrians can also result in severe injuries. Moreover, the study found that the risk of serious injury is higher for pedestrians and cyclists who are aged over 65 years [6]. An additional research effort tried to assess the risk of injury or fatality resulting from pedestrian–bicycle collisions; the findings indicate that the probability of death is negligible, while the likelihood of injury is as infrequent as the probability of death in a plane crash [7].

A study conducted in Melbourne, Australia, examined pedestrian injuries resulting from collisions with bicycles from 2006 to 2016 [8]. The study found that there was no increase in such injuries during this period and that the frequency of these injuries was significantly lower than the frequency of pedestrian injuries resulting from collisions with motor vehicles [8]. Varnild et al. conducted a comprehensive analysis of pedestrian and cyclist injuries in Sweden from 2003 to 2017, which was during the time that the "Vision Zero" road safety policy was implemented [9]. The authors found that injuries to both pedestrians and serious injuries to cyclists were significantly less common outside of the road where there was no interaction with traffic, and they recommended the separation of unprotected road users from motorized traffic. Aligned with the above conclusions are the results of the study by Soleil Cloutier et al. (2022), which set up a pilot project for allowing bike riding in a pedestrian street and monitored users' behavior and conflicts; the authors stated that the co-existence of pedestrians and cyclists only poses a few risks to users' safety [10].

Except for the abovementioned studies that use data about injuries and collisions, important indications about how safe the co-existence of pedestrians and cyclists is can be provided by questionnaire surveys that examine users' perceptions and attitudes. Kang and Fricker (2016) analyzed responses with respect to some recorded videos from shared infrastructures in China and concluded that the opinion of pedestrians about infrastructures where co-existence with cyclists is required tends to be negative; however, this negative attitude can be moderated if sufficient space is allocated to pedestrians and if speed restrictions for cyclists are set [11]. The issue of space allocation and cyclists' speed moderation has also been discussed by other studies. A study that was conducted in Thessaloniki, Greece, identified that over-dimensioning a bicycle lane in pedestrians–cyclists shared infrastructure has a negative impact on pedestrians' perceptions regardless of the cyclists' flows [12]; meanwhile, a study in Bristol concluded by stating that moderating cyclists' speed is essential for improving pedestrians' perceptions [13]. Some interesting points have been raised by the study of Hatfield and Prabhakharan (2016). Their analysis identified that an important issue for cyclists is that they fail to adequately supervise children or animals, while an issue that pedestrians face is that cyclists do not warn when they are about to pass [14]. A more recent study by Gkekas et al. (2020) highlighted that a high number of pedestrians and cyclists have experienced traffic conflict in shared infrastructure, which leads to cyclists avoiding areas where increased interactions with pedestrians exist [15].

Despite the limited number of serious injuries that have occurred from cyclist and pedestrian collisions, it has become understood that safety issues still exist, and the design of spaces where these two categories of users co-exist continues to be challenging [16,17]. To improve safety in pedestrian–cyclist shared spaces, several measures can be implemented. These measures mainly aim to reduce the conflicts and events between the users which, according to previous studies, have been shown to have an important effect on users' perceptions and result in a deteriorated perceived quality of service [18–20]; they are mostly experienced in cases of high traffic volumes and traffic complexity [21]. One of the most effective measures is to create clearly marked paths for each mode of transportation. This can include separate bike lanes and pedestrian walkways, each with their own unique markings and signage. Additionally, speed limits can be lowered to reduce the risk of accidents, and appropriate lighting can be installed to improve visibility, especially during nighttime hours. Other measures include the use of warning signs and pavement markings to alert cyclists to the presence of pedestrians. In cases where users have limited experience with pedestrian–cyclist co-existence, such measures are even more important, as the limited experience can result in a negative assessment of the infrastructure and confrontations between the users, therefore leading to an unwillingness to use the specific infrastructure [22,23].

The present paper aims to assess the performance of a measure that has rarely been applied internationally in practice. More specifically, this paper assesses how crosswalks in the form of zebra crossings within a bicycle lane affect not only the perceptions, but also the behavior of both cyclists and pedestrians in shared spaces. Because such a measure has only been used in limited cases worldwide, its assessment is considered essential; moreover, the assessment can provide evidence about whether this measure can be efficient for improving safety conditions in areas where high flows of cyclists and pedestrians are being concentrated. Focusing on the implementation of the pedestrian crossings as the main point of interest, we attempt to unravel the behavior and attitude of users regarding this measure, specifically with respect to their attitudes towards the other group of users and with respect to their perception about the shared space overall, including preferences for additional interventions. To this end, we have developed a conceptual framework for investigating the aforementioned issues. This framework was specified and estimated in the form of latent variable and path models for both pedestrians and cyclists. For the purposes of our research, we collected data via a questionnaire survey that was administered online. To the best of our knowledge, there is no other quantitative study that has aimed to holistically assess pedestrian crossings in bicycle lanes; therefore, the results of this paper can not only guide researchers, but also practitioners and authorities.

The remainder of the paper is organized as follows; Section 2 presents the area of our case study and the details of the questionnaire survey data collection. This is followed by Sections 3 and 4, which present some preliminary descriptive and inferential statistics analysis for providing a better understanding of the collected data. Section 5 presents our conceptual framework; this is followed by Section 6, which focuses on the modelling approach and results. This paper concludes with a Section 7.

2. Materials and Methods

2.1. Case Study

Thessaloniki is located in northern Greece. According to the latest data from the Hellenic Statistical Authority, the city's permanent population stands at 1,091,424; out of the permanent population, women comprise 52.5% (573,228) and men comprise 47.5%

(518,196). The transport system in Thessaloniki includes a comprehensive bus network, and a metro system is expected to start its operation in the beginning of 2024. Furthermore, some micromobility services exist [24,25]. Thessaloniki's Sustainable Urban Mobility Plan states that 41.3% of trips in the city are made using private cars, while public transport and motorcycles make up 33.7% and 11% of the modal share, respectively. However, the use of active modes of transportation such as walking and cycling is relatively low, accounting for only 9.2% and 1.7% of the modal share, respectively [26].

The most popular place for cycling in the city of Thessaloniki is the seafront area, where a bicycle lane runs parallel to the waterfront; this offers cyclists a comfortable and scenic route. The bicycle lane is well-marked and separated from vehicular traffic, making it an enjoyable experience for cyclists of all skill levels. The lane is also wide enough and mainly accommodates recreational cyclists. The seafront area also provides a great pedestrian-friendly environment for citizens and visitors of Thessaloniki. A spacious pedestrian promenade runs parallel to the bicycle lane and offers a great place for a leisurely walk or jog.

However, in the seafront area, the high levels of pedestrian and cycling flows, especially during weekends, result in conflicts between pedestrians and cyclists, which raises safety concerns. These concerns became even greater since the growth of micromobility vehicles. Following these issues and for the further designation of the areas that are provided for walking compared with those that are provided for cycling, horizontal signage has been added by the local authorities. Actually, pedestrian crossings, as shown in Figure 1, were painted for reducing the conflict points between cyclists and pedestrians and for alerting both user categories.

Figure 1. Pedestrian crossing in the bicycle lane of the seafront area.

According to Hamilton-Baillie [1,2], shared spaces are characterized by two main elements: the minimization of segregation between users and the reliance on mutual respect. These two elements are applicable in the seafront area, as segregation between cyclists and pedestrians exists; however, this does not include any physical elements, and the harmonious co-existence of users heavily relies on mutual respect. Moreover, the specific area can be considered as a destination place that mostly attracts people for recreational purposes, which is an additional attribute of shared spaces. In this sense, we characterize the seafront area as a shared space, even if an indication about space allocation exists.

2.2. Data Collection

To investigate the behavior of the users in the bicycle lanes' crossings and assess the performance of this measure, a questionnaire survey was carried out. At first, the questionnaire was designed; the questionnaire included four different sections. The first section includes questions that are related to both the socioeconomic profile of the respondent as well as the experience of the respondent with respect to the use of the seafront area, i.e., the main travel purpose when using the seafront and frequency of using it as a pedestrian and cyclist. The second section follows the point of view of a pedestrian and includes questions regarding the choice to use the crossings, the perceptions of it, and the assessment of cyclists' behavior when a pedestrian is using the crossings. The third section includes similar questions to the second one, but follows the point of view of a cyclist. In the fourth section, the respondents are asked to provide an assessment of the crossings with respect to safety and flow improvement. It should be noted that if the respondent was more frequently using the seafront as a pedestrian, they were asked to answer the second section of the questionnaire but not the third one; if they were mostly using it as a cyclist, they were asked to answer the third section but not the second one. This way, each respondent participated from the perspective in which they mostly experienced the infrastructure.

The questionnaire was designed in an electronic format and was distributed within the community of the Aristotle University of Thessaloniki by email to all registered members (i.e., both students and employees). This approach was selected due to the COVID-19 constraints and to reach a greater sample, as the university community has been found to be keener in responding to such surveys. A total of 1194 questionnaires were successfully completed, and they were used for the statistical analysis; from the responses, 1059 were from people that mostly used the seafront as pedestrians and 135 were from people that mostly used it for cycling.

3. Descriptive Statistics

3.1. Participants

Out of the overall sample (1194 respondents), 60.2% of respondents were female, 37.6% were male, and 2.2% did not wish to state their gender. With respect to age, 71.3% of respondents were 18–24 years while 21.2% were 25–39 years old. Regarding the other age groups, 6.4% of respondents were 40–54 years old and 1.2% were 55–64 years old. These findings suggest that our sample consisted of younger individuals. This potential bias towards younger respondents was potentially due to the fact that the survey was administered online and was also circulated within university-related means. The latter is further supported by the respondents' occupation, as 80.7% of respondents were university students. In the question regarding what type of user the respondents would identify themselves as, 88.7% stated that they were pedestrians and 11.3% identified as cyclists.

The study area itself is mainly a leisure infrastructure; this fact was also reflected in the listed trip purpose of the users, where 81% of respondents reported that they used the shared space for leisure/walking. This was followed by physical activity (12.7%), while other activities represented the remaining responses. More than half of the respondents (51.26%) never cycle at the infrastructure, and 72.9% walked at the area at least once a week (14.7% stated daily use). However, when focusing on respondents identifying themselves as cyclists, approximately 30% of respondents use the infrastructure daily, whereas approximately 41% use it more than once a week. That is, the cyclists of the sample are on average very familiar with the infrastructure.

Overall, the respondents were satisfied with the quality of the infrastructure as expressed in terms of cleanliness, safety, comfort, and aesthetics. Moreover, the implementation of pedestrian crossings was in general well-received as it was more perceived as helping with user interactions rather than interrupting their flows. Following this positive perception, it is very interesting that 72.11% of respondents stated that they would use the infrastructure more after the implementation of interventions. All suggested interventions (e.g., physical separation, better lighting conditions, different coloring of the bike lane, different surface material for the bike lane, additional signs at the pedestrian crossings) were considered to be very important or important, with the exception of moving the bicycle lane to a different location.

3.2. Pedestrians and Infrastructure

Out of the 1059 pedestrian respondents, 23.9% responded as having not ever noticed the existence of pedestrian crossings on the cycling lane. Additionally, although only 3.3% of respondents reported a collision with a cyclist, 46.2% mentioned a near miss, which is an expected finding because of the former being less likely to occur. There was not a clear trend regarding the perception towards sharing the space with cyclists, as the pedestrians' responses were split across different levels of agreement; however, there was a higher tendency to agree with the arrangement.

Regarding the pedestrian crossings, there was not a clear outcome observed regarding their choice for crossing the cycle path. However, when using the crossings, approximately 75% of respondents stated that they checked for oncoming bicycles before crossing. Moreover, pedestrians were positively inclined with respect to the perception of safety when using the crossings, and most pedestrians stated that they do not react with anger when a cyclist does not give them priority to cross.

Most pedestrians agreed that cyclists respect the boundaries of the bicycle lane. However, their perceptions mostly range from neutral to negative regarding cyclists decreasing their speed when they attempt to cross using a pedestrian crossing. The same also applies regarding cyclists giving priority to pedestrians and reacting with anger when pedestrians attempt to cross without the presence of a pedestrian crossing. However, lower levels of cyclist anger were reported for interactions that took place at a pedestrian crossing.

3.3. Cyclists and Infrastructure

Out of the 135 cyclists' responses, 90.4% reported that they observed the pedestrian crossings, which is a higher proportion compared with pedestrians. Moreover, 12.6% reported a crash with a pedestrian, while 71.9% reported a near miss. These are also higher levels, compared with the pedestrians' responses. Like pedestrians, there was not a distinct pattern with respect to the perception of cyclists for sharing space with the former user type.

The majority of cyclists (57.8%) totally agreed with respect to the statement of reducing their speed when a pedestrian was at a pedestrian crossing. Similarly, most cyclists reported giving priority to pedestrians and higher levels of reacting with anger when pedestrians crossed the bike lane without the presence of a pedestrian crossing. There was not a clear trend in the level of agreement regarding the additional safety by the pedestrian crossings. Cyclists agree that pedestrians do not respect the boundaries of the bicycle lane, and the cyclists do not agree that pedestrians choose the pedestrian crossings or check for oncoming bicycles while crossing.

4. Inferential Statistics

Before the modeling exercise, the responses of pedestrians and cyclists were compared to obtain a better understanding of the potential differences in how the two groups perceive each other. Given the ordered nature of the examined questions, the Wilcoxon rank sum test with continuity correction was used. The test only provides the p-value that is related to the significance; hence, the direction of the differences in responses was examined via the mean and median values. Some of the most notable findings are presented in the following subsections.

4.1. Users and Infrastructure

The responses of pedestrians and cyclists did not significantly differ with respect to sharing the infrastructure with each other; in general, there is a neutral opinion from both sides. Cyclists stated that they feel significantly less safe by the presence of pedestrian crossings ($W = 63{,}662$, $p = 0.034$), which may indicate that the infrastructure is more well-received by pedestrians. On the other hand, pedestrians are of the opinion that cyclists move outside the bike lane to a greater extent compared with cyclists perceiving pedestrians walking in the bike lane ($W = 45{,}158$, $p < 0.001$). It is interesting that despite the majority of pedestrians stating that they are carefully checking before crossing the bike lane, the opinion of cyclists is significantly different regarding the matter ($W = 33,200$, $p < 0.001$). Moreover, cyclists disagree to a greater extent that pedestrians choose the pedestrian crossings to walk through the bike lane ($W = 36,999$, $p < 0.001$). Cyclists also perceive pedestrians as reacting with anger more compared with what pedestrians reported about themselves ($W = 84,614$, $p < 0.001$). No significant difference was found regarding the reaction of cyclists in the event of a pedestrian crossing the bike lane. On the other hand, pedestrians disagree significantly more that cyclists reduce their speed when the former group is trying to cross the bike lane by using the pedestrian crossing ($W = 106,788$, $p < 0.001$). The same pattern was observed with respect to cyclists giving priority to pedestrians from the perspective of when the latter group is waiting at the pedestrian crossing ($W = 98,677$, $p < 0.001$).

4.2. Attitudes towards Infrastructure

The respondents were asked about a series of quality-related elements regarding the infrastructure, namely comfort, safety, aesthetic of the environment, and cleanness. No significant differences were found for these indicators except for the latter ($W = 79,822$, $p = 0.022$). Pedestrians disagree more regarding the cleanliness of the infrastructure; however, this observation is unlikely to be related to their interactions with cyclists. With respect to the implementation of the pedestrian crossings, pedestrians agree to a greater extent on the crossings provide more safety ($W = 56,085$, $p < 0.001$) compared with cyclists. Similarly, the former group agrees more that pedestrian crossings allow for easier interactions between the users ($W = 61,042$, $p = 0.004$). Pedestrians also had higher levels of agreement regarding the improvement of cyclists' flows due to the pedestrian crossings ($W = 63,398$, $p = 0.025$), whereas the opposite was found with respect to the disruption of flow ($W = 82,101$, $p = 0.003$). No significant differences were observed, however, between the two groups of users regarding the improvement or decline of pedestrians' smooth flow. Finally, the respondents were asked about a series of potential interventions that could be implemented to the infrastructure. No significant differences were observed regarding any of the interventions, except for the change in surface material for the bike lane; cyclists agreed more regarding this intervention ($W = 91,056$, $p < 0.001$).

5. Conceptual Framework

5.1. Variables

The variables of the survey were arranged in a number of groups that we considered in our analysis. These groups were:

- User's background: this group of variables refer to the background of the respondents and their past experience with the other group of users, particularly regarding any occurrence of a crash or near miss.
- Behavior of the other group: this concept refers to how a group of users perceives the behavior of the other group in their interactions at the infrastructure under study.
- Perceived quality of the infrastructure: this concept represents the general perception of a group of users regarding the overall quality of the infrastructure.
- Perception about pedestrian crossing: this group of variables captures the overall perception of a group of users regarding the efficiency of the pedestrian crossing.
- Interventions: this group represents the opinion of the respondents towards the implementation of specific interventions.
- Behavior of a group of users regarding the pedestrian crossings.

These groups of variables were considered in the development of a conceptual framework that would unravel the relationships between the attitudes of users about the infrastructure, the pedestrian crossings, their opinion about the other users and, ultimately, their behavior.

5.2. Factor Analysis and Latent Constructs

Prior to the estimation of the models that reflect our conceptual framework, a series of factor analyses was performed to examine the validity of the constructs used. In particular, we examined every group of questions using an exploratory factor analysis (EFA) to confirm whether the examined items (questions) were part of the same factor or if they were part of multiple factors. The results, including the expected factor and outcome after performing the EFA, are presented in the Appendix A in Tables A1 and A2 for the pedestrians and cyclists, respectively. In both tables, the factor loading values refer to the values along with the factor that each item was most related unless otherwise indicated. It should be mentioned that the variables in the two tables were not all examined simultaneously. The EFA was performed separately to each of the original groups in order to confirm our a priori expectations regarding the relevance among the questionnaire items of specific blocks of questions. Additional factors were then generated based on the results of the EFA for each group of questions.

Among the most interesting outcomes regarding the EFA analysis that was applied on the pedestrian sample, it is worth mentioning that our questions in the theme of Pedestrian and pedestrian crossings infrastructure were grouped into two categories: behavior at the pedestrian crossings, which includes items regarding pedestrians' behavior and reactions when using the pedestrian crossings; and attitudes towards using the pedestrian crossings. In the next general concept of the perceived cyclists' behavior when interacting with pedestrians at the bicycle lane, the items were grouped in two concepts, namely, cyclists' behavior and cyclists' anger. Moreover, the items related towards the perception of the pedestrian crossings in general were grouped into those related to the positive effects of pedestrian crossings in interactions and negative effects of pedestrian crossings in interactions. With respect to the implementation of interventions that could increase the use of the shared space infrastructure, two themes occurred: Soft interventions, which refer to measures that should be taken on the existing infrastructure/cycle lane; and Hard interventions, which refer to the implementation of physical separations or transportation of the bicycle lane to a different location. Finally, the general quality indicators of the infrastructure resulted in only one factor.

The cyclists' sample was considerably smaller; however, the EFA that was applied in the same manner as the pedestrians' sample suggested comparable results. In particular, regarding the pedestrians' behavior, two factors were extracted of which one could be related to pedestrians' anger. One difference compared with the pedestrians' sample regarding the perception towards the pedestrian crossings was found: although one factor was about the positive impact of pedestrian crossings, the second was related to the impact of pedestrian crossings on the flow of bicycles rather than the negative impact of the measure on both users' flows. Although the perception regarding the improvement of cyclists' flows had a higher loading on the latter factor, it also had a comparable value with respect to the positive impact factor. Another difference was related to one of the intervention items: although the hard intervention-related items were part of the same factor (as in the pedestrians' sample), one additional item was also related to this factor, specifically the different coloring of the bicycle lane. Finally, all items that were related to the perceived general quality of the infrastructure were part of the same factor.

5.3. Proposed Conceptual Framework

Our conceptual model was based on the idea of unravelling the stated behavior of pedestrians and cyclists with respect to the use of the pedestrian crossings. Moreover, by considering the stated behavior and overall attitudes towards the shared infrastructure, we aim to understand whether it would be possible to increase the use of the infrastructure. Based on the initial groups of variables and the constructs generated as part of the EFA analyses, the conceptual framework was developed as follows:

- Level 1: the perceptions of users that may drive their opinion about the infrastructure, the other group of users, and their interactions. An example is the perception about sharing spaces in general.
- Level 2: in this level, we considered the overall perception about the quality of the infrastructure under investigation.

- Level 3: in the first layer of this level, we considered the perception about the behavior of the other group of users under the presence of pedestrian crossings. We assume that this perception, together with the variables from the two previous levels, is what then affects the opinion about the usefulness of the pedestrian crossings in facilitating interactions.
- Level 4: here, we included the preference for the implementation of potential new interventions as a function of the variables from the previous levels.
- Level 5: in this final level, we examined the behavior of a group of users around the pedestrian crossings, their general feelings towards pedestrian crossings, and the potential of increasing the use of the infrastructure as a function of these two factors combined with the variables from the previous levels.

6. Results

6.1. Model Specification

Our conceptual framework was implemented for the pedestrians' sample in the form of a structural equation model. This approach was selected given that the EFA analysis suggested several factors that were ultimately implemented as latent variables. Structural equation models (SEMs) are composed of two main parts. The latent variable model captures the relationship between endogenous (dependent) and exogenous (independent) latent variables. The measurement model reflects the impact of latent variables on the observed variables (indicators). The main formula (Equation (1)) of a SEM is

$$\eta = B\eta + \Gamma\xi + \zeta \tag{1}$$

where η is an ($m \times 1$) vector of the endogenous variables, ξ is an ($n \times 1$) vector of the exogenous latent variables, and ζ is an ($m \times 1$) vector of the random disturbance. The m and n indicators indicate the number of the endogenous and exogenous latent variables. The elements of the B and Γ matrices are the parameters of the model. The B matrix is an ($m \times m$) parameter matrix of the latent endogenous variables, and the Γ matrix is an ($m \times n$) parameter matrix for the latent exogenous variables. The main formulae of the measurement model are

$$x = \Lambda x\xi + \delta \tag{2}$$

for the exogenous variables (Equation (2)) and

$$y = \Lambda y\eta + \varepsilon \tag{3}$$

for the endogenous variables (Equation (3)), where the observed variables are indicated by the vectors y ($p \times 1$) and x ($q \times 1$). The p and q indicators denote the number of the endogenous and exogenous indicator (observed) variables, respectively. The matrix Λy ($p \times m$) reflects the parameters of the y elements, while the matrix Λx ($q \times n$) indicates the parameters of the x elements. The measurement errors for y are indicated by the ($p \times 1$) vector ε and for x by the ($q \times 1$) vector δ.

Although the same sets of variables were available both for the pedestrians' and cyclists' samples, the latter had a considerably smaller sample size. Because of this, we followed a more simplified approach. That is, rather than considering latent variables, we implemented a path model where the items' values that composed each factor of the cyclists' sample were averaged. These simpler average values were then used as the model input. Both models were estimated using the diagonally weighted least square method, which is more robust for categorical or non-normal variables, using the lavaan (v 0.6.12) [27] package of R software [28].

6.2. Results for Pedestrians

The results of the pedestrians' model are presented in Table 1. The measurement equation model results are presented in Table A3 of the Appendix A. The item codes in

Table A3 correspond to those presented in Table A1. The goodness-of-fit indices presented in Table A4 of the Appendix A suggest a moderate but acceptable fit. Starting from the more generic aspects of pedestrians' perceptions, the general attitude towards sharing the infrastructure with cyclists had a negative association with reporting a near miss with a cyclist. That is, having a negative experience with a cyclist was negatively related to sharing space with the latter. The attitude towards sharing was positively related to the perceived overall quality of the infrastructure. Hence, pedestrians who do not agree with shared space are also less likely to attribute qualities such as comfort or safety to the infrastructure. Older respondents were more likely to perceive the infrastructure as of lower quality. Older individuals as well as those who reported a near miss with a cyclist were also less likely to hold a positive opinion towards the positive behavior of cyclists when interacting with pedestrians at the bicycle lane. The perception about the positive impact of pedestrian crossings was related to the general perception about the quality of infrastructure and the perception towards cyclists' behavior. In the next level of the model, the variables related to the general perception towards the infrastructure, cyclists, and pedestrian crossings were used to investigate their relationship with preferred interventions that could increase the infrastructure's use. The preference for soft interventions were positively related to the positive perception about the pedestrian crossings and attitude towards sharing the space with cyclists. On the other hand, a positive perception regarding the general quality of the infrastructure was negatively related to the preference for the implementation of soft interventions. With respect to the preference for the implementation of hard interventions, the same outcomes are observed with respect to the perception of the pedestrian crossings, the overall quality of the infrastructure, and the perceived efficiency of the pedestrian crossings. Moreover, the preference for hard interventions was negatively associated with the perceived positive behavior of cyclists in interactions at the bike lane. Pedestrians who perceive such behavior to a lesser extent are more likely to prefer the implementation of hard interventions, i.e., physical separation or moving the infrastructure to a different location. The final level of the model regards the impact of both the general perception and the specific perception about the pedestrian crossings of pedestrians on their behavior and preference for the latter. Pedestrians were more likely to report negative behavior (indicated by not checking for bicycles before crossing or expressing anger) if they did not share a positive attitude towards the behavior of cyclists in their interactions or towards the efficiency of the pedestrian crossings, preferred the implementation of hard interventions, or did not agree with the shared space. Pedestrians' preference for pedestrian crossings was related to the perception about cyclists' behavior and the positive effects of implementing the pedestrian crossings. Finally, the positive effects of implementing the pedestrian crossings and the preference for the implementation of both soft and hard interventions was positively associated with the increased use of the infrastructure.

Table 1. Structural model parameter estimates—Pedestrians' model.

Paths			Estimate	z-Value	p-Value
Perception towards sharing space	←	Near miss with a cyclist	−0.25	−3.60	0.000
Overall quality	←	Perception towards sharing space	0.26	12.31	0.000
Overall quality	←	Age: 18–24	−0.17	−3.52	0.000
Perceived cyclists' (positive) behavior	←	Near miss with a cyclist	−0.21	−4.14	0.000
Perceived cyclists' (positive) behavior	←	Age: 18–24	−0.28	−4.77	0.000
Positive effect of pedestrian crossings	←	Perceived cyclists' (positive) behavior	0.18	8.11	0.000
Positive effect of pedestrian crossings	←	Overall quality	0.17	9.87	0.000
Soft interventions	←	Positive effect of pedestrian crossings	0.39	15.63	0.000
Soft interventions	←	Overall quality	−0.20	−8.48	0.000
Soft interventions	←	Perception towards sharing space	0.09	4.24	0.000
Hard interventions	←	Perceived cyclists' (positive) behavior	−0.87	−4.10	0.000
Hard interventions	←	Positive effect of pedestrian crossings	0.88	3.94	0.000

Table 1. *Cont.*

	Paths		Estimate	z-Value	p-Value
Hard interventions	←	Overall quality	−1.11	−4.07	0.000
Pedestrians' (negative) behavior	←	Perceived cyclists' (positive) behavior	−0.50	−7.16	0.000
Pedestrians' (negative) behavior	←	Hard interventions	−0.32	−3.83	0.000
Pedestrians' (negative) behavior	←	Positive effect of pedestrian crossings	0.30	5.64	0.000
Pedestrians' (negative) behavior	←	Perception towards sharing space	0.24	6.72	0.000
Attitudes towards using the pedestrian crossing	←	Perceived cyclists' (positive) behavior	0.18	6.05	0.000
Attitudes towards using the pedestrian crossing	←	Positive effect of pedestrian crossings	0.38	12.16	0.000
Increased use after interventions	←	Positive effect of pedestrian crossings	0.03	2.10	0.036
Increased use after interventions	←	Soft interventions	0.09	7.68	0.000
Increased use after interventions	←	Hard interventions	0.05	4.06	0.000

6.3. Results for Cyclists

The results of the pedestrians' model are presented in Table 2. The goodness-of-fit indices presented in Table A4 of the Appendix A suggest a good fit. With respect to the overall perceived quality of the infrastructure, cyclists who did not agree with sharing the space with pedestrians were also more likely to perceive lower levels of quality. Moreover, female respondents were more likely to perceive an overall higher quality. A near miss with a pedestrian was negatively related to the positive perceived behavior of pedestrians when the latter cross the bike lane. The same observation was also made in the model of pedestrians. Another similar trend in the pedestrians' model was that the perceived overall quality was positively related to positive effects in interactions due to the presence of pedestrian crossings. The latter was negatively associated with the preference for the implementation of hard interventions. It should be mentioned that among the tested model specifications, the overall perception about the quality of the infrastructure was also positively related to the perception about the implementation of hard interventions. However, it was not possible to include in the model specification both the perception about the positive impact of pedestrian crossings and the overall perceived quality, as the direct effect of the latter was losing significance. It is likely that both variables are important, but our model was not able to capture these effects due to limitations in the sample size. The reported behavior of cyclists was negatively associated with the perceived positive behavior of pedestrians. That is, cyclists who perceived a negative behavior from pedestrians were also more likely to express behavior such as speed reduction or giving priority. Positive cyclist behavior was also related to the perception about sharing the road with pedestrians. These findings are consistent with the pedestrians' model. Among the various model specifications tested, the overall perceived quality of the infrastructure was positively related to the positive cyclists' behavior. However, this variable was losing significance if both this and the perception towards sharing were included (possibly given that their relation was already included in the model specification); hence, the overall perceived quality was dropped in the final model specification. It is possible that infrastructure that is not perceived to be of high quality (with sharing the space being one of the reasons for this) could trigger aggressive behavior to a greater extent. Regarding the preference for soft interventions, unlike the pedestrians' model, no significant associations were found. The inclination towards having a positive perception for using the infrastructure (reflected in the perception of safety alone) was related to the overall perceived benefits from the presence of pedestrian crossings. This finding is the same as the one obtained in the pedestrians' model. In the latter, however, the behavior of the other group of users also had a significant impact. Finally, the willingness to use the infrastructure more was related to the implementation of hard interventions. Overall, the two models are consistent; however, fewer significant relationships were found in the pedestrians' model, which could be attributed to sample size limitations.

Table 2. Structural model parameter estimates—Cyclists' model.

Paths			Estimate	z-Value	p-Value
Overall quality	←	Perception towards sharing space	0.22	4.17	0.000
Overall quality	←	Gender: Female	−0.40	−2.69	0.007
Perceived pedestrians' (positive) behavior	←	Near miss with a pedestrian	−0.67	−3.39	0.001
Positive effect of pedestrian crossings	←	Overall quality	0.29	3.77	0.000
Hard interventions	←	Positive effect of pedestrian crossings	−0.29	−2.76	0.006
Cyclists' (negative) behavior	←	Perceived pedestrians' (positive) behavior	−0.55	−3.99	0.000
Cyclists' (negative) behavior	←	Perception towards sharing space	0.23	3.21	0.001
Attitudes towards interacting with the pedestrian crossing	←	Positive effect of pedestrian crossings	0.57	3.42	0.001
Increased use after interventions	←	Hard interventions	0.09	2.04	0.041

6.4. Sample Size and Power Analysis

As reported in Table A4 of the Appendix A, the goodness-of-fit indices suggested a good fit of the models to the data. An additional test focuses on the sample size of the models, which is relevant especially for the cyclists' model due to the lower sample size. In general, there are no strict rules for sample sizes in SEM models so much as rules of thumb [29]. Another approach is to investigate the overall statistical power of a model with respect to a specific goodness-of-fit indicator (for instance, RMSEA) and with respect to Type II errors. For instance, a power analysis can be conducted to determine the sample size for an SEM model, which can detect the misspecification expressed as RMSEA ≥ 0.05; this is usually considered as the maximum acceptable value for a great model fit (or any other critical value) [30]. Except for this a priori power analysis, it is also possible to examine the achieved power of a model with respect to a given sample size and the degrees of freedom (post hoc power analysis). We conducted the aforementioned types of power analyses using the R package 'semPower' (v. 2.0.1) [31]. The results are presented in Table 3 where, for the given degrees of freedom (taken from our models' results), we conducted a priori and post hoc analyses for different levels of RMSEA accuracy. For the pedestrians' model, the required sample size was always smaller than what was available in the data sample size. Moreover, the post hoc model could be able to detect a misspecification expressed as RMSEA with a probability higher than >0.999 for all of the critical values of RMSEA tested. On the other hand, the cyclists' model would require a sample of 225 individuals to detect a power of 80% RMSEA ≥ 0.05. However, relaxing this to higher values (either 0.08 or 0.1) results in the required sample sizes being smaller than what is available in the data. This is further reflected in the post hoc analysis, where the power for detecting RMSEA ≥ 0.05 was only 0.49; however, this probability was above 0.95 for critical RMSEA values of 0.08 or 0.1. Hence, the cyclists' model could be benefited from an additional sample as it would help to detect even more significant relationships via the model; however, the current sample size is sufficient to ensure the detection of misspecification under a moderate fit.

Table 3. Results of the power analyses.

	Pedestrians' Model		Cyclists' Model	
RMSEA	Sample Size (A Priori)	Power (Post Hoc)	Sample Size (A Priori)	Power (Post Hoc)
0.05	93	>0.999	225	0.49
0.08	37	>0.999	89	0.967
0.1	24	>0.999	58	>0.999
	N = 1059		N = 135	
	df = 281		df = 57	

7. Discussion

Whether it is for active or sustainable mobility, walking and cycling represent two very important means of transportation. It is essential that we better understand the drives for selecting such alternatives in everyday life. In this study, we followed the implementation of a number of pedestrian crossings along the bike lane of a shared space infrastructure for pedestrians and cyclists. We then conducted a quantitative analysis of exploratory nature to unravel the behavior and perceptions of users, taking into account their overall experience and perceptions. Previous interactions with other road users (e.g., near misses) are of essential importance in the formation of perceptions and attitudes, and these interactions can also affect the overall opinion about the infrastructure, as has been recognized by previous studies [15,32]. Although the pedestrian crossings do not physically separate the two groups of users, they may create a sense of "order" in the interactions between users. In the case of pedestrians, the pedestrian crossings were not perceived for facilitating only interactions that were related to the behavior of cyclists, but they were also related to the perceived quality of the infrastructure. That is, although shared spaces may be more optimal solutions with respect to saving space, they may not be as well-received by users who may prefer some form of structure in the interactions. This is further validated in the pedestrians' model, where there was a relationship between the positive impact in the interactions from the presence of the pedestrian crossings and the preference for the implementation of further interventions, comprising both soft and hard interventions. The latter components were also negatively related to the overall perceived quality, which is a factor that is also related to the perception of the pedestrian crossings. A very interesting finding was that for the pedestrians, a preference towards the implementation of hard interventions was positively related to higher levels of stated aggressive behavior. It should be noted that previous studies identify a clear preference among pedestrians for a clear indication of the cycling lane [12,13]; meanwhile, the preference among cyclists opinions are split; some studies identify that cyclists also prefer a clear indication of the cycling lane [23,33], while another study has identified that cyclists prefer mixed traffic conditions [13]. Moreover, perceptions about the contribution of the pedestrian crossings in facilitating interactions was also significant, along with the perceived behavior and perceptions towards sharing the infrastructure. For cyclists, the relationship between the perception of aggressive behavior about pedestrian crossings (and hence the current state of things) is indirect, as the latter item was associated with the implementation of hard interventions. For both pedestrians and cyclists, the perception that the pedestrian crossings could facilitate interactions also influences the sense of safety. We can hence observe a pattern that is similar to the existing situation in the interactions between pedestrians and motorized traffic; the pedestrian crossings provide some sort of structure in the interactions, determining priorities and potentially raising the perceived safety and overall perceived reliability of the system. Although the implementation of pedestrian crossings along a bicycle lane may have less defined rules in practice, it also denotes some sort of order in the execution of interactions. Of course, this is most likely the case for the users who agree with this specific measure. Unravelling our conceptual framework backwards, the preference for using the pedestrian crossings or perceiving additional safety by their presence is ultimately related to the initial perception about shared spaces. Regardless of if they are users who perceive the pedestrian crossings as useful and prefer to use them or users who simply exhibit aggressive behavior in interactions that take place in the bicycle lane, the common element is the distinct separation and definition of rules in the interactions. This is an expecting finding, as the difference in how a group of users perceives the other was evident in the inferential statistics analysis. The competitiveness of how the groups perceive each other was reflected in them having similar variables regarding the attitudes towards the pedestrian crossings and their extensions in defining rules for interactions, which are eventually linked to negative behavior and the preference for interventions that would separate the users.

Of course, while interpreting our results, one needs to consider that all responses were collected via an online survey; no field measurements were performed to confirm the validity of the trends found in our data. This has two main potential negative extensions that could induce bias to our results: (a) our findings only represent the views of those familiar with online technologies and which could access the survey; and (b) there is no validation regarding the actual situation. On top of these issues, one must consider that our sample mostly consisted of university students; this was mainly due to the channels that we used for the circulation of the survey, which was web-based and was not performed in person due to the COVID restrictions at the time. Another major issue that potentially affected our capability in observing significant results was the low sample size for cyclists, which can be attributed to the same reasons. Although it is very likely that the proportion between pedestrians and cyclists would be comparable in real-life observations at the shared infrastructure, additional data are still required to deduce more robust conclusions with respect to cyclists. Hence, subjectivity is an issue here, which was made obvious when the perceptions of the two groups of users were compared for similar matters and when significant differences were found. Moreover, our results are potentially more representative for specific groups of pedestrians and cyclists rather than the total population of these users. Potential extensions of the current research could focus on collecting the same data from a broader sample via an in-person data collection rather than online data collection. Moreover, more observations from cyclists are required. Additionally, field observations at crossings and bike lanes in general locations could provide very useful insights with respect to the existing situation and interactions between users. It must be mentioned that the initial objective of the current research study was to collect field observations; however, due to circumstances related to the COVID-19 pandemic, the original research direction was changed. However, it must be highlighted that when investigating factors behind use or intention to use, survey data are more crucial in order to understand the underlying factors. Field observations, on the other hand, can have a supplementary role in providing insights with respect to the discrepancies between how one perceives themselves and the actual situation.

Nevertheless, the patterns observed in our conceptual framework, which was applied to a great extent in both of our samples, suggest that some level of segregation is desirable. Even though our sample considered the physical separation as important, such an intervention could reduce the flexibility and comfort for all users. Hence, measures such as different coloring or surface material and additional vertical signs could enhance the sense of separation and improve user interactions. Given the preferences towards some sort of separation, it is very likely that completely unmarked shared spaces could be confusing for users and could lead to considerably lower cycling speeds. This seems to be the case at least in cities, where cycling and space sharing experience is limited, as such an experience is essential for the harmonious co-existence of users and can highly affect users' perceptions [22]. At the same time, cycling lanes must be sufficiently wide enough to allow for a good level of service for the bicycle flows. Despite the implementation of any potential intervention, a major issue, as derived from our inferential statistics analysis, is based on the beliefs that the two groups of users have for each other, which could suggest a mutual perceived lack of respect. For instance, it is very likely that cyclists do not feel additional safety due to the presence of crossings, as they also believe that pedestrians do not cross only from these areas. On the other hand, pedestrians perceive cyclists as more aggressive compared with how the latter group perceives themselves. Although these issues are overlooked due to the small number of physical conflicts and their low severity, it must be ensured that users understand how to behave when using specific parts of the infrastructure. This issue, combined with the previously mentioned challenges, must be kept in mind both during the development of new infrastructure and the implementation of interventions in existing infrastructure by the respective authorities.

Author Contributions: Conceptualization, S.B.; methodology, E.P., A.N. and S.B.; software, E.P.; validation, E.P.; formal analysis, C.M. and E.P.; investigation, C.M. and S.B.; resources, S.B.; data curation, C.M. and E.P.; writing—original draft preparation, E.P. and A.N.; writing—review and editing, E.P., A.N. and S.B.; visualization, E.P. and A.N.; supervision, S.B. All authors have read and agreed to the published version of the manuscript.

Funding: This research received no external funding.

Institutional Review Board Statement: The study was conducted in accordance with the Declaration of Helsinki and was approved by the Institutional Ethics Committee of Aristotle University of Thessaloniki (protocol code 86794/2021; date of approval 8 April 2021).

Informed Consent Statement: Informed consent was obtained from all subjects involved in the study.

Data Availability Statement: The data are not publicly available, but they can be provided upon request to the corresponding author.

Conflicts of Interest: The authors declare no conflict of interest.

Appendix A

Table A1. Factor loadings of the pedestrians' sample EFA analysis.

Expected Factor	Derived Factor	Items	Item Code	Factor Loading
Pedestrian and pedestrian crossing	Behavior at pedestrian crossing	Check for oncoming bicycles before crossing	I_1	0.756
		React with anger when cyclist does not give priority	I_2	−0.748
	Attitudes towards using the pedestrian crossing	Choice of pedestrian crossings to cross the bike lane	I_3	0.773
		Increased sense of safety when choosing the pedestrian crossing	I_4	0.74
Perception about cyclists' behavior	Cyclists' behavior	Cyclists reduce their speed at pedestrian crossings	I_5	0.95
		Cyclists give priority at pedestrian crossings	I_6	0.659
		Cyclists are moving within the bike lane boundaries	I_7	0.361
	Cyclists' anger	Cyclists react with anger when pedestrian crosses from a pedestrian crossing	I_8	0.997
		Cyclists react with anger when pedestrian does not cross from a pedestrian crossing	I_9	0.461
Perception about the pedestrian crossing	Positive effect of pedestrian crossing	The pedestrian crossings facilitate interactions	I_{10}	0.783
		The pedestrian crossings provide additional safety	I_{11}	0.72
		The pedestrian crossings facilitate the pedestrians' flows	I_{12}	0.585
		The pedestrian crossings facilitate the cyclists' flows	I_{13}	0.498
	Negative effect of pedestrian crossing	The pedestrian crossings interrupt the cyclists' flows	I_{14}	0.678
		The pedestrian crossings interrupt the pedestrians' flows	I_{15}	0.494
Quality indicators of the shared infrastructure (Overall quality)		Aesthetics/environment	I_{16}	I_{16}
		Comfort	I_{17}	I_{17}
		Safety	I_{18}	I_{18}
		Cleanness	I_{19}	I_{19}
Interventions	Soft interventions	Different material for the pedestrian crossings	I_{20}	0.662
		Additional signs at the pedestrian crossings	I_{21}	0.659
		Additional lighting at the pedestrian crossings	I_{22}	0.649
		Different surface material for the bike lane	I_{23}	0.615
		Different coloring of the bike lane	I_{24}	0.577
	Hard interventions	Different location for the bike lane	I_{25}	0.655
		Physical separation of the bike lane	I_{26}	0.362

Table A2. Factor loadings of the cyclists' sample EFA analysis.

Expected Factor	Derived Factor	Items	Factor Loading
Cyclists and pedestrian crossings		Speed reduction at pedestrian crossings	0.756
		Give priority to pedestrians at pedestrian crossings	0.765
Perception about pedestrians' behavior	Pedestrians' behavior	Pedestrians are respecting the bike lane boundaries	0.86
		Pedestrians are checking for oncoming bicycles when crossing	0.73
		Pedestrians are choosing the pedestrian crossings to cross	0.604
	Pedestrians' anger	Pedestrians are reacting with anger with cyclists do not give them priority	0.581
Perception about the pedestrian crossings	Positive effect of pedestrian crossings	The pedestrian crossings provide additional safety	0.795
		The pedestrian crossings facilitate interactions	0.753
		The pedestrian crossings facilitate the cyclists' flows	0.488 (−0.53 *)
		The pedestrian crossings facilitate the pedestrians' flows	0.455
	Cyclists' flow *	The pedestrian crossings interrupt the cyclists' flows	0.7 *
Quality indicators of the shared infrastructure (overall quality)		Aesthetics/environment	0.803
		Comfort	0.603
		Safety	0.597
		Cleanness	0.713
Interventions	Soft interventions	Additional signs at the pedestrian crossings	0.761
		Additional lighting at the pedestrian crossings	0.545
		Different material for the pedestrian crossings	0.446
		Different coloring of the bike lane	0.328
	Hard interventions	Different location for the bike lane	0.62
		Physical separation of the bike lane	0.542
		Different surface material for the bike lane	0.417

* Loadings associated to the Cyclists' flow factor.

Table A3. Measurement model results of pedestrians' model.

Paths			Estimate	z-Value	P (>|z|)
Perceived cyclists' (positive) behavior	←	I_7	0.43	12.75	0.000
Perceived cyclists' (positive) behavior	←	I_5	0.90	17.69	0.000
Perceived cyclists' (positive) behavior	←	I_6	0.93	17.65	0.000
Overall quality	←	I_{17}	0.81	27.17	0.000
Overall quality	←	I_{18}	0.60	24.03	0.000
Overall quality	←	I_{16}	0.96	28.27	0.000
Overall quality	←	I_{19}	0.68	25.25	0.000
Soft interventions	←	I_{22}	0.62	21.48	0.000
Soft interventions	←	I_{23}	0.65	21.10	0.000
Soft interventions	←	I_{20}	0.73	22.20	0.000
Soft interventions	←	I_{21}	0.65	21.84	0.000
Soft interventions	←	I_{24}	0.57	19.81	0.000
Hard interventions	←	I_{25}	0.17	4.23	0.000
Hard interventions	←	I_{26}	0.25	4.46	0.000
Positive effect of pedestrian crossings	←	I_{11}	0.76	23.20	0.000
Positive effect of pedestrian crossings	←	I_{10}	0.75	22.87	0.000
Positive effect of pedestrian crossings	←	I_{13}	0.49	18.11	0.000
Positive effect of pedestrian crossings	←	I_{12}	0.58	19.87	0.000
Pedestrians' (negative) behavior	←	I_1	1.13	13.73	0.000
Pedestrians' (negative) behavior	←	I_2	−0.71	−15.52	0.000
Attitudes towards using the pedestrian crossings	←	I_3	0.77	17.09	0.000
Attitudes towards using the pedestrian crossings	←	I_4	1.19	15.54	0.000

Table A4. Goodness-of-fit indices.

Measure	Value	
	Pedestrians' Model	Cyclists' Model
Absolute fit		
RMSEA	0.059	0.020
SRMR	0.064	0.077
GFI	0.98	0.979
Incremental fit		
AGFI	0.976	0.971
CFI	0.850	0.966
TLI	0.828	0.961

References

1. Hamilton-Baillie, B. Shared Space: Reconciling People, Places and Traffic. *Built Environ.* **2008**, *34*, 161–181. [CrossRef]
2. Hamilton-Baillie, B. Towards shared space. *Urban Des. Int.* **2008**, *13*, 130–138. [CrossRef]
3. Stepan, O.; Rotaru, I. Street design, streetscape and traffic calming. *Transp. Learn. Proj. Train. Modul.* **2012**, *5*, 1–57.
4. Rupprecht Consult (Ed.) *Guidelines for Developing and Implementing a Sustainable Urban Mobility Plan*, 2nd ed. 2019. Available online: https://www.eltis.org/sites/default/files/sump_guidelines_2019_interactive_document_1.pdf (accessed on 1 April 2023).
5. Aultman-Hall, L.; LaMondia, J. Evaluating the Safety of Shared-Use Paths: Results from Three Corridors in Connecticut. *Transp. Res. Rec. J. Transp. Res. Board* **2005**, *1939*, 99–106. [CrossRef]
6. Chong, S.; Poulos, R.; Olivier, J.; Watson, W.; Grzebieta, R. Relative injury severity among vulnerable non-motorised road users: Comparative analysis of injury arising from bicycle-motor vehicle and bicycle-pedestrian collisions. *Accid. Anal. Prev.* **2010**, *42*, 290–296. [CrossRef]
7. Grzebieta, R.; McIntosh, A.; Chong, S. Pedestrian-Cyclist Collisions: Issues and Risk. In Proceedings of the Australasian College of Road Safety Conference, Melbourne, Australia, 1–2 September 2011.
8. O'Hern, S.; Oxley, J. Pedestrian injuries due to collisions with cyclists Melbourne, Australia. *Accid. Anal. Prev.* **2019**, *122*, 295–300. [CrossRef]
9. Varnild, A.; Tillgren, P.; Larm, P. What types of injuries did seriously injured pedestrians and cyclists receive in a Swedish urban region in the time period 2003–2017 when Vision Zero was implemented? *Public Health* **2020**, *181*, 59–64. [CrossRef]
10. Soleil Cloutier, M.; Brodeur-Ouimet, P.; Lamarche, A.A.; Pierre-Maxime, L. Making the Most of a Pedestrian Street: Insights from a Pilot-Project Allowing Cyclist to Ride Along Pedestrians in Montreal, Canada. *J. Transp. Health* **2022**, *25*, 101438. [CrossRef]
11. Kang, L.; Fricker, J. Sharing urban sidewalks with bicyclists? An exploratory analysis of pedestrian perceptions and attitudes. *Transp. Policy* **2016**, *49*, 216–225. [CrossRef]
12. Nikiforiadis, A.; Basbas, S. Can pedestrians and cyclists share the same space? The case of a city with low cycling levels and experience. *Sustain. Cities Soc.* **2019**, *46*, 101453. [CrossRef]
13. Delaney, H.; Melia, S.; Parkhurst, G. Walking and cycling on shared-use paths: The user perspective. In Proceedings of the Institution of Civil Engineers-Municipal Engineer, London, Canada, 1–4 June 2016; Volume 170, pp. 175–184.
14. Hatfield, J.; Prabhakharan, P. An investigation of behaviour and attitudes relevant to the user safety of pedestrian/cyclist shared paths. *Transp. Res. Part F Traffic Psychol. Behav.* **2016**, *40*, 35–47. [CrossRef]
15. Gkekas, F.; Bigazzi, A.; Gill, G. Perceived safety and experienced incidents between pedestrians and cyclists in a high-volume non-motorized shared space. *Transp. Res. Interdiscip. Perspect.* **2020**, *4*, 100094. [CrossRef]
16. Nikiforiadis, A.; Basbas, S.; Mikiki, F.; Oikonomou, A.; Polymeroudi, E. Pedestrians-Cyclists Shared Spaces Level of Service: Comparison of Methodologies and Critical Discussion. *Sustainability* **2021**, *13*, 361. [CrossRef]
17. Zhang, C.; Du, B.; Zheng, Z.; Shen, J. Space sharing between pedestrians and micro-mobility vehicles: A systematic review. *Transp. Res. Part D Transp. Environ.* **2023**, *116*, 103629. [CrossRef]
18. Botma, H. Method to determine level of service for bicycle paths and pedestrian-bicycle paths. *Transp. Res. Rec.* **1995**, *1502*, 38–44.
19. Nikiforiadis, A.; Basbas, S.; Garyfalou, M.I. A methodology for the assessment of pedestrians-cyclists shared space level of service. *J. Clean. Prod.* **2020**, *254*, 120172. [CrossRef]
20. Kazemzadeh, K.; Laureshyn, A.; Winslott Hiselius, L.; Ronchi, E. Expanding the Scope of the Bicycle Level-of-Service Concept: A Review of the Literature. *Sustainability* **2020**, *12*, 2944. [CrossRef]
21. Liang, X.; Meng, X.; Zheng, L. Investigating conflict behaviours and characteristics in shared space for pedestrians, conventional bicycles and e-bikes. *Accid. Anal. Prev.* **2021**, *158*, 106167. [CrossRef] [PubMed]
22. Nikiforiadis, A.; Chatzali, E.; Ioannidis, V.; Kalogiros, K.; Paipai, M.; Basbas, S. Investigating factors that affect perceived quality of service on pedestrians-cyclists shared infrastructure. *Travel Behav. Soc.* **2023**, *31*, 323–332. [CrossRef]
23. Paschalidis, E.; Basbas, S.; Politis, I.; Prodromou, M. "Put the blame on . . . others!": The battle of cyclists against pedestrians and car drivers at the urban environment. A cyclists' perception study. *Transp. Res. Part F Traffic Psychol. Behav.* **2016**, *41*, 243–260. [CrossRef]

24. Ayfantopoulou, G.; Salanova Grau, J.M.; Maleas, Z.; Siomos, A. Micro-Mobility User Pattern Analysis and Station Location in Thessaloniki. *Sustainability* **2022**, *14*, 6715. [CrossRef]
25. Nikiforiadis, A.; Paschalidis, E.; Stamatiadis, N.; Paloka, N.; Tsekoura, E.; Basbas, S. E-scooters and other mode trip chaining: Preferences and attitudes of university students. *Transp. Res. Part A Policy Pract.* **2023**, *170*, 103636. [CrossRef]
26. Centre for Research and Technology Hellas—Hellenic Institute of Transport. Thessaloniki's Sustainable Urban Mobility Plan, 1st ed. 2019. Available online: https://www.svakthess.imet.gr/Portals/0/Diavoulefseis/Diavoulefsi03/SVAK_Thessalonikis.pdf (accessed on 1 April 2023).
27. Rosseel, Y. lavaan: An R package for structural equation modeling. *J. Stat. Softw.* **2012**, *48*, 1–36. [CrossRef]
28. R Core Team. R: A language and environment for statistical computing. In *R Foundation for Statistical Computing*; R Core Team: Vienna, Austria, 2022; Available online: http://www.R-project.org/ (accessed on 1 April 2023).
29. Kyriazos, T.A. Applied psychometrics: Sample size and sample power considerations in factor analysis (EFA, CFA) and SEM in general. *Psychology* **2018**, *9*, 2207. [CrossRef]
30. Jobst, L.J.; Bader, M.; Moshagen, M. A tutorial on assessing statistical power and determining sample size for structural equation models. *Psychol. Methods* **2021**, *28*, 207–221. [CrossRef]
31. Moshagen, M. Power Analysis for Structural Equation Models: SemPower Manual. 2021. Available online: https://cran.r-project.org/web/packages/semPower/semPower.pdf (accessed on 1 April 2023).
32. Poulos, R.G.; Hatfield, J.; Rissel, C.; Flack, L.K.; Grzebieta, R.; McIntosh, A.S. Cyclists' self-reported experiences of, and attributions about, perceived aggressive behaviour while sharing roads and paths in New South Wales, Australia. *Transp. Res. Part F Traffic Psychol. Behav.* **2019**, *64*, 14–24. [CrossRef]
33. Nazemi, M.; van Eggermond, M.; Erath, A.; Axhausen, K. Studying Cyclists' Behavior in a Non-naturalistic Experiment Utilizing Cycling Simulator with Immersive Virtual Reality. In Proceedings of the 98th Annual Meeting of the Transportation Research Board, Washington, DC, USA, 13–17 January 2019.

 *sustainability*

Article

Traffic Circle—An Example of Sustainable Home Zone Design

Stanisław Majer and Alicja Sołowczuk *

Department of Construction and Road Engineering, West Pomeranian University of Technology in Szczecin, 71-311 Szczecin, Poland; stanislaw.majer@zut.edu.pl
* Correspondence: alicja.solowczuk@zut.edu.pl; Tel.: +48-91-449-40-36

Abstract: A significant number of new metered parking systems have been introduced in recent years by the local authorities of various spa towns in Poland in connection with home zone conversion projects. The traffic signs posted in these locations were limited to the beginning and end of the demarcated parking area. Traffic circle (TC) is an example of a traffic calming measure (TCM) used in home zones to slow down the traffic (case study—home zone in a small spa village). This article presents the results of a study investigating the speed reductions obtained within a home zone and a traffic circle used as traffic calming measure. The indispensable speed surveys were carried out in relation to this study in two periods: in summer when the streets are crowded with tourists and in September with little pedestrian traffic. Two research hypotheses were formulated as part of the speed data analysis to verify the slowing effect of the traffic circle and the relevance of the traffic circle's design parameters and location, road function and the surrounding streetscape. For each hypothesis, statistical analyses were carried out using two nonparametric tests: two-sample Kolmogorov–Smirnov test and median test. The third research hypothesis formulated in this study was related to sustainable development factors related to fuel consumption and traffic-related air pollution, including carbon dioxide, carbon monoxide, nitrogen oxide and hydrocarbons. This hypothesis was verified by estimating the amount of air pollution in the home zone under analysis in three different situations (scenarios): in summer with the travel speed reduced by pedestrian traffic to ca. 8–10 km/h, in September with a small number of pedestrians and 20–25 km/h resulting speed between traffic circles, reduced at the traffic circle, and in a theoretical 30 km/h zone with 25–30 km/h assumed speed between traffic circles, dropping at the traffic circle. These analyses confirmed the appropriateness of the traffic circle as a home zone traffic calming measure, as long as its design is based on a detailed analysis of the relevant factors, including location, road function and the surrounding streetscape.

Keywords: traffic calming; traffic circle; reduce speed; home zone; sustainability; air pollution; streetscape character

Citation: Majer, S.; Sołowczuk, A. Traffic Circle—An Example of Sustainable Home Zone Design. *Sustainability* **2023**, *15*, 16751. https://doi.org/10.3390/su152416751

Academic Editors: Luca D'Acierno, Salvatore Leonardi and Natalia Distefano

Received: 23 October 2023
Revised: 6 December 2023
Accepted: 7 December 2023
Published: 12 December 2023

1. Introduction

In the field of road construction, sustainability is understood as construction or reconstruction of road components intended to serve the needs of the current and future generations, while maintaining a balance between economic growth on the one hand and environmental protection and community well-being on the other. Four sustainability types (or pillars) have been distinguished: human, social, economic and environmental. Sustainable transport systems are distinguished by environmental and cost awareness, including the cost of land purchase, construction and future operation, which belong to the environmental and economic sustainability pillars. This article tackles, in particular, the issue of the effective redesign of urban streets and public spaces from the viewpoint of sustainable mobility in urban areas. According to Horn and Jansson [1], redesigning of urban streets and public spaces always involves considerable environmental, economic and social costs in the long run. Therefore, it is so important to assess the effectiveness of any contemplated projects, which should include reference to the social dimension, i.e., the desired coincidence and integration of environmental and traffic safety benefits.

The ever growing or flourishing economy has brought a massive increase in the number of motor vehicles, mainly private cars, which has affected public space quality [2]. This is particularly true of small towns and spas, where public spaces play a specific role by defining the prestige and attractiveness of the place. However, this must not affect the well-being of the local community or compromise the safety of traffic or the availability of transport means. Bearing in mind the sustainable development of the urban environment, these requirements should be considered and implemented in line with accepted urban planning principles. As the first step in the process, the existing transport system of the spa town should be analysed, paying attention to the designated functions [3,4]. This should include determination of local residents' and tourists' needs, bearing in mind the intended use and function of a given public space. Towns can have one of the following spatial structures (Figures 1 and 2):

— monocentric, with a clear-cut centre both functionally and spatially and a generally oblong, elliptical, square, rectangular or semi-circular shape (Figure 2a,b),
— bipolar, made up of two or more distinctly bordered urban entities that may merge into a rectangular or tubular system (Figure 2c),
— polycentric, formed through development and merging of smaller entities (Figure 2d).

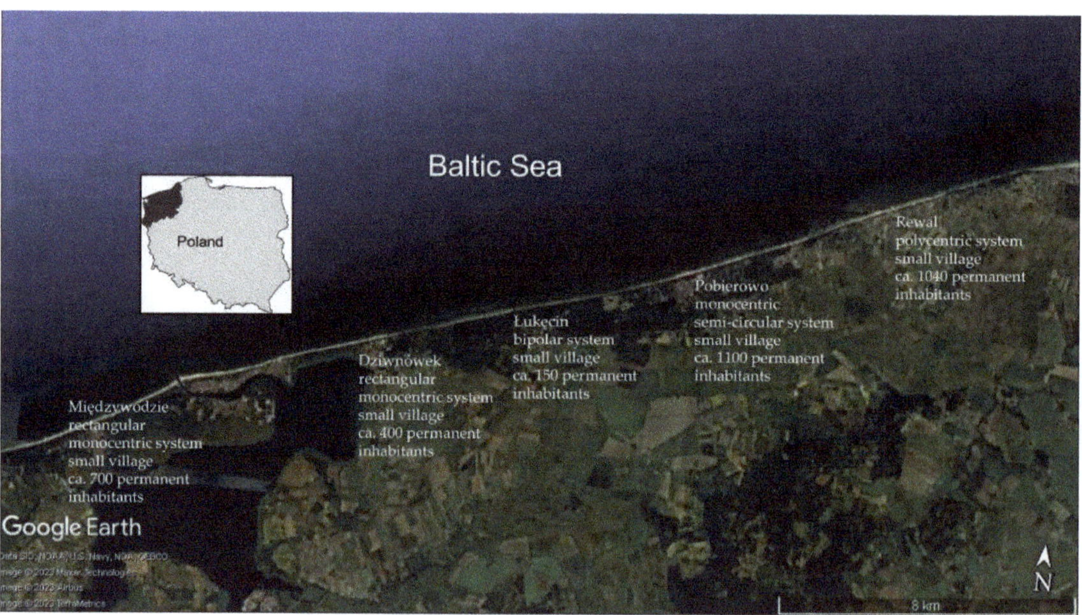

Figure 1. Small seaside spa villages in the West Pomerania region of Poland showing the spatial structure type (shown in Figure 2 below) and indicative number of inhabitants. Source: own picture drawn on Google Earth satellite image [5].

Figure 2 shows some Polish seaside settlements, where brighter orange areas represent the historic centres and indigenous residents' homes and paler orange areas are occupied by recently built resorts and homes of the younger generations. The route of the provincial road running through these illustrative spas is also marked.

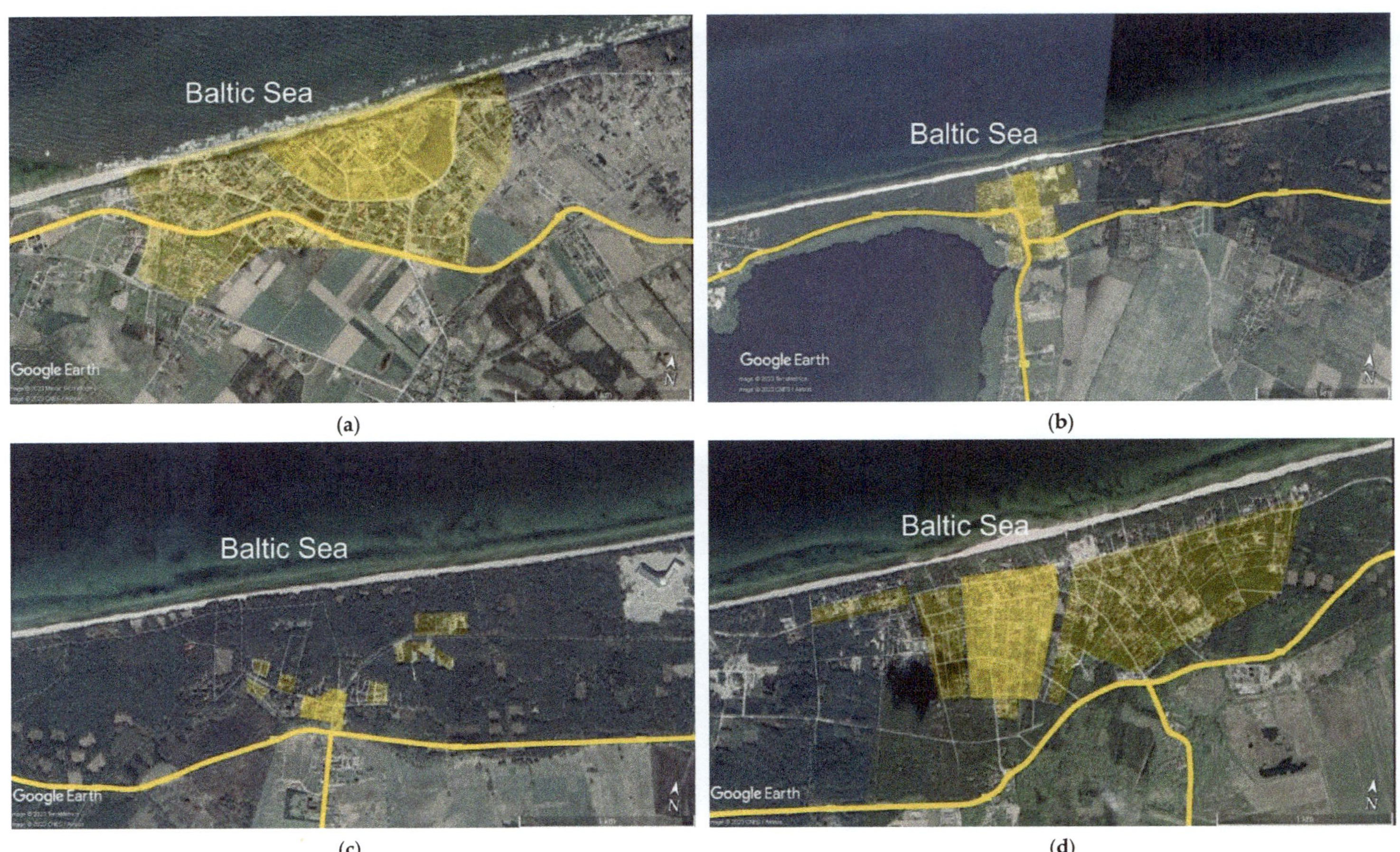

Figure 2. Spatial structures of selected seaside settlements in Poland: (**a**) monocentric semi-circular system of Pobierowo; (**b**) rectangular monocentric system of Dziwnówek; (**c**) bipolar system of Łukęcin; (**d**) polycentric system of Rewal. Source: own work against the background of a satellite image from Google Earth [5].

When contemplating projects intended to address transport issues and revive smaller settlements and spas in line with sustainable development principles, it is indispensable to consider the four above-mentioned sustainability pillars: human, social, economic and environmental. The first wave of urban sustainability projects aimed at public space, social or economic revival involved regeneration of urban green areas, planning of green streets and green infrastructure and care for the natural environment [6–9]. These factors should be taken collectively as the basis for the preparation of traffic calming projects. In the case of public space revival projects, it is also necessary to consider a factor related to the safety of pedestrians, cyclists and drivers or passengers, because pedestrian fatalities constitute a high percentage of all urban road accidents [10–12] (ca. 70% according to [12]). Broken down by age, 9% of these fatalities are people up to 24 years of age, 42% are between 25 and 65 years old and 49% are older than 65 years [11]. As reported in [12], pedestrian fatalities constitute 36% of all urban road accident fatalities in urban areas, as compared to 14% cyclists, 19% motorised two wheelers, 26% car occupants, etc.

Other challenges to public space revival planning are air pollution issues (related to the ever-increasing number of motor vehicles) and traffic noise pollution caused by road traffic in general and with consideration of effective traffic management strategies (TMS) [13–16]. The air pollution and traffic noise pollution issues are the main considerations in the selection of appropriate traffic calming measures and the planning of traffic calming systems in urban areas [17,18]. These considerations may be addressed through appropriate modelling [19] or field monitoring of existing traffic calming measures [13]. A number of studies have investigated the issue of air pollution in the vicinity of existing traffic calming measures as part of traffic calming studies [13,20–22]. Traffic calming measures generally cause sudden slowing of traffic, making vertical and horizontal deflections an undesired option when considered from the air pollution and traffic noise pollution angles [23]. Noteworthy, the resistance to this kind of traffic calming measure, as shown by locals, road authorities, road and landscape engineers and urban planners, is directed more towards vertical deflections and less towards horizontal deflections.

Traffic calming may involve urban traffic management, functional classification of streets and/or introduction of home or 30 km/h zones in the area concerned. Other than in larger towns where 30 km/h zones may be applied in combination with home zones, in smaller towns and spa towns home zones are the preferred option, bearing in mind pedestrian amenity improvements. This issue is particularly evident in spa villages and recreational resorts. Various traffic calming measures, i.e., horizontal and vertical deflections, can be used to slow down the traffic as part of traffic calming projects [24–33]. These include chicanes, road narrowing, speed tables or speed humps and mini roundabouts. Two-way to one-way conversions may also be implemented as part of a traffic calming project [34].

Raised junctions, mini roundabouts or traffic circles are often used in home or 30 km/h zones. The differences between roundabouts and traffic circles can be found in different publications [35,36]. However, they generally refer to older designs of these road components, covering larger surface areas [36]. However, in traffic calming areas, home zones in particular, there is a tendency to design traffic circles without reconstructing the approach legs, realigning of kerb-lines, etc. [37] to cut the project cost. Then, it is appropriate to refer to them with the term proposed in [38], i.e., "mini traffic calming circle". Figure 3 shows some examples of such traffic circles that can be found in traffic calming areas in Poland. The traffic circle's central island may be elevated above (Figure 3a) or installed flush with the surrounding road surface (Figure 3b). The central island may, but not necessarily, promote circular traffic. It may be imposed by appropriately used traffic signs and pavement markings. An example of a traffic circle with traffic signs imposing circular traffic around a raised central island is shown in Figure 3a. In Figure 3b, the traffic circle extends over the entire area between the kerb-lines, without imposing circular traffic as a result.

(a)

(b)

Figure 3. Examples of traffic circles used in Poland on two-way streets: (**a**) an example of a sectioned-off central island on a raised junction in Mierzyn, including traffic signs directing the traffic around the island. Source: photo by Alicja Sołowczuk; (**b**) example of coloured/textured surface used on the central island of a raised junction in Puławy. Source: Google Earth [5].

The effectiveness of traffic circles in traffic calming applications has not, as yet, been reported in the literature, as opposed to extensively covered experiments and evaluations to verify the effectiveness of speed tables. Where the central island does not impose circular traffic, traffic circles are similar to speed tables in terms of traversability. Among the key benefits of traffic circle retrofitting projects are calming of traffic, reduced travel delay, compact size making it possible to keep within the existing right-of-way, low project cost and improved traffic safety.

This motivated the authors to undertake the research described in this article, i.e., an evaluation of the effectiveness of traffic circles as a traffic calming measure in home zones. The initial assumptions included a defined transverse profile of the traffic circles' central island and the relevance of pedestrian traffic volume in the home zone area. A small seaside spa town featuring a grid street pattern located on the Baltic coast in Poland was chosen as study area.

The following research hypotheses were defined:

Hypothesis H1. *"A traffic circle has a significant traffic calming effect when located in a home zone of a spa village".*

Hypothesis H2. *"The central island should have its transverse profile appropriate to the street function and location and the surrounding streetscape's character".*

Hypothesis H3. *"Provision of traffic circles in home zones of small spa towns should be considered as part of the urban street and public space redesign projects contemplated in these locations".*

Since Hypothesis H2 consists of three independent parts, three auxiliary hypotheses have been formulated:

Hypothesis 2A. *"Are the "before" and "after" speeds and speed reductions influenced by pedestrian traffic?"—speed data from two traffic surveys carried out at two different times of the year were considered to test this hypothesis.*

Hypothesis 2B. *"Are the "before" and "after" speeds and speed reductions influenced by the TC lo-cation and its place in the sequence along the streets, or by the surrounding streetscape?"—only traffic circles located on the same street were considered.*

Hypothesis 2C. *"Are the "before" and "after" speeds and speed reductions influenced by the street function and surrounding streetscape?"—traffic circles with parallel locations on two analysed streets with different importance, function and streetscape were considered.*

Section 2 of this article describes the object of study, which are seven traffic circles built about 20 years ago in a small seaside spa town in Poland, and the applied research methods. The study results are presented in Section 3. The results are discussed in Section 4 and the final conclusions of the study are given in Section 5. Figure 3 presents the stages of the study on traffic circle effectiveness as a traffic calming measure for home zone areas.

2. Materials and Methodology

2.1. Study Area

The study area was a home zone of Międzywodzie—a small seaside spa town located in Poland (Figure 4). With almost 700 permanent inhabitants, Międzywodzie is considered as a small village. However, in summer over a 12,000 visitors come to the village, turning it into a spa village. The village is divided into two independent parts by the DW102 through road that runs through it. The village is constantly growing with more and more B&Bs, health resorts, small holiday apartments, food outlets, shops, etc., being built all the time. The study area, including the analysed traffic circles, is located in the centre of Między-wodzie and extends over three streets, including a home zone (Figure 5—Zwycięstwa St., Kasztanowa St. and Wojska Polskiego St.).

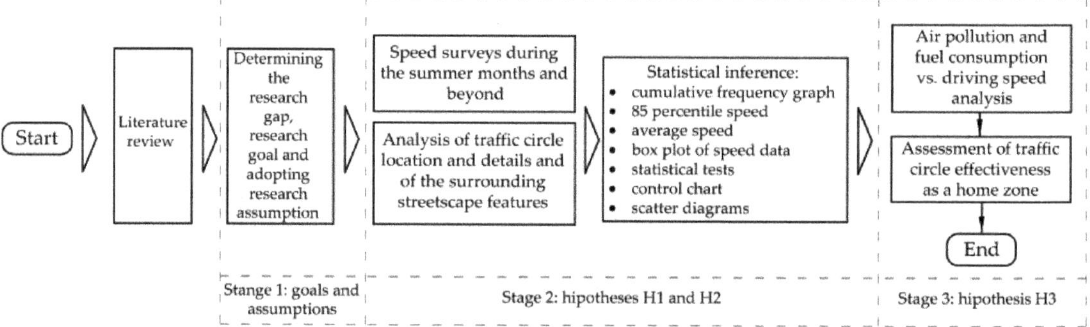

Figure 4. Adopted stages of the research. Source: own work.

With a growing number of tourists in summer, this small spa town has had to deal with serious traffic-related issues. The main challenge was the high volume of pedestrian traffic on the way to and from the beach and a high demand for parking spaces there. Metered Parking Systems with designated and appropriately marked parking places are introduced in such places, similarly to those in home zones, to cope with these growing parking problems. In this situation, a home zone was implemented at the beginning of the 21st century in the central part of Międzywodzie (Figure 5). This involved two-way to one-way conversion of a few streets, constructing a few traffic circles, making the paved paths run flush with the carriageway surface, and demarcating of parking spaces. Improved traffic safety was another benefit of the implemented home zone. However, with no accidents recorded for the period 1995–2023 in the central road accident register SEWIK [39], we cannot give a poor traffic safety record as the grounds for home zone implementation. In fact, it was intended to cope with the parking problems and improve mobility amenities for tourists making their way through the main streets of this spa village.

Figure 5. Both the study area and the home zone (marked in orange) are located to the north of the DW102 road (the analysed traffic circle locations are marked with small red circles). Source: own work against the background of a satellite image from Google Earth [5].

Considering the cost of reconstruction and available land constraints, the cheapest project option was chosen with no changes to kerb-lines. Footpaths were brought flush with the road surface and the old asphalt pavement was replaced with block paving on a few street sections and permeable concrete grid paving in the parallel on-street parking areas. Raised traffic circles were built at a few junctions, yet without changing the approach leg widths. All the main streets were 5.0–6.0 m wide. After reconstruction and installation of flush kerbs separating the path from the carriageway, the carriageway width changed to 5.0 m in all cases. Currently, each street includes ca. 2.0 m wide parallel on-street parking spaces. These parking spaces are demarcated by a different surfacing material and with pavement markings applied thereon. Appropriate traffic signs have also been placed to indicate parking locations. Without demarcated pedestrian crossings, pedestrians may walk all over the carriageway and footpath width.

The locations of the seven analysed traffic circles are shown in Figure 6. This number includes two traffic circles located on Zwycięstwa St. (No. 1 and No. 2 in Figure 6). These junctions have three entry legs and one exit leg each. The remaining five traffic circles included three located on four-leg junctions (No. 3, No. 4 and No. 5 in Figure 6) and two located on staggered junctions on Wojska Polskiego St. (No. 6 and No. 7 in Figure 6). Traffic circles No. 5, No. 6 and 7 are located on Wojska Polskiego St., the promenade of Międzywodzie, lined with small restaurants, fishermen's houses, ice cream parlours, pastry shops, small local markets, boutiques, etc. Various events take place during summer weekends along the whole of Wojska Polskiego St. between the DW102 provincial road and traffic circle No. 5. In summer too, further on to the north, a summer fair is held during which stalls are placed over the whole carriageway width. The northern part of Wojska Polskiego St. is blocked as a result. In turn, during the above-mentioned weekend events, stalls selling various merchandise are placed on the footpaths and on the carriageway. At all intersections, the side streets have 5 m wide carriageways, and have no footpaths running along the road or demarcated on-street parking spaces. Figure 5 shows as additional information the traffic directions on the analysed one-way streets and the height Δh of the raised central island.

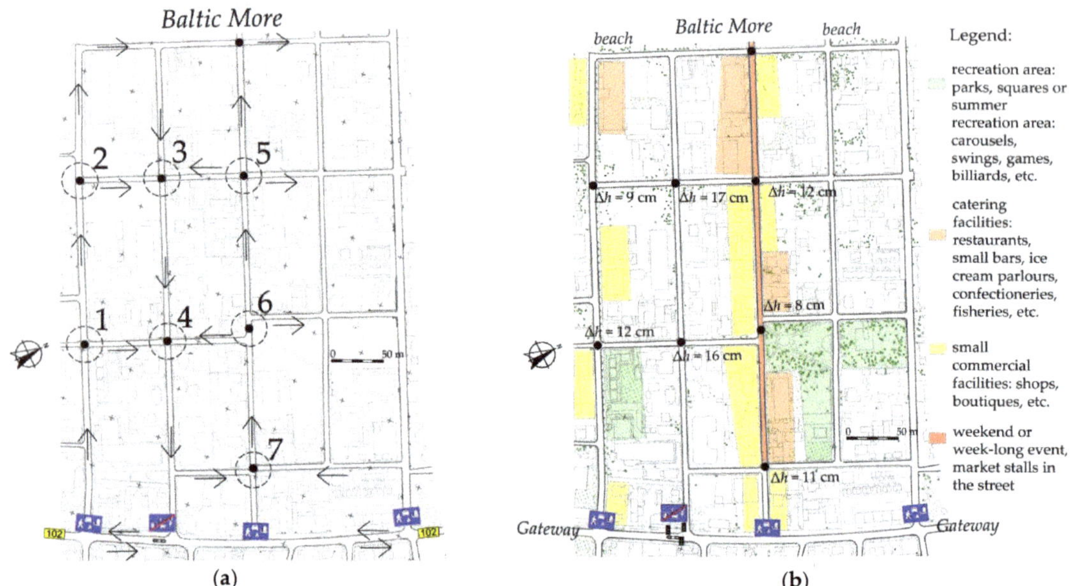

Figure 6. Study area details: (**a**) traffic circles' locations and numbering; (**b**) home zone divided into functional sub-areas. Source: own work.

2.2. Traffic Volume and Speed Surveys

The object of this case study is a spa village. SR4 traffic detection devices [40] were used to simultaneously measure traffic speeds and volumes as part of this study. The siting of the survey stations is shown in Figure 7. The devices were mounted on the existing traffic delineator posts on the way to and past the analysed traffic circles (before and after), at the one-way street entries and exits and between the junctions. Due to low traffic volumes (up to 35 veh./h) noted in the area under analysis, the surveys were discontinued when the number of logged vehicles exceeded 100. Considering the speed logging characteristics and 0.01 s logging time accuracy, we can assume free traffic flow conditions, allowing each observed vehicle and the following vehicles to move freely without an obstacle vehicle ahead (according to [41]). In the summer, free traffic flow may be arbitrarily related to logged vehicles, as the main obstacles on the road were the pedestrians walking over the whole carriageway width and thus making faster driving impracticable. Information on the gear the drivers used between the junctions was randomly gathered from those who pulled over to park. In September, they generally drove in second gear when making their way through the traffic circle, shifting to third gear on the section between the junctions. This information was then used to calculate fuel consumption.

Taking account of crowded streets in summer and the low number of pedestrians beyond this season, the speed surveys were carried out in two representative periods: in summer and in the last week of September. Considering these low speeds and occupied parking spaces, it was justified to limit the comparative analyses to the values of v_{85} and v_{av}.

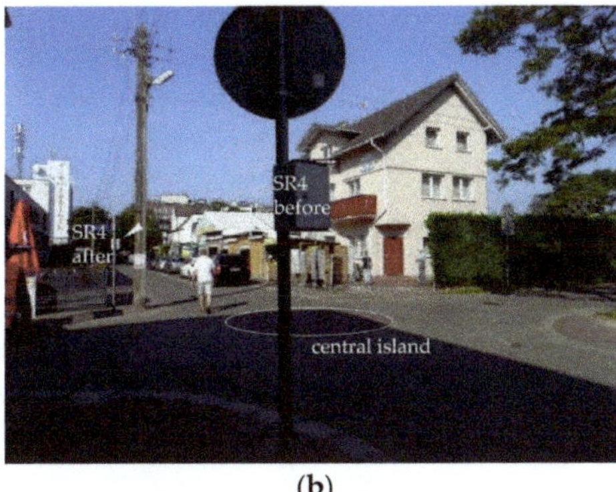

(a)   (b)

Figure 7. Survey station equipment (SR4): (**a**) traffic detection device; (**b**) SR4 sited before and after traffic circle No. 6. Source: photo by Alicja Sołowczuk.

2.3. Methodology

The statistical analysis sequence is shown in Figure 8 below.

For the research hypothesis H3, it was decided to use the conclusions and the data given by Merkisz et al. [22]. The study area included streets with speed humps and zone 30 streets for comparison. Air pollution was measured during a few passes of a test vehicle carrying special gas detection equipment. The air pollution values are the means of the data logged during the respective passes, including vehicle speeds and exhaust emission levels. Based on our analysis, we determined that the traffic conditions during the survey of Merkisz et al. were similar to the conditions prevailing at the traffic circles under analysis.

As regards fuel consumption, it was decided to separately consider two driving scenarios: scenario 1—when the car drives in second gear all the way, and scenario 2—when the car drives in second gear through the traffic circle, shifting to third gear upon entering an in-between section.

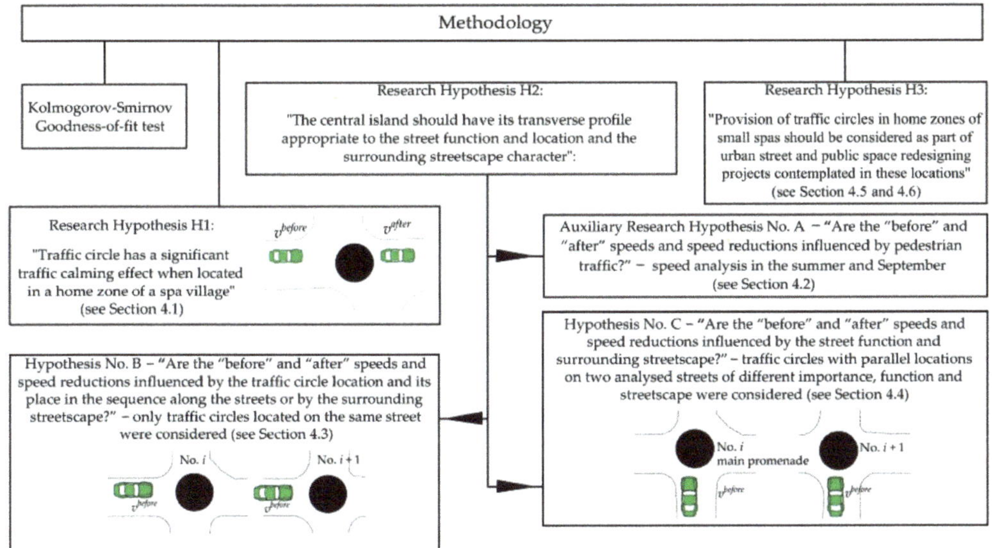

Figure 8. Sequence of statistical analysis. Source: own work.

3. Results

3.1. Characteristics of the Traffic Calming Measurement TCM

In the study area, traffic circles were built as part of the project to convert the existing public spaces to a home zone. The traffic circles' central islands were surfaced with red concrete paving bricks. The junction approach legs were not widened as part of the project. In the whole home zone area, the footpaths were brought flush with the carriageway surface, separated by a kerb-line made of kerb units laid flat. The design allowed leaving the existing drainage system unchanged. On the demarcated parallel on-street parking spaces, a permeable concrete grid pavement was laid. The travel lanes were, in turn, surfaced with grey paving bricks. The parallel on-street parking spaces on Zwycięstwa St. and Kasztanowa St. start and end ca. 7–10 m from the junction edge, thus creating apparent bulb-outs, yet with no kerb-line or markings applied on the pavement surface. On the Międzywodzie promenade, they start and end ca. 3–4 m from the junction edge. The parameters of the analysed traffic circles are given in Table 1 below.

Table 1. Parameters of the analysed traffic circles. Source: own work.

Traffic Circle	Δh [1], m	Street Function	l_1, m [2]	l_2, m [3]	Streetscape Characteristics
No. 1	0.12	footstreet	190	150	summer recreation area
No. 2	0.09	footstreet	150	125	summer recreation area, small businesses open in summer
No. 3	0.17	vehicular street	125	150	private properties
No. 4	0.16	vehicular street	150	125	private properties
No. 5	0.11	main promenade	140	125	small catering and small commercial facilities open in summer
No. 6	0.08	main promenade	135	140	year-round recreation areas and catering businesses
No. 7	0.12	main promenade	60	135	year-round shopping centre, post office, bank, etc.

[1] Δh—difference of level between the central island centre and edge when passing through the traffic circle; [2] l_1—length of traffic circle approach section; [3] l_2—length of road section from the traffic circle end and the next junction.

For streetscape beautification reasons, two-colour paving slabs were used on the footpath and small grafted trees were planted alongside. Kerbing was installed around the trees to prevent oil-contaminated water from penetrating into the plant bed. On the side facing the road, each tree was protected with metal guard posts on either side. In a few places, low-height tree boxes or plant beds were placed within the street width. Benches and litter bins were placed between the newly planted trees. Wider footpath portions are designated to be occupied by stalls, counters and A-frame ads for boutiques and small food serving businesses. The streetscape character of the selected streets for the installed traffic calming measurement TCM is visualised in Figure 9.

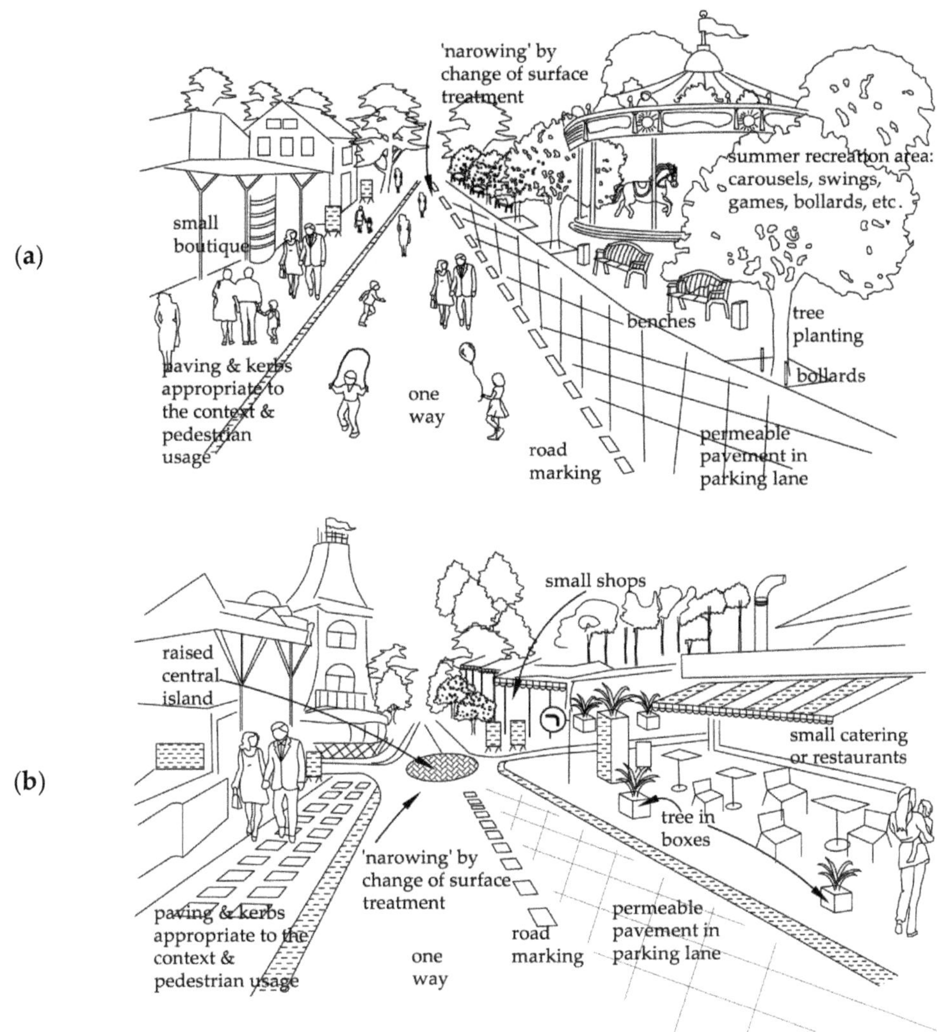

Figure 9. Streetscape and traffic calming measurements and visualisation of the analysed streets: (**a**) one-way street with parallel on-street parking on the right-hand side (Zwycięstwa St.); (**b**) one-way main promenade hosting various weekend events (Wojska Polskiego St.). Source: own drawings.

3.2. Plan and Cross-Section of Selected Traffic Circles

Figure 10 shows transverse cross-sections and plan views of traffic circles located at selected junctions. The island diameter was 4.00 m in all cases. The varying parameter was the difference in level between the island centre and its perimeter. The one-way streets had 5.00 m wide carriageways, including a 2.00 m width demarcated for parallel on-street parking. The side streets had footpaths on some parts and no demarcated parking spaces. The footpaths were surfaced by concrete paving slabs in two contrasting colours. The footpaths were brought flush with the carriageway and separated from it by white kerb units laid flat (Figures 9 and 11).

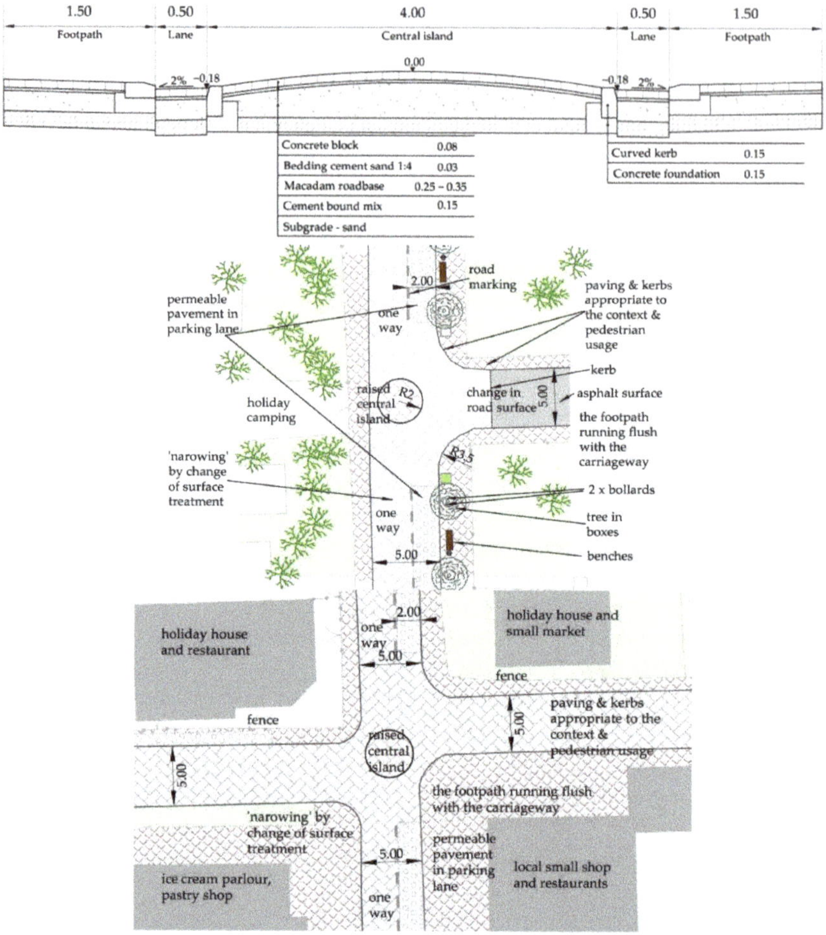

Figure 10. Visualization of the analysed traffic circles: (**a**) transverse cross-section; (**b**) plan view of traffic circle No. 2 located on a T-junction; (**c**) plan view of traffic circle No. 7 located on the main promenade (all dimensions in metres). Source: own work.

Figure 11. Analysed traffic circles: (**a**) traffic circle No. 3—Δh = 0.17 m; (**b**) traffic circle No. 6—Δh = 0.08 m (all dimensions in metres). Source: photo by Alicja Sołowczuk.

3.3. Speed Survey Data Processing

Figures 12 and 13 show the speed ranges calculated from the summer and September survey data, respectively. The red dashed line represents averaged v_{85} values calculated using the summer and September survey data, respectively, for all the survey stations.

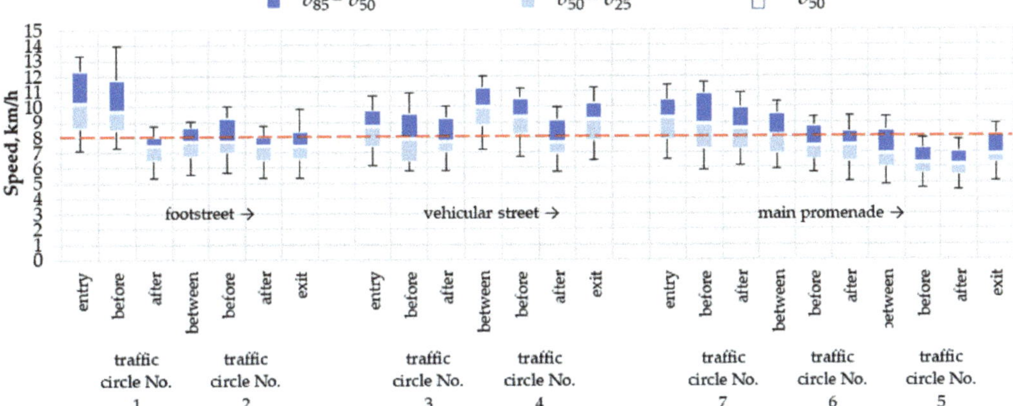

Figure 12. Box plot of speed data measured in summer. The whiskers represent the minimum and maximum values; lower and upper edges of the boxes determine the first and third quartiles; the bold white line designates the median value (the arrow indicates the direction of movement). Source: own work.

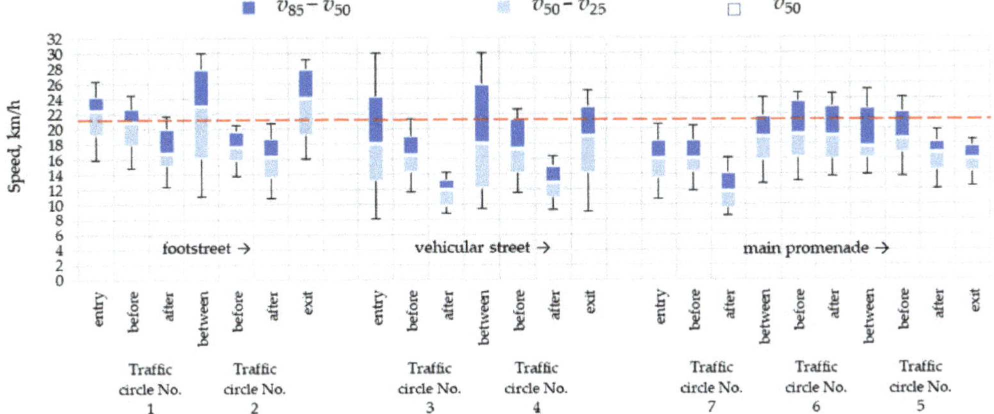

Figure 13. Box plot of speed data measured in September. The whiskers represent the minimum and maximum values; lower and upper edges of the boxes determine the first and third quartiles; the bold white line designates the median value (the arrow indicates the direction of movement). Source: own work.

However, in summer, the main factor slowing the traffic are the tourists walking all over the one-way carriageways and footpaths (Figure 14). During this period, the parallel parking spaces were occupied for most of the time, making the carriageway apparently, and actually, narrower.

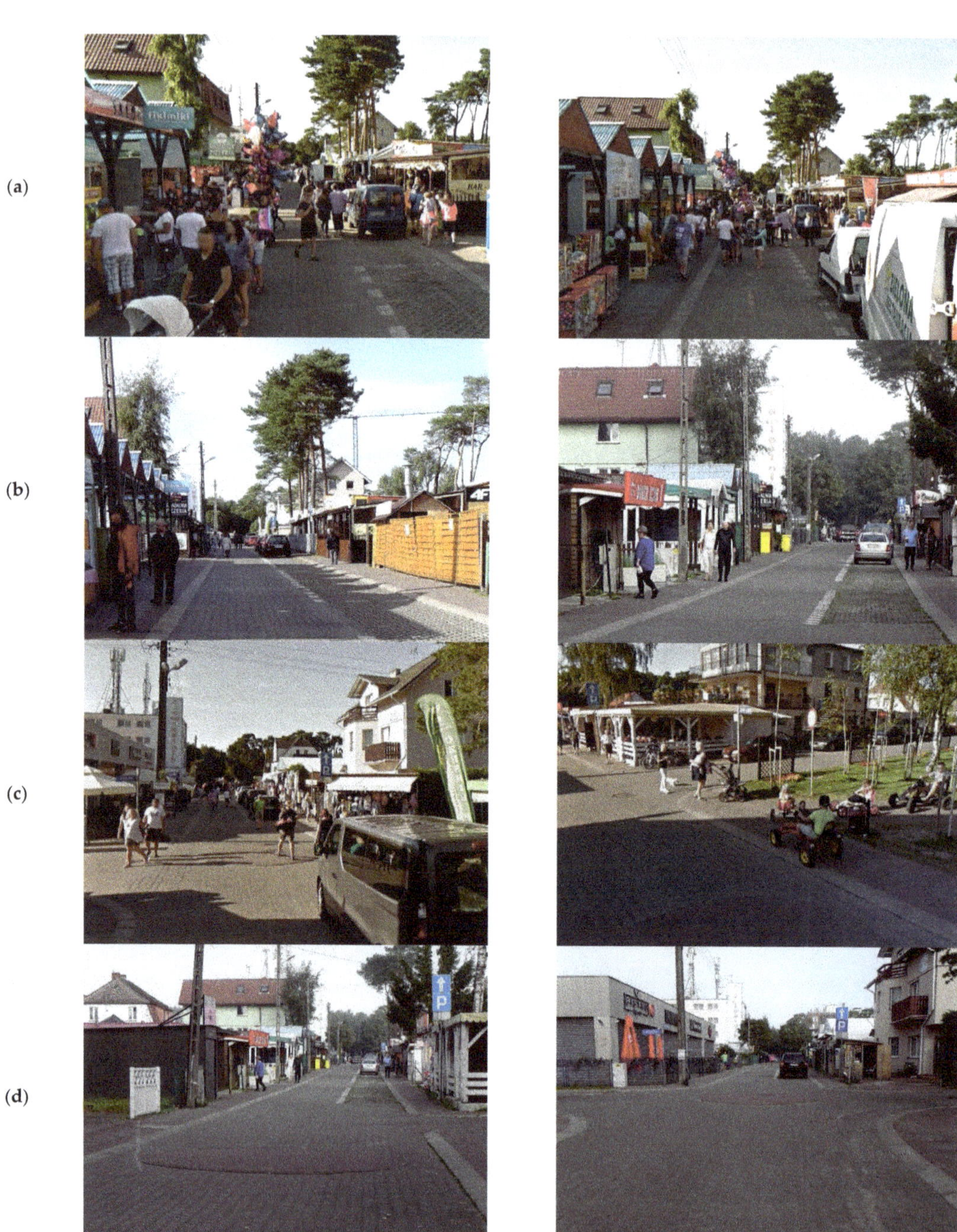

Figure 14. Differences in pedestrian traffic on the main promenade in the home zone area: (**a**) summer season—main promenade (Source: Google Earth [5]); (**b**) September—main promenade (Source: photo by Alicja Sołowczuk); (**c**) summer season—traffic circle No. 5 and No. 6 (Source: Google Earth [5]); (**d**) September—traffic circle No. 5 and No. 6 (Source: photo by Alicja Sołowczuk).

In the off-season period, the tourist business is limited to health and spa facilities, a few shops and some boutiques. Other businesses are closed. In the off-season period, only some of the parking spaces were used by owners of the local properties in the area. With less obstacles on the road ahead, traffic can be handled more efficiently and at higher speeds, both between and within the junctions (Figures 13 and 14).

The results given in Figure 13 allow us to conclude that, even without summer visitors, traffic circles effectively slow down the traffic within junctions. In order to traverse the raised central island, the drivers reduce their driving speed by about 12–15 km/h. However, the slowing effect is limited to a max. 10 m distance. In summer, the speeds of travel on the sections between the junctions ranged from 7 to 10 km/h depending on the number of pedestrians and vehicles driving in and out of the parking spaces. In September, this range increased on the sections between the traffic circles to 20–25 km/h, the exact speed depending most probably on the number of parked vehicles.

The cumulative frequency graph in Figure 15 shows the cumulative distribution function (CDF) representing the situation at four traffic circles. The cumulative distribution function and 85th percentile speed differences on these traffic circles stem from different street functions, their place in the sequence and the surrounding and varying streetscape characteristics. All these traffic circles featured a similar difference of level in the range 0.09–0.12 m. The cumulative density functions and values of v_{85} and v_{av} indicate smaller speeds immediately before and past traffic circle No. 2, as well as a smaller approach speed. This may be due to the placing of the sequence of traffic circles (traffic circle No. 2 is passed as the second traffic circle when driving down Zwycięstwa St.), and about 150 m spacing between the subsequent traffic circles. Other relevant factors may include pedestrian traffic, in the summer season, using the whole carriageway width and the surrounding streetscape features. Traffic circle No. 2 and the holiday camping area featured more parked cars and street businesses, including boutiques and food outlets, compared to traffic circle No. 1, with the surrounding developments limited to homes and small resorts.

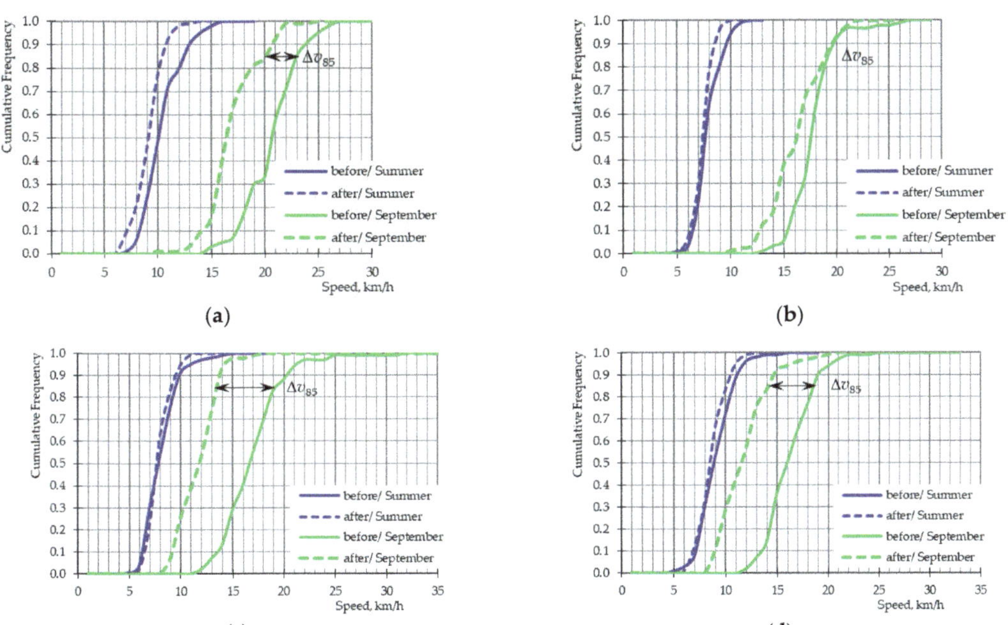

Figure 15. Illustrative cumulative frequency graph of traffic circle crossing speeds: (**a**) traffic circle No. 1; (**b**) traffic circle No. 2; (**c**) traffic circle No. 3; (**d**) traffic circle No. 7. Source: own work.

In turn, traffic circle No. 3 is the first when driving down Kasztanowa St., located about 120 m from the street entry and featuring the greatest transverse slope of the central island with a $\Delta h = 0.17$ m level difference. On the street section leading to traffic circle No. 3, vehicles are parked year-round due to the open spas and health resorts. In addition, the first in the sequence is traffic circle No. 7, located on Wojska Polskiego St. about 60 m from the entry to this one-way street. It has a smaller transverse slope of the central island, with $\Delta h = 0.11$ m difference in level. On the way to traffic circle No. 7, vehicles are present at all times due to the central location, with many year-round open shops and public amenities. The cumulative density functions and the calculated values of v_{85} and v_{av} indicate slightly smaller speeds of vehicles passing traffic circle No. 7, attributed to the surrounding streetscape features and more pedestrians.

4. Discussion

The statistical inference method was adopted for processing of the speed data obtained for all the traffic circles under analysis. The process started with the Kolmogorov–Smirnov Goodness-of-fit test, carried out to verify normality of distribution of the analysed speed populations (Equation (1)). Normality of speed distribution was confirmed for all the analysed traffic circles in both traffic survey periods.

Kolmogorov–Smirnov Goodness-of-fit test

$$\text{Null hypothesis } H_0 \text{ and alternative hypothesis } H_1: \begin{cases} H_0 = F(v) = F_0(v) \\ H_1 = F(v) \neq F_0(v) \end{cases}, \lambda_\alpha = 1.36, \ \lambda = 0.05. \tag{1}$$

where: H_0—null hypothesis, H_1—alternative hypothesis, $F(v)$—empirical cumulative frequency curve, $F_0(v)$—theoretical cumulative frequency curve, λ_α—critical values, α—adopted significance level.

Next, two of the three research hypotheses were verified using the obtained speed results research hypotheses: Research Hypothesis H1 and Research Hypothesis H2 (see Section 1). Dealing with a non-measurable characteristic, nonparametric tests (two-sample Kolmogorov–Smirnov test and median test) were chosen to verify the research hypotheses (H1 and H2).

4.1. Research Hypothesis H1—"A Traffic Circle Has a Significant Traffic Calming Effect When Located in a Home Zone of a Spa Village"

These were the two-sample K–S test (Equation (2)) and the median test (Equation (3)). In the case of Research Hypothesis H1, both statistical tests were performed for the summer and September "before" and "after" speed parameters for each of the analysed traffic circles. These tests revealed a significant difference between the "before" and "after" speed parameters for all the traffic circles except for traffic circle No. 6, thus confirming the effect of traffic circles on speed reduction for all the analysed traffic circles (Research Hypothesis H1). Some statistical analysis results obtained for traffic circle No. 1 and No. 2 are given in Table 2 below.

Two-sample Kolmogorov–Smirnov test :

$$\text{Null hypothesis } H_0 \text{ and alternative hypothesis } H_1: \begin{cases} H_0 : F\left(v^{before}\right) = F\left(v^{after}\right), \\ H_1 : F\left(v^{before}\right) \neq F\left(v^{after}\right), \end{cases} \lambda_\alpha = 1.36, \ \alpha = 0.05. \tag{2}$$

where: H_0—null hypothesis, H_1—alternative hypothesis, $F(v^{before})$—before speed cumulative distribution function, $F(v^{after})$—after speed cumulative distribution function, λ_α—critical values, α—adopted significance level.

Median test :

$$\text{Null hypothesis } H_0 \text{ and alternative hypothesis } H_1: \begin{cases} H_0 : F_1(v_{50}) = F_2(v_{50}), \\ H_1 : F_1(v_{50}) \neq F_2(v_{50}), \end{cases} \chi^2_\alpha = 3.84, \ \alpha = 0.05. \tag{3}$$

where: H_0—null hypothesis, H_1—alternative hypothesis, $F_1(v_{50})$—number of results below v_{50} from both populations, $F_2(v_{50})$—number of results above v_{50} from both populations, $\chi^2{}_\alpha$—critical values, α—adopted significance level.

Table 2. Sample results of statistical tests—traffic circle No. 1 and No. 2. Source: own work.

	K–S Goodness-of-Fit Test [1]		Two-Sample K–S Test [2]	Median Test [3]
	Before	After		
		Data from the Summer		
Traffic circle No. 1	$\lambda = 0.76 < \lambda_\alpha = 1.36$	$\lambda = 0.34 < \lambda_\alpha = 1.36$	$\lambda = 4.05 > \lambda_\alpha = 1.36$	$\chi^2 = 30.3 > \chi_\alpha{}^2 = 3.84$
Traffic circle No. 2	$\lambda = 0.73 < \lambda_\alpha = 1.36$	$\lambda = 0.58 < \lambda_\alpha = 1.36$	$\lambda = 4.45 > \lambda_\alpha = 1.36$	$\chi^2 = 38.2 > \chi_\alpha{}^2 = 3.84$
		Data from the September		
Traffic circle No. 1	$\lambda = 0.92 < \lambda_\alpha = 1.36$	$\lambda = 1.00 < \lambda_\alpha = 1.36$	$\lambda = 4.03 > \lambda_\alpha = 1.36$	$\chi^2 = 83.8 > \chi_\alpha{}^2 = 3.84$
Traffic circle No. 2	$\lambda = 0.90 < \lambda_\alpha = 1.36$	$\lambda = 0.75 < \lambda_\alpha = 1.36$	$\lambda = 2.28 > \lambda_\alpha = 1.36$	$\chi^2 = 27.2 > \chi_\alpha{}^2 = 3.84$

[1] Kolmogorov–Smirnov Goodness-of-fit test (Equation (1)): $\lambda_\alpha = 1.36$, $\alpha = 0.05$. [2] Two-sample Kolmogorov–Smirnov test (Equation (2)): $\lambda_\alpha = 1.36$, $\alpha = 0.05$. [3] Median test (Equation (3)): $\chi_\alpha{}^2 = 3.84$, $\alpha = 0.05$.

4.2. Research Hypothesis H2—"The Central Island Should Have Its Transverse Profile Appropriate to the Street Function and Location and the Surrounding Streetscape Character"

4.2.1. Auxiliary Hypothesis 2A—"Are the "Before" and "After" Speeds and Speed Reductions Influenced by Pedestrian Traffic?"

Since the speed variation analysis for the summer and September surveys (Figures 12–14) showed a considerable traffic slowing effect of the pedestrian crowds during the summer season, this factor was also subjected to the statistical tests comparing the summer and September data populations (Equations (4) and (5)). The results of both tests are given in Table 3. Based on these results, showing a difference between the "before" and "after" speed data populations in almost all cases, we can conclude that the analysed factor has a statistically significant traffic slowing effect. This effect was not confirmed only for traffic circle No. 6, for which a result close to the critical value was obtained for the "before" speed in one test only. Traffic circle No. 6 is located on a staggered T-junction (Figures 6 and 11b) without demarcated on-street parallel parking and no parked cars (Figures 6 and 11b).

Two-sample Kolmogorov–Smirnov test :

$$\text{Null hypothesis } H_0 \text{ and alternative hypothesis } H_1 : \begin{cases} H_0 : F\left(v^{Summer}\right) = F\left(v^{September}\right), \\ H_1 : F\left(v^{Summer}\right) \neq F\left(v^{September}\right), \end{cases} \lambda_\alpha = 1.36, \ \alpha = 0.05. \tag{4}$$

where: H_0—null hypothesis, H_1—alternative hypothesis, $F(v^{Summer})$—summer speed cumulative distribution function, $F(v^{September})$—September speed cumulative distribution function, λ_α—critical values, α—adopted significance level.

Median test :

$$\text{Null hypothesis } H_0 \text{ and alternative hypothesis } H_1 : \begin{cases} H_0 : F_1(v_{50}) = F_2(v_{50}), \\ H_1 : F_1(v_{50}) \neq F_2(v_{50}), \end{cases} \chi^2{}_\alpha = 3.84, \ \alpha = 0.05. \tag{5}$$

where: H_0—null hypothesis, H_1—alternative hypothesis, $F_1(v_{50})$—number of results below v_{50} from both populations, $F_2(v_{50})$—number of results above v_{50} from both populations, $\chi^2{}_\alpha$—critical values, α—adopted significance level.

Table 3. Results of statistical tests related to Hypothesis 2A—effect of pedestrians on the "before" and "after" traffic circle speeds measured in the summer season and in September. Source: own work.

Test	Traffic Circle						
	No. 1	No. 2	No. 3	No. 4	No. 5	No. 6	No. 7
Data from Test K–S test [1]							
Before (Summer and September)	4.05	6.9	5.7	4.5	2.3	1.35	4.9
After (Summer and September)	7.00	7.2	7.5	7.7	7.1	7.3	7.1
Data from Median test [2]							
Before (Summer and September)	26.6	167.9	165.5	174.4	179.7	166.0	164.5
After (Summer and September)	97.9	153.8	139.3	152.4	180.8	197.8	56.1

[1] Two-sample Kolmogorov–Smirnov test λ (Equation (4)): $\lambda_\alpha = 1.36$, $\alpha = 0.05$. [2] Median test (Equation (5)): $\chi_\alpha^2 = 3.84$, $\alpha = 0.05$.

4.2.2. Auxiliary Hypothesis 2B—"Are the "Before" and "After" Speeds and Speed Reductions Influenced by the Traffic Circle Location and Its Place in the Sequence along the Streets or by the Surrounding Streetscape?"

In order to verify Hypothesis 2B concerning the effect of traffic circle location and place in the sequence on a given street and the effect of the immediate surroundings, two statistical tests were conducted for the combined data for paired traffic circles (Equations (6) and (7)). The pairs were made up of consecutive traffic circles located on the same street (Table 4—Equations (6) and (7)). The test results for "before" traffic circle speeds in September are given in Table 4. The obtained statistics show that populations of different traffic circles are different, i.e., must not be combined in one set. This confirms hypothesis B on the statistically significant effect of traffic circle location (i.e., place in the traffic circle sequence when driving down the street), and of the surrounding streetscape character on the obtained speed reduction results.

Two-sample Kolmogorov–Smirnov test :

$$\text{Nullhypothesis } H_0 \text{ and alternative hypothesis } H_1 : \begin{cases} H_0 : F\left(v^{No.\ i}\right) = F\left(v^{No.\ i+1}\right), \\ H_1 : F\left(v^{No.\ i}\right) \neq F\left(v^{No.\ i+1}\right), \end{cases} \lambda_\alpha = 1.36,\ \alpha = 0.05. \tag{6}$$

where: H_0—null hypothesis, H_1—alternative hypothesis, $F(v^{No.\ i})$—speed cumulative distribution function on the traffic circle No. i, $F(v^{No.\ i+1})$—speed cumulative distribution function on the traffic circle No. $i + 1$, i—traffic circle preceding, $i + 1$ traffic circle following, λ_α—critical values, α—adopted significance level.

Median test :

$$\text{Null hypothesis } H_0 \text{ and alternative hypothesis } H_1 : \begin{cases} H_0 : F_1(v_{50}) = F_2(v_{50}), \\ H_1 : F_1(v_{50}) \neq F_2(v_{50}), \end{cases} \chi^{2}_{\alpha} = 3.84,\ \alpha = 0.05. \tag{7}$$

where: H_0—null hypothesis, H_1—alternative hypothesis, $F_1(v_{50})$—number of results below v_{50} from both populations, $F_2(v_{50})$—number of results above v_{50} from both populations, χ^2_α—critical values, α—adopted significance level.

Table 4. Statistics related to Hypothesis 2B to verify the effect of the traffic circle location, place in the sequence and the surrounding streetscape features on the "before" speed measured in the September survey. Source: own work.

Test	Analysis of Traffic Circles Located along the Main Streets			
	No. 1 and 2	No. 3 and 4	No. 5 and 6	No. 6 and 7
Data from Test K–S test [1]	4.1	1.1	1.6	3.3
Data from Median test [2]	48.4	3.6	8.0	67.6

[1] Two-sample Kolmogorov–Smirnov test λ (Equation (6)): $\lambda_\alpha = 1.36$, $\alpha = 0.05$. [2] Median test (Equation (7)): $\chi_\alpha^2 = 3.84$, $\alpha = 0.05$.

4.2.3. Auxiliary Hypothesis 2C—"Are the "before" and "after" Speeds and Speed Reductions Influenced by the Street Function and Surrounding Streetscape?"

In order to verify Hypothesis 2C, two statistical tests were conducted for combined data for paired traffic circles (Equations (7) and (8)). The pairs were made up of traffic circles located on neighbouring streets at parallel locations (Table 5—Equations (7) and (8)). The test results for "before" traffic circle speeds in September are given in Table 5. The obtained statistics show that populations of different traffic circles are different, i.e., must not be combined in one set. This confirms hypothesis C on the statistically significant effect of traffic circle location, street function and of the surrounding streetscape character on the obtained speed reduction results.

Two-sample Kolmogorov–Smirnov test :

$$\text{Null hypothesis } H_0 \text{ and alternative hypothesis } H_1 : \begin{cases} H_0 : F\left(v^{No.\ i}\right) = F\left(v^{No.\ i+1}\right), \\ H_1 : F\left(v^{No.\ i}\right) \neq F\left(v^{No.\ i+1}\right), \end{cases} \lambda_\alpha = 1.36, \ \alpha = 0.05. \tag{8}$$

where: H_0—null hypothesis, H_1—alternative hypothesis, $F(v^{No.\ i})$—speed cumulative distribution function on the traffic circle No. i, $F(v^{No.\ i+1})$—speed cumulative distribution function on the traffic circle No. $i + 1$, i—traffic circle on the analysed street, $i + 1$—traffic circle on the adjacent street at a parallel location, λ_α—critical values, α—adopted significance level.

Table 5. Statistics related to Hypothesis 2C to verify the effect of the traffic circle location and the surrounding streetscape features at a parallel location on the "before" speed measured in the September survey. Source: own work.

Test	Analysis of Traffic Circle Located on Parallel Side Streets			
	No. 1 and 7	No. 2 and 5	No. 3 and 5	No. 4 and 7
Data from Test K–S test [1]	4.3	1.32	2.1	1.6
Data from Median test [2]	83.8	8.6	8.7	15.7

[1] Two-sample Kolmogorov–Smirnov test λ (Equation (8)): $\lambda_\alpha = 1.36$, $\alpha = 0.05$. [2] Median test (Equation (7)): $\chi_\alpha^2 = 3.84$, $\alpha = 0.05$.

Now we can conclude that both hypotheses (Research Hypothesis H1 and Research Hypothesis H2) were confirmed in the statistical inference process. Thus, whenever traffic circles are designed in home zones, the location, street function and the surrounding streetscape features should be taken into account in the process and the traffic circle design parameters should be implemented accordingly.

4.3. Trajectory and Speed Profile Analysis

As mentioned, a traffic circle may (but not necessarily) promote circular movement. In this case, the only 3 m wide carriageway, along the one-way street with demarcated on-street parallel parking, passing the traffic circle makes keeping to the right-hand side impracticable. Instead, the drivers tended to pull left, navigating past the raised central island on its left-hand side. This constitutes the main difference between a traffic circle located on a one-way street and a mini roundabout located on a two-way carriageway. Figure 16 shows examples of different travel paths noted during the surveys. Driving through a traffic circle depended on the driver's skills and habits, differences in level of the central island, vehicle ground clearance, and the surrounding features, including street stalls, buildings, various fences and dense shrubs that could obscure the view of the side road junctions (Figures 14 and 16c,d).

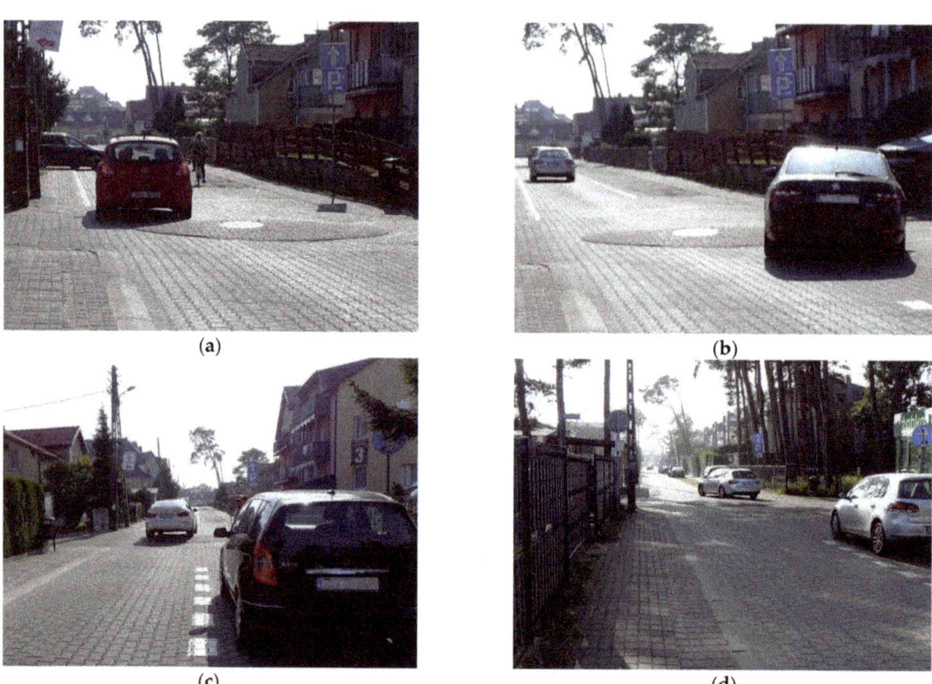

Figure 16. Examples of passenger cars driving through traffic circles: (**a**) traversing the island on the left-hand side; (**b**) traversing the island on the right-hand side; (**c**) traversing through the central part of the island; (**d**) bypassing the island by pulling right. Source: photo by Alicja Sołowczuk.

The v_{85} and v_{av} profiles are presented in Figure 17 for different slopes of the traffic circle central islands, different longitudinal slopes on the way to and past the traffic circle, and different surrounding streetscapes. In addition, these profiles include the speeds on the way to, immediately before, within and past the traffic circle, in order to expose the actual speed variation. The results of both surveys are included.

Firstly, the speed profile data reveal completely different traffic conditions at the times of the two surveys, i.e., in summer and in September (Figure 17). In the former case, with crowds of tourists walking on the footpath and on the carriageway, the traffic circles had some, though a rather small, slowing effect on the road traffic. Much higher speeds and a much more pronounced slowing effect of the traffic circles was noted in September when pedestrian traffic is limited to the guests at the health resorts and spas of Międzywodzie. The greatest speed reductions were recorded at traffic circles No. 3 and No. 4 located on Kasztanowa St. The speed reduction differences between these two are attributed to different longitudinal slopes on the junction approach sections. The next in order of speed reduction amount were the traffic circles located on Wojska Polskiego St. and Zwycięstwa St., with central island level difference of $\Delta h = 0.11$–0.12 m. Speed reductions of about 4 km/h were obtained on traffic circles No. 5 and No. 7, depending on the approach and departure speeds and the surrounding streetscape features. For example, in the case of traffic circle No. 5, there are local markets and boutiques at the main legs and the demarcated on-street parking spaces start as close as 3–4 m from the secondary leg kerb-line. These parking spaces are used by the spa house and are occupied also in autumn, prominently narrowing the travel lane. These are the main factors contributing to the obtained speed reductions. Traffic circle No. 7 is located 60 m from the home zone entry and this location defines the observed approach speeds. The home zone entry area and the junction corners are occupied by year-round open markets and public amenities, including post office, bank, pharmacy, etc., generating pedestrian traffic and frequent driving in and out of the parking spaces. The situation is different at traffic circles No. 1 and No. 2, with very few pedestrians or parked vehicles in autumn. This resulted in a smaller amount of speed reduction in autumn, in the order of 2 km/h.

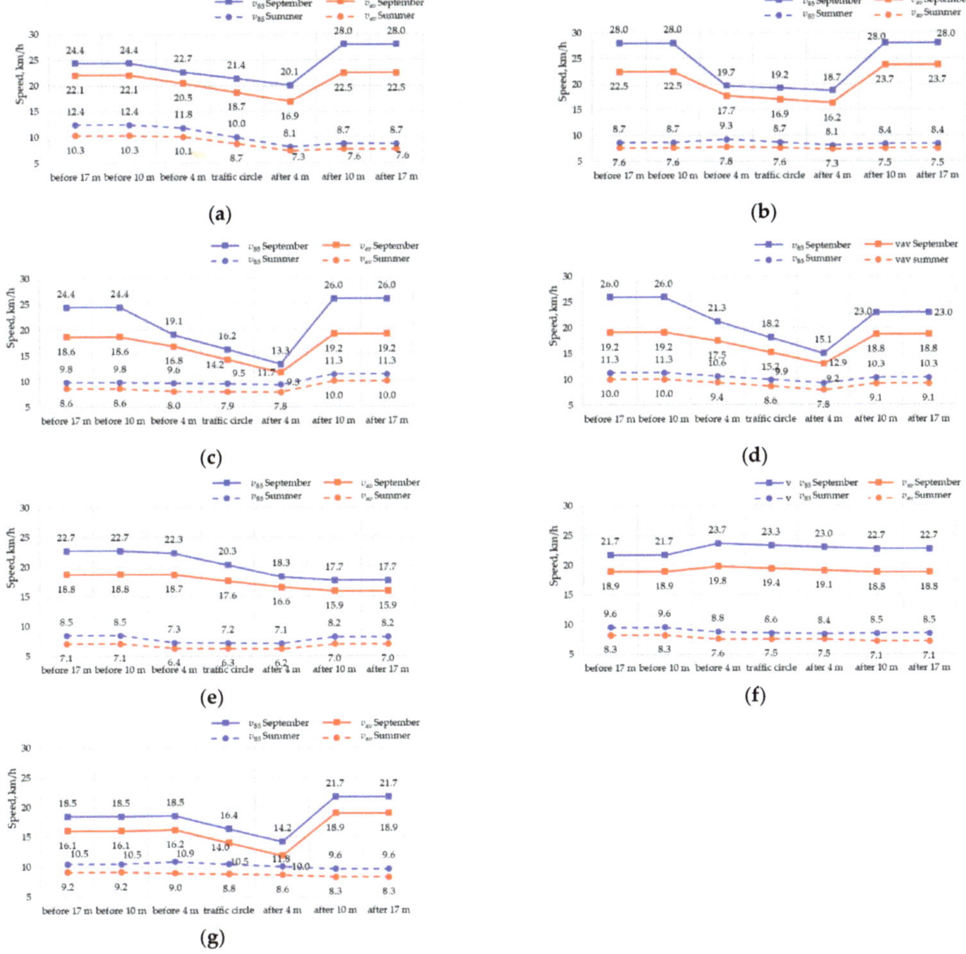

Figure 17. Traffic circle driving speed profiles: (**a**) No. 1—Δh = 0.12 m; (**b**) No. 2—Δh = 0.09 m; (**c**) No. 3—Δh = 0.17 m; (**d**) No. 4—Δh = 0.16 m; (**e**) No. 5—Δh = 0.12 m; (**f**) No. 6—Δh = 0.08 m; (**g**) No. 7—Δh = 0.11 m. Source: own work.

The least traffic slowing effect was noted for traffic circle No. 6, which featured a small difference in level of $\Delta h = 0.07$ m and no demarcated on-street parking next to it. To the right of the one-way street, behind the footpath, there is an urban park, and on the left-hand side there are year-round open small food outlets. The above factors mean that drivers practically have at their disposal the entire width of the road, equal to 5 m, and do not slow down when crossing traffic circle No. 6. Driving speeds become steady on the section between traffic circle No. 7 and traffic circle No. 5 due to the low number of pedestrians in September and a small number of parked vehicles (Figures 11b and 16).

4.4. Regression Analysis

Summing up, we can say that the speed reduction obtained with traffic circles depends on the central island difference in level, longitudinal slope of the approach section, effective carriageway width (whether or not limited by vehicles parked in the demarcated parking spaces), presence of pedestrians, traffic circle location and the surrounding streetscape features. Figure 18 represents the relationship between the obtained speed reductions and the difference in level of the central island, based on the September survey data.

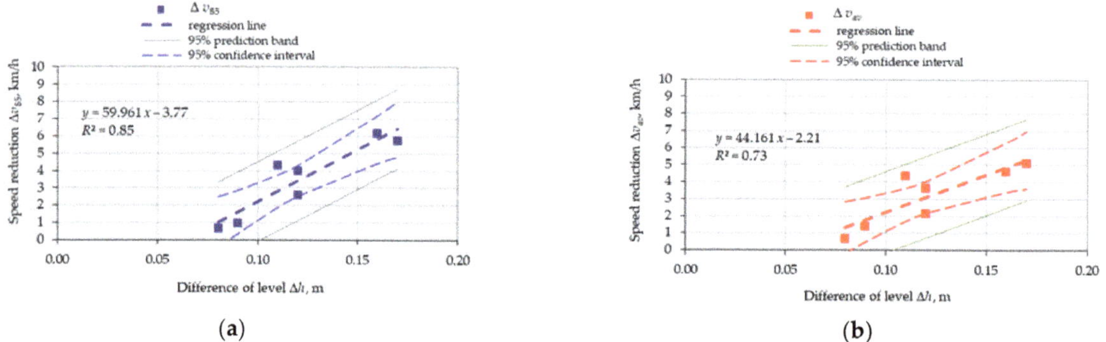

Figure 18. Speed reduction vs. central island difference of level Δh: (a) v_{85}; (b) v_{av}. Source: own work.

Therefore, in sustainable redesigning of urban streets and public spaces, which may include setting up home zones, it is important to select the appropriate traffic calming measures and use appropriate design parameters, matching the existing streetscape features, as the resulting speed reductions have a considerable bearing on the noise and air pollution in the area [42]. These dependencies are related to the third of the research hypotheses, i.e., research Hypothesis H3. For sustainable home zone projects, it is therefore important to avoid sudden speed drops [42] and consider the range in impact of traffic calming measures when planning their locations [26,43–45]. Bearing this in mind, when analysing the speed reductions obtained for the analysed traffic circles, references were made to the results obtained by other researchers reporting similar speed reductions obtained with speed tables [22,43]. The different traffic calming measures (speed table and traffic circle) have different applications and, for example, speed tables are an option for places with a 30 km/h or 40 km/h desired speed, and thus they are not appropriate for home zones requiring reduction to lower speeds. With similar climbing phase characteristics, driving with these two traffic calming measures differs due to the inclined top surface in the case of traffic circles and the level surface in the case of speed tables. The literature does not, as yet give any experimental results comparing these two traffic calming measures in home zone applications. This being so, this article presents the results obtained on three one-way streets, including seven traffic circles having different central island transverse slopes.

4.5. Air Pollution in Three Traffic Scenarios—Research Hypothesis H3

The home zone speed recommendations differ between countries and guidelines, ranging from walking speed, as recommended by the Dutch guidelines [46], to 16 km/h in England [47], to 20 km/h in several other countries [30,33,48–51]. Therefore, air and traffic noise pollution estimations were based on the data obtained for driving through speed tables at speeds in the range of 10–30 km/h [22,43]. Considering the driving speeds measured between the junctions, which also did not exceed the specified speed range (Figures 15 and 16), we can conclude that the analysed traffic circles did not cause abrupt accelerations and decelerations that could be considered undesired due to the impact on the surrounding environment, as per [18,22,43,52]. The level of air pollution was estimated by comparing the three scenarios analysed in this article, using for comparison the research data published by Merkisz et al. [22]. In the first scenario, a vehicle drove through a home zone traffic circle in summer at an almost steady speed of 8–10 km/h, as shown in Figure 15. The second scenario concerned the speeds logged in the home zone September survey, as shown in Figure 16, varying over a considerably wider range of 10–25 km/h. The third scenario was added for comparison, in order to demonstrate the traffic circle's effectiveness as a home zone traffic calming measure TCM. The simulated air pollution in the 30 km/h zone was based on the assumption that the raised central island of the traffic circle causes a speed reduction corresponding to that determined during the home zone traffic survey of September. On the sections between the traffic circles, driving speeds were assumed to vary between 25 and 30 km/h. The air pollution results for these three scenarios are given in Figure 19 below, broken down into carbon dioxide (CO_2), carbon monoxide (CO), nitrogen oxide (NO_x) and hydrocarbons (HC).

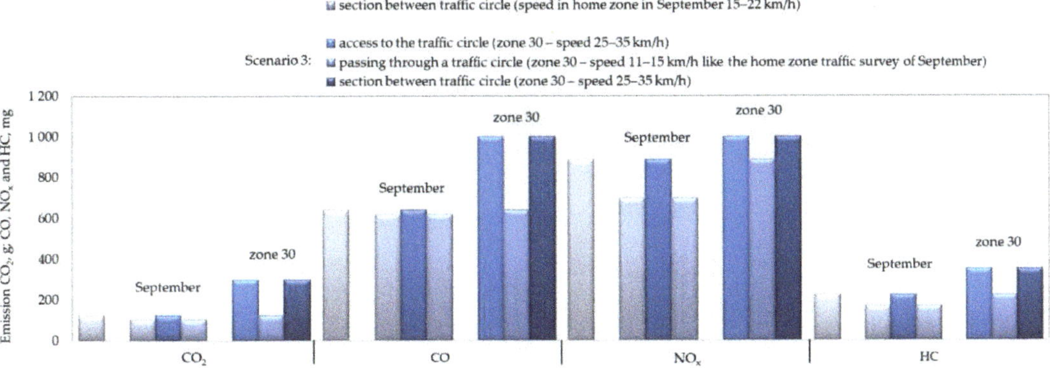

Figure 19. Air pollution in the three main street driving scenarios. Source: own work.

The analysis of the home zone data given in Figure 19 and the 30 km/h zone simulated data showed that the traffic circle related air pollution may be higher in the 30 km/h zone than in a home zone. Comparing the home zone air pollution data in summer and in September, it is justified to conclude that carbon dioxide and carbon monoxide pollution along the street does not differ between the summer and September despite different driving speeds recorded in these periods, due to an almost steady speed in summer of about 8–10 km/h (Figures 12 and 15). The levels of nitrogen oxides and hydrocarbons were, in turn, different in the home zone under analysis. That said, the level of air pollution will be definitely lower in summer due to steadier driving speeds.

4.6. Fuel Consumption in Three Traffic Scenarios

Sudden speed variations would also influence fuel consumption. However, with the short distances between the subsequent traffic circles and small driving speed variations on the way, and with steady speeds of 8–10 km/h due to pedestrian traffic, we can expect drivers to shift to second gear. In September, the driving speeds ranged from 20 to 25 km/h, depending on the number of pedestrians, decreasing only right at the traffic circle. Some drivers drove in third gear, thus reducing fuel consumption. However, almost all the interviewed drivers confirmed shifting to second gear on the approach to the traffic circle. Considering the short distances of travel through the analysed streets and using the mean fuel consumption data in litres/100 km depending on gear, as given in [53], we may expect only very slight differences in fuel consumption. For example, travelling at 8–10 km/h in second gear in summer, the car would use about 0.043 litres of petrol to reach the end of the street, and in September this amount would drop to 0.033 litres due to using third gear, with 20–25 km/h travel speeds practicable on this section. However, if a 30 km/h zone were implemented in this street, a car would use 0.035 litres of petrol driving between the traffic circles in third gear at 25–30 km/h.

The comparisons performed as part of this study, covering pedestrian safety (especially in summer), reduction in driving speed, air pollution and fuel consumption, indicate the suitability of the traffic circle as an effective and economic traffic calming measure for use in home zone applications.

5. Conclusions

The following conclusions may be drawn based on the results of the above-described analyses:

- The traffic calming effect and the amount of speed reduction due to traffic circles depend, to a large extent, on the height of the raised central island Δh. The resulting values of R^2 (v_{85}) = 0.85 and R^2 (v_{av}) = 0.72 indicate that 85% or 72% of the dependent variable variation (v_{85}, v_{av}) may be explained by a relationship with the independent variable (Δh). Now, the remaining 15% or 28% of the variability should be attributed to the effect of other relevant factors (traffic circle location, place in the sequence, street function and the surrounding streetscape features), and other random factors.
- The transverse slope of the central island should be determined in a prior analytical study and implemented in the home zone design, taking into account the following factors:
  - travel lane width,
  - distance between the start and end of the on-street parallel parking spaces and the side street edge,
  - spacing distance between subsequent traffic circles,
  - the surrounding features, such as the locations and opening hours of markets, restaurants and public amenities throughout all seasons of the year.
- The research findings and verification of the formulated research hypotheses show that, for main promenades lined with many retail outlets (seasonal, generating high pedestrian traffic in summer) in home zones located in spa villages, Δh values should be moderate, i.e., max. 8–10 cm. This value may be increased to max. 11–12 cm in other streets with smaller pedestrian traffic and a smaller number of retail businesses and other outlets. In turn, much greater Δh values should be applied in primarily vehicular streets that are not lined with retail businesses or other outlets and have much lighter pedestrian traffic. These higher values of Δh recommended for the above-described type of street may, for example, ranging from 17–19 cm when, past the traffic circle, the street runs for another 150 m or more. For shorter remaining street lengths, such as 50–100 m, Δh should preferably range between 14–16 cm.
- When the traffic calming areas are designed in line with sustainability principles, allowing for extensive use of the carriageway space by pedestrian traffic, as is the case

in this article, one-way traffic should be the first option, and green street/infrastructure components should be used, as far as practicable, for beautification reasons.

- In order to prevent exceeding of the desired speed range on the sections between traffic circles, encouraged by a lack of vehicles parked on the street, fixed-type side obstacles should be designed at the beginning and end of such sections. These obstacles include flowerbeds, planters, and concrete or wooden tree boxes, as shown in Figure A1 in Appendix A.
- Finally, the authors believe that the issue of increased fuel consumption due to driving in lower gears in traffic calmed areas, such as home zones, may be effectively resolved by the global transition to electric vehicles and sustainable design of traffic calming projects.

There are a few limitations that affect this study. One example concerns the speeds in the September traffic survey, which are defined by driving habits typical of Polish drivers and Polish traffic rules. Thus, slightly different speeds may be obtained in research projects carried out in other countries. Fuel consumption estimates in this project were also based on geographically specific French data and different estimates may be obtained with input data typical of other locations.

Author Contributions: Conceptualization, S.M. and A.S.; methodology, S.M. and A.S.; software, S.M.; formal analysis, S.M. and A.S.; data curation, S.M. and A.S.; writing—original draft preparation, S.M. and A.S.; writing—review and editing, S.M. and A.S.; visualization, S.M. and A.S.; supervision, S.M.; project administration, S.M. All authors have read and agreed to the published version of the manuscript.

Funding: This research received no external funding.

Institutional Review Board Statement: Not applicable.

Informed Consent Statement: Not applicable.

Data Availability Statement: The data presented in this study are available in the article.

Acknowledgments: The authors would like to thank Marcin Jóźwiak and the employees of the Public Roads Authority in Kamień Pomorski for their assistance during the traffic surveys carried out as part of this study and for providing us with the information on the home zone conversion project in Międzywodzie.

Conflicts of Interest: The authors declare no conflict of interest.

Appendix A

(a)

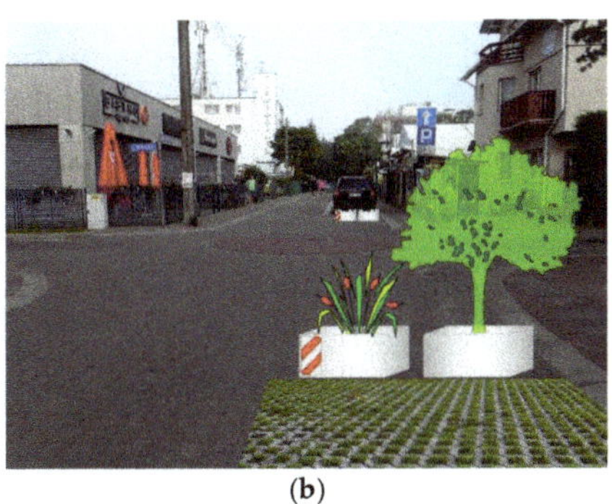

(b)

Figure A1. Visualization of the proposed traffic safety improvement measures using green infrastructure components suitable for home zone applications: (**a**) tree boxes with appropriately selected species used as side obstacles (greenery resistant to drought, frost, exhaust emissions); (**b**) side obstacles with appropriate grafted tree species. Source: own work.

References

1. Horn, B.E.; Jansson, A.H.H. Traffic safety and environment: Conflict or integration. *Int. Assoc. Traffic Saf. Sci. IATSS Res.* **2000**, *24*, 21–29.
2. Boglietti, S.; Tiboni, M. Analyzing the criticalities of public spaces to promote sustainable mobility, XXV International Conference Living and Walking in Cities—New scenarios for safe mobility in urban areas (LWC 2021), 9-10 September 2021, Brescia, Italy. *Transp. Res. Procedia* **2022**, *60*, 172–179. [CrossRef]
3. Paszkowski, Z. *Miasto Idealne w Perspektywie Europejskiej z jego Związki z Urbanistyką Współczesną*; Towarzystwo Autorów i Wydawców Prac Naukowych UNIVERSITAS: Kraków, Poland, 2011. (In Polish)
4. Paszkowski, Z. *Historia Idei Miasta od Antyku do Renesansu*; Zachodniopomorski Uniwersytet Technologiczny w Szczecinie: Szczecin, Poland, 2015. (In Polish)
5. Google Earth. Available online: http://www.earth.google.com (accessed on 2 September 2023).
6. Mell, I.; Clement, S. Progressing Green Infrastructure planning: Understanding its scalar, temporal, geo-spatial and disciplinary evolution. *Impact Assess. Proj. Apprais.* **2020**, *38*, 449–463. [CrossRef]
7. Meerow, S. A green infrastructure spatial planning model for evaluating ecosystem service tradeoffs and synergies across three coastal megacities. *Environ. Res. Lett.* **2019**, *14*, 125011. [CrossRef]
8. Learn about Green Streets. Available online: https://www.epa.gov/G3/learn-about-green-streets (accessed on 18 July 2023).
9. *Green Streets, Residential Streets Commercial Streets Arterial Streets Alleys. A Conceptual Guide to Effective Design Solutions*; United States Enviromental Ptotection Agency US EPA Region 3: Philadelphia, PA, USA, 2009. Available online: https://nepis.epa.gov/Exe/ZyPDF.cgi/P10059Y4.PDF?Dockey=P10059Y4.PDF (accessed on 18 July 2023).
10. WHO. *Pedestrian Safety: A Road Safety Manual for Decision-Makers and Practitioners*; WHO: Geneva, Switzerland, 2013.
11. Slootmans, F. *European Road Safety Observatory Facts and Figures—Pedestrians*; European Commission, Directorate General for Transport: Brussels, Belgium, 2021. Available online: https://road-safety.transport.ec.europa.eu/system/files/2023-02/ff_pedestrians_20230213.pdf (accessed on 18 July 2023).
12. Slootmans, F. *European Road Safety Observatory Facts and Figures—Urban Areas*; European Commission, Directorate General for Transport: Brussels, Belgium, 2022. Available online: https://road-safety.transport.ec.europa.eu/system/files/2022-07/ff_roads_inside_urban_areas_20220707.pdf (accessed on 18 July 2023).
13. Bigazzi, A.Y.; Rouleau, M. Can traffic management strategies improve urban air quality? A review of the evidence. *J. Transp. Health* **2017**, *7*, 111–124. [CrossRef]
14. Harris, G.J.; Stait, R.E.; Abbott, P.G.; Watts, G.R. *Traffic Calming: Vehicle Generated Noise and Ground-Borne Vibration Alongside Sinusoidal, Round-Top and Flat-Top Road Humps*; TRL Report 416; Transport Research Laboratory: Crowthorne, Berkshire, UK, 1999. Available online: https://trl.co.uk/uploads/trl/documents/TRL416.pdf (accessed on 18 July 2023).
15. Lu, X.; Kang, J.; Zhu, P.; Cai, J.; Fei Guo, F.; Zhang, Y. Influence of urban road characteristics on traffic noise. *Transp. Res. Part D Transp. Environ.* **2019**, *75*, 136–155. [CrossRef]
16. McAlexander, T.P.; Gershon, R.R.; Neitzel, R.L. Street-level noise in an urban setting: Assessment and contribution to personal exposure. *Environ. Health* **2015**, *14*, 18. [CrossRef]
17. Ahn, K.; Rakha, H. A field evaluation case study of the environmental and energy impacts of traffic calming. *Transp. Res. Part D Transp. Environ.* **2009**, *14*, 411–424. [CrossRef]
18. Yang, X.; McCoy, E.; Hough, K.; de Nazelle, A. Evaluation of low traffic neighbourhood (LTN) impacts on NO_2 and traffic. *Transp. Res. Part D Transp. Environ.* **2022**, *113*, 103536. [CrossRef]
19. Cloke, J.; Boulter, P.; Davies, G.P.; Hickman, A.J.; Layfield, R.E.; McCrae, I.S.; Nelson, P.M. *Traffic Management and Air Quality Research Programme*; TRL Report 327;Transport Research Laboratory: Crowthorne, Berkshire, UK, 1998.
20. Vardoulakis, S.; Kettle, R.; Cosford, P.; Lincoln, P.; Holgate, S.; Grigg, J.; Kelly, F.; Pencheon, D. Local action on outdoor air pollution to improve public health. *Int. J. Public Health* **2018**, *63*, 557–565. [CrossRef] [PubMed]
21. Burns, J.; Boogaard, H.; Polus, S.; Pfadenhauer, L.M.; Rohwer, A.C.; van Erp, A.M.; Turley, R.; Rehfues, E.A. Interventions to reduce ambient air pollution and their effects on health: An abridged Cochrane systematic review. *Environ. Int.* **2020**, *135*, 105400. [CrossRef]
22. Merkisz, J.; Pielucha, J.; Fuć, P.; Nowak, M. Assessment of vehicle emission indicators for diverse urban microinfrastructure. *Combust. Engines* **2013**, *154*, 787–793.
23. Kuss, P.; Nicholas, K.A. A dozen effective interventions to reduce car use in European cities: Lessons learned from a meta-analysis and transition management. *Case Stud. Transp. Policy* **2022**, *10*, 1494–1513. [CrossRef]
24. Mass Hingway. *Traffic Calming and Traffic Management, Chapter 16*; Mass Hingway: Winchester, MA, USA, 2006. Available online: https://www.winchester.us/DocumentCenter/View/4073/MassDOT-Traffic-Calming-and-Traffic-Management (accessed on 18 July 2023).
25. *Traffic Calming Guidelines Chapter 6*; Devon County Council, Engineering and Planning Department: Devon, UK, 1995.
26. *Roads Development Guide*; East Ayrshire, Strathclyde Regional Council London: Kilmarnock, UK, 2010.
27. *Traffic Calming, Local Transport Note 1/07*; Department for Regional Development (Northern Ireland): Newry, Ireland; Scottish Executives, Welsh Assembly Government: London, UK, 2007.
28. Webster, P. *Traffic Calming—Design Guide, Nottinghamshire County Council*; Department for Transport: Nottingham, UK, 2004.

29. OMGEVING cvba. *IPOD, Richtlijnen, Dimensioneringen, Statuten en Checklist Voor het Openbaar Domein in Gent*; OMGEVING cvba: Berchem-Antwerpia, The Neterlands, 2011.
30. *Vejdirektoratet, Urban Traffic Areas—Part 7—Speed Reducers*; Vejdirektoratet-Vejregeludvalget: Copenhagen, Denmark, 1991.
31. Lárus, Á.L. *Techniques of Speed Reduction—Danish Experiences*; ICTCT International Co-operation on Theories and Concepts in Traffic Safety No. 277/99; ICTCT: Copenhagen, Denmark, 1999.
32. *Traffic Calming Guidelines TCG*; Devon County Council, Engineering and Planning Department: Devon, UK, 1991.
33. *Directives for the Design of Urban Roads*; RASt 06; Road and Transportation Research Association (FGSV): Köln, Germany, 2006.
34. Ignaccolo, M.; Zampino, S.; Maternini, G.; Tiboni, M.; Leonardi, S.; Inturri, G.; Le Pira, M.; Elena Cocuzza, E.; Distefano, N.; Giuffrida, N.; et al. How to redesign urbanized arterial roads? The case of Italian small cities. In Proceedings of the XXV International Conference Living and Walking in Cities—New scenarios for Safe Mobility in Urban Areas (LWC 2021), Brescia, Italy, 9–10 September 2021. [CrossRef]
35. Are Roundabouts and Traffic Circles the Same? The Answer Might Surprise You. Available online: https://www.motorbiscuit.com/roundabouts-traffic-circles-same-answer-surprise (accessed on 18 July 2023).
36. Traffic Circles and Roundabouts. Available online: https://www.surrey.ca/services-payments/parking-streets-transportation/roads-in-surrey/traffic-circles-and-roundabouts (accessed on 18 July 2023).
37. Melton, L.; Shumard, D. *Mini Roundabouts and Neighborhood Traffic Circles*; NCTCOG Public Works Roundup, City Council Regular Meeting: Burleson, TX, USA, 21 May 2019. Available online: https://www.nctcog.org/getmedia/57bdd772-1d6b-4d1f-a344-94ab249ec392/2019PWR-MiniRAB-FINAL.pdf (accessed on 18 July 2022).
38. Rao, P.S.N.; Bhagwati, S.; Satish Khanna, S. *City Level Projects: Street Design Guidelines*; Delhi Urban Art Commission: Delhi, India, 2020.
39. System Ewidencji Wypadków i Kolizji. Available online: https://sewik.pl/ (accessed on 18 July 2023).
40. *Speed Displays Traffic Detection, Radar, Detection, Software*; Vitronic: Kędzierzyn Koźle, Poland, 2015.
41. Transportation Research Board Highway. *Capacity Manual HCM 2000*; Transportation Research Board TRB: Washington, DC, USA, 2000.
42. Künzler, P.; Dietiker, J.; Steiner, R. *Nachhaltige Gestaltung von Verkehrsräumen im Siedlungsbereich, Grundlagen für Planung, Bau und Reparatur von Verkehrsräumen*; Herausgegeben vom Bundesamt für Umwelt BAFU: Bern, Switzerland, 2011. Available online: https://www.bafu.admin.ch/dam/bafu/de/dokumente/luft/uw-umwelt-wissen/nachhaltige_gestaltungvonverkehrsraeumenimsiedlungsbereich.pdf (accessed on 12 August 2023).
43. Faheem, H. Suitability of existing traffic calming measures for use on some highways in Egypt. In Proceedings of the 9th International Conference on Civil and Architecture Engineering, Cairo, Egypt, 29–31 May 2012. Available online: https://iccae.journals.ekb.eg/article_44388.html (accessed on 2 September 2023). [CrossRef]
44. Wirksamkeit Geschwindigkeitsdämpfender Maßnahmen Außerorts; Hessisches Landesamt für Straßen-und Verkehrswesen: Hessen, Germany. 1997. Available online: https://docplayer.org/57666796-Wirksamkeit-geschwindigkeitsdaempfender-massnahmen.html (accessed on 2 August 2020).
45. Distefano, N.; Leonardi, S. Evaluation of the benefits of traffic calming on vehicle speed reduction. *Civ. Eng. Archit.* **2019**, *7*, 200–214. [CrossRef]
46. Centre for Research and Contract Standardization in Civil Engineering (CROW). *Recommendations for Traffic Provisions in Built-Up Areas: ASVV*; CROW: Ede, The Netherlands, 1998.
47. Department for Transport (DfT). *Home Zones: Challenging the Future of Our Street*; Department for Transport (DfT): London, UK, 2005.
48. Gharehbaglou, M.; Khajeh-Saeed, F. Woonerf: A Study of Urban Landscape Components on Living Streets. *MANZAR* **2018**, *10*, 40–49. [CrossRef]
49. Van den Boomen, T. Wed Met de Regels! Het Nieuwe Woonerf (Away with the Rules! he New Woonerf). *NCR Handelsblad* **2002**, *5*.
50. Collarte, N. The Woonerf Concept. Rethinking a Residential Street in Somerville, Master of Arts in Urban and Environmental Policy and Planning MA (857), Tufts University, 7 December 2012. Available online: https://docplayer.net/35878575-The-woonerf-concept-rethinking-a-residential-street-in-somerville-natalia-collarte-60-wadsworth-street-apt-12h-cambridge-ma-857.html (accessed on 18 July 2023).
51. Clayden, A.; McKay, K.; Wild, A. Improving residential liveability in the UK: Home zones and alternative approaches. *J. Urban Des.* **2006**, *11*, 55–71. [CrossRef]
52. Pérez-Acebo, H.; Ziółkowski, R.; Linares-Unamunzaga, A.; Gonzalo-Orden, H. A Series of Vertical Deflections, a Promising Traffic Calming Measure: Analysis and Recommendations for Spacing. *Appl. Sci.* **2020**, *10*, 3368. [CrossRef]
53. Nina67. Consommation D'essence en Fonction de Vitesse et Rapport, Astuces-Pratiques; Article of 23 July 2015. Available online: https://www.astuces-pratiques.fr/auto-moto/consommation-d-essence-en-fonction-de-vitesse-et-rapport (accessed on 12 August 2023).

 *sustainability*

Article

Turbo-Roundabouts as an Instrument for Improving the Efficiency and Safety in Urban Area: An Italian Case Study

Salvatore Leonardi *[ID] and Natalia Distefano [ID]

Department of Civil Engineering and Architectural, University of Catania, Viale Andrea Doria, 6, 95125 Catania, Italy
* Correspondence: salvatore.leonardi@unict.it; Tel.: +39-09-5738-2202

Abstract: In recent years, numerous turbo-roundabouts have been built in many European countries. To date, there are no turbo-roundabouts in Italy and even the regulations do not provide for their implementation. Turbo-roundabouts are considered the ideal alternative to multi-lane roundabouts as they have numerous advantages. However, they offer better operational performance only for specific traffic flow distributions. This research used the case study of an important and complex urban arterial road in eastern Sicily, Italy, to compare the operational and safety performance between multi-lane roundabouts and turbo-roundabouts. The evaluations were carried out with two simulation software: (1) AIMSUN Next 20.0.1 (operational performance); (2) SSAM 3.0 (safety performance). The results show that at medium/low traffic volumes, multi-lane roundabouts are significantly superior to turbo-roundabouts in terms of operational performance. At high traffic volumes, the operational performance of turbo-roundabouts improves significantly. As regards the safety parameters, for turbo-roundabouts there is always an increase in the TTC and PET, a reduction in maximum speeds and decelerations. There is also a significant decrease in conflict points. Ultimately, the safety and efficiency performance of turbo-roundabouts should: (1) Encourage administrations to replace the multi-lane roundabouts (illegal in Italy) with turbo-roundabouts; (2) encourage Italian legislators to revise intersection design legislation to include turbo-roundabouts among possible design solutions.

Keywords: urban road infrastructures; turbo-roundabouts; road safety; congestion; traffic scenarios

Citation: Leonardi, S.; Distefano, N. Turbo-Roundabouts as an Instrument for Improving the Efficiency and Safety in Urban Area: An Italian Case Study. *Sustainability* **2023**, *15*, 3223. https://doi.org/10.3390/su15043223

Academic Editors: Armando Cartenì and Filomena Mauriello

Received: 30 December 2022
Revised: 20 January 2023
Accepted: 7 February 2023
Published: 10 February 2023

1. Introduction

The problem of road congestion is relatively new, although strongly felt. The roads that were built until the first half of the last century, especially in urban and peri-urban areas, were not designed for high traffic volumes. Intersections were (and still are) crossings between two or more different streets, and the few motor vehicles that used them initially did not cause congestion problems.

With the increase in circulating vehicles, roads in many urban contexts have become "in crisis", unable to accommodate and manage the growing flows of urban traffic. It has therefore been necessary to regulate the busiest intersections by more complex means than simple "Stop" or "Yield" signs.

Hence the proliferation of signals that allow you to give the green light to all traffic flows in turn, also based on mutual consistency in terms of traffic volume. This device has evolved over time, so that today we have very complex traffic light systems that, for example, change the green time depending on the length of the queues to be reduced, or give the green light when a bus is approaching (to speed up its journey), or are activated only when vehicles are approaching, all automatically.

However, in numerous cases, the use of roundabouts has been shown to improve intersection functionality, in terms of the number of vehicles that can be handled in the reference time unit. Roundabouts do not require traffic signal control; the traffic rule states that before entering the roundabout, arriving vehicles must give priority to vehicles already in the same roundabout. This type of regulation generally reduces the average waiting times of vehicles compared to traffic light regulation, and thus reduces the queues along the roads converging to the intersection.

In addition to improving vehicular traffic, roundabouts can also provide important safety and environmental benefits [1]. In terms of road safety, it should be noted that the presence of roundabouts significantly slows down approaching vehicles in terms of their travel speed, so that collisions between vehicles at a roundabout are much less frequent and less violent than collisions at signalized intersections (where the intersecting roads are straight ahead). In addition, conflict points are significantly reduced compared to standard intersections, and in particular, roundabouts do not have intersection conflict points. Roundabouts can therefore be considered traffic calming measures in all respects, and as such are particularly useful in urban contexts where they are appreciated by all categories of road users [2,3].

However, as far as the environmental aspect is concerned, it should be borne in mind that roundabouts allow a reduction in pollutant emissions and noise caused by road traffic due to the smoother movement of vehicles (less braking, less time spent standing with the engine running, etc.) [4].

Therefore, roundabouts are design solutions that contribute to the revitalization of the urban environment [5]. It should be noted, however, that the above benefits are mainly realized at single-lane roundabouts. The main international regulations, including the Italian one, prohibit the construction of roundabouts with more than one lane on the circulatory roadway. Nevertheless, a great many two-lane (and even three-lane) roundabouts are still used in infrastructural contexts around the world today. The main disadvantage of the above configurations is the creation of dangerous intersection conflict points on the circulatory roadway, which do not exist in the single-lane configurations (Figure 1).

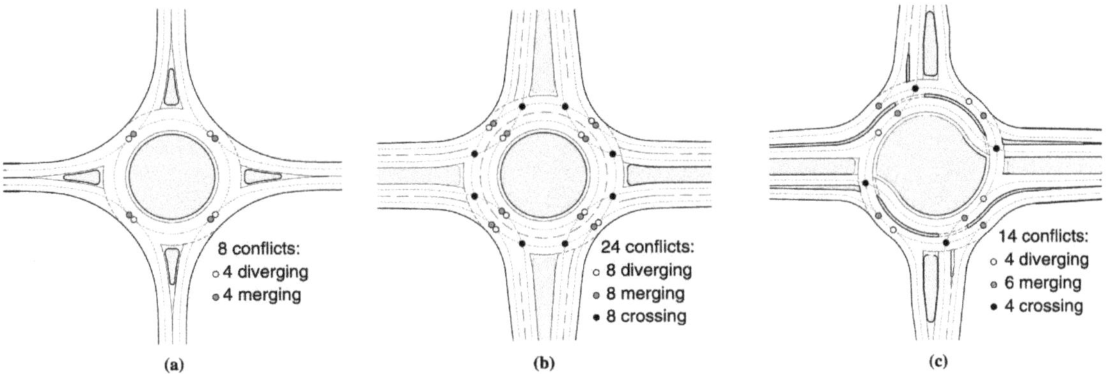

Figure 1. Conflict points (a) single-lane roundabout; (b) double-lane roundabout; (c) turbo-roundabout.

For these reasons, many countries are looking for a solution as to what to do with their existing "standard" multi-lane roundabouts in order to improve the level of traffic safety and capacity. In the last years, many countries of northern Europe (such as the Netherlands, Poland, Czech Republic, Germany, Slovenia, and the UK) have solved the problems of low traffic safety and capacities of existing "standard" multi-lane roundabouts by adopting some alternative types of roundabouts, which decrease the number of conflict points. One of them is the turbo-roundabout [6].

The turbo-roundabout is an innovative scheme of the two-lane roundabout. Professor L.G.H. Fortuijn first introduced it the late 1990s as a safer and more efficient alternative to the standard multi-lane roundabouts. The first turbo-roundabout was built in the Netherlands in 1998. At the turbo-roundabout the traffic flows run separately before entry into the roundabout, they occupy separate lanes within circulatory roadway, and traffic flows are also separate at the exit from the roundabout. Physical separation is obtained through delineators, that is specially-shaped elements, which hinder change of traffic lanes in the various elements making up the roundabout. Therefore, turbo-roundabouts' defining characteristic are physical barriers between circular lanes. Drivers have to choose an entry lane on the approach leg to the roundabout, based on opportune lane markings. This brings an undeniable advantage for user safety, i.e., the number of conflict points on turbo-roundabouts is lower than on multi-lane roundabouts (Figure 1). Two-lane roundabouts have 24 conflict points, whereas turbo-roundabout have only 14 (4 crossing, 4 diverging, and 6 merging conflict points) [6,7].

This study compared the performance of multi-lane roundabouts and turbo-roundabouts in terms of efficiency and safety. For this purpose, two pieces of microsimulation software (AIMSUN Next and SSAM) provided by the Department of Civil Engineering and Architecture of the University of Catania were used and their specificities are explained in Section 3.3.

Roundabouts characterized by the presence of more than one lane on the circulatory roadway are not allowed by Italian law. However, many roundabouts from before 2006 (the year in which the regulations for road intersections were published) that are still in operation have a double lane on the circulatory roadway, and it is desirable that they be adapted with more suitable design solutions to ensure, above all, a higher level of safety. In this study, therefore, a case study is proposed referring to a design configuration in the urban context of the city of Catania (Italy), characterized by the presence of a series of multi-lane roundabouts. The peculiarity of this study is that we did not want to perform a performance comparison with respect to single roundabouts, but considered a whole road section where two roundabouts ("multi-lane" in the existing configuration and "turbo" in the design hypothesis) were the main intersections of a road infrastructure, called "Circonvallazione", characterized by a high traffic volume.

2. Literature Review

In the scientific literature, in addition to the aforementioned reduction of conflict points compared to multi-lane roundabouts, turbo-roundabouts are consistently attributed with various advantages in terms of safety [6–10] even if specific accidents occur there [11]. In particular: (1) the number of conflict points on turbo-roundabouts is lower than on multi-lane roundabouts (Figure 1). Two-lanes roundabouts have 24 conflict points, whereas turbo-roundabout have only 14 (4 crossing, 4 diverging, and 6 merging conflict points); (2) improper lane changes and illegal turns can be significantly avoided; (3) the physical separation of lanes leads to optimum utilization of entry and circulatory lanes, which, in turn, can boost capacity; (4) the spiral road markings in conjunction with the raised lane dividers promote low driving speed; (5) major-road vehicles are limited to cross one circulating lane, as opposed to two circulating lanes in case of minor-road vehicles. Thus, more major-road vehicles can be processed through available critical gaps within the circulatory lanes.

On the contrary, there are some disadvantages that could have a negative impact on operational performance: (a) U-turns are not allowed on minor approaches of basic turbo-roundabouts; (b) passing vehicles on minor approaches are forced to use only the left entry lane, resulting in less flexibility than offered to arriving drivers on main approaches. The impact of the aforementioned disadvantages on traffic operation is mainly dependent on volumes and proportions of vehicular traffic movements of minor legs [6,7].

Other research shows that turbo-roundabouts have a greater capacity than single-lane roundabouts [12], while, depending on the geometric organization and the traffic flows, it could be less than that of multi-lane roundabouts [13]. Choosing the best design solution may sometimes require the use of simulation software that represents, analyzes, and predicts the behavior of vehicle traffic at the site under study. These software offer different levels of detail for analysis:

- macroscopic; treat stationary and aggregated values,
- mesoscopic; study the temporal evolution of all the variables and,

- microscopic; intermediate level between the two previously-mentioned levels.

There are various, mainly commercial, microsimulation software for evaluating the operational performance of the individual components of the road network and the road network as a whole. Among the main microsimulation software tested in research conducted around the world, which have proven reliable and are now widely used and consolidated in the context of roundabout performance characterization, are the following:

- VISSIM, developed by Planung Transport Verkehr (PTV), a German company, is a microscopic simulation program for modeling multimodal transport operations. VISSIM is characterized by a discrete, stochastic, and time step-based model in which vehicle units are represented as individual entities.
- AIMSUN Next, developed by Siemens (one of the largest manufacturers of signal control systems), is capable of generating various traffic conditions based on either stochastic route choice or dynamic user equilibrium.
- PARAMICS, developed by Quadstone Limited, a Scottish company, is a software for modeling the movement and behavior of individual vehicles and transit on local and regional freeway networks.
- TRITONE, developed by University of Calabria, is an open-source platform that also allows the evaluation of traffic safety performance through a set of indicators that represent the interactions in real time between different pairs of vehicles of the traffic flow.

Table 1 shows the advantages and disadvantages of each of the above-mentioned software, based on the results of the studies conducted by different researchers [14–29].

Table 1. Advantages and disadvantages of the main micro-simulation software.

	VISSIM	AIMSUN Next	PARAMICS	TRITONE
Advantages				
User-defined algorithms for vehicle movement control.		X		X
Appropriate for traffic simulation, traffic data analysis, planning, etc.	X	X	X	X
Intersection type is not predefined.	X			
The duration of traffic analysis can be defined by the user.	X			
Includes psycho-physical model for car-following.				
Includes car-following models and other calibration methods.				X
Suitable for safety analysis based on vehicle trajectory.		X	X	X
Disadvantages				
Developing a complete algorithm for safety analysis is difficult, especially for new users				X
There are few options for modeling accidents.			X	
Coding the inputs and outputs is very time consuming and labor intensive.	X	X		
The modeled trajectories are not realistic.	X			
Traffic volume is determined using origin-destination matrices.			X	

In one study, PARAMICS microsimulation software was used to investigate the operational performance of a two-lane turbo-roundabout and a three-lane traditional roundabout. The authors concluded that the turbo-roundabout had a 12–20% higher capacity [30]. Other authors evaluated the performance of multi-lane roundabouts using KREISEL 7.0 software. The results showed that turbo-roundabouts were able to achieve higher capacity in most cases regardless of the saturation level [31].

Analysis of a turbo-roundabout and an existing two-lane roundabout in Bogotà, Colombia, using VISSIM showed that converting a two-lane roundabout to a turbo-roundabout resulted in a 7% increase in overall capacity [32]. Similarly, a study of the effects of converting an existing two-lane roundabout to a basic turbo-roundabout in Portugal concluded that the total capacity of conventional roundabouts was nearly 3% lower [33].

By using a Dutch capacity model, the Multilane Roundabout Explorer, it was shown that turbo-roundabouts were superior to traditional roundabouts whenever traffic flow did not exceed the threshold of 3500 pcph. The authors indicated that the capacity improvement of turbo-roundabouts compared with conventional roundabouts ranged from 25% to 35% [34].

Another study found that conventional roundabouts always performed better as long as circulating vehicles did not exceed 700 pcph. However, when circulating traffic exceeded this value and reached up to 3000 pcph, basic turbo-roundabouts became superior and consistently outperformed conventional roundabouts [35].

In contrast, other studies have shown that conventional two-lane roundabouts have 20–30% higher capacity compared to basic turbo-roundabouts. The authors also added that minor approaches of basic turbo-roundabouts perform better in very rare scenarios where more than 60% of the main traffic flow turns right [30,36,37].

Comparison of four alternatives to conventional roundabouts, one of which is a basic turbo-roundabout, showed that turbo-roundabouts perform best when 70% of vehicles turn right on each approach [38].

One study examined three turbo-roundabouts that were designed to replace three existing two-lane roundabouts in Portugal. Results from the AIMSUN microsimulation model showed that turbo-roundabouts were superior to conventional roundabouts when the saturation level was below 70% [39]. The results of a study conducted in Ghana using VISSIM software showed that turbo-roundabouts provided 19% higher capacity than conventional roundabouts [40]. Nevertheless, their minor approaches were always operationally inferior. In a study conducted in Italy, an analytical capacity model was used to evaluate an existing two-lane roundabout with congestion and a turbo-roundabout. The results of the study stated that each approach of a turbo-roundabout can ultimately handle 15–84% more vehicles, depending on traffic volumes and traffic distribution on site [41].

Even in this case, the results still seem unclear and sometimes contradictory. Two studies showed that drivers experience significantly less delay at turbo-roundabouts than at multi-lane roundabouts when most of the traffic flow is through the main approaches [42,43]. However, according to another study, turbo-roundabouts have shorter delays than multi-lane roundabouts when traffic volumes are balanced, although the differences are minimal [31].

The outputs of micro-simulation models can also be used to evaluate the safety level of road networks. This approach can be achieved using SSAM software developed by FHWA, which automates conflict analysis by processing vehicle trajectories (vehicle position, speed and acceleration profiles) generated during simulation.

Therefore, by coupling the micro-simulation performed with software such as VISSIM, AIMSUN Next, PARAMICS, and TRITONE with SSAM, it is possible to study different scenarios from a traffic safety point of view and to quickly evaluate possible measures to improve the safety level of the infrastructure elements of a road network.

Some studies have highlighted some concerns about SSAM, namely: The ability of a simulated trajectory to reflect complex real-world driving behavior, calibration efforts to obtain reliable safety results, or the inability of SSAM to determine the probability that each estimated conflict will result in an accident [44–47]. Although the SSAM required accurate calibration of the traffic model, some authors nevertheless found reasonable relationships between the conflicts estimated by the SSAM and real accidents [48–51].

Regarding specifically the application of SSAM to roundabouts, some studies have shown that the values of surrogate safety indicators obtained for the configurations of two-lane roundabouts and turbo-roundabouts are particularly reliable [51,52]. In contrast, the safety indicators obtained for single-lane roundabouts are less reliable [24,51–54]. A study conducted in Italy has shown that regardless of the traffic micro-simulation model used, the safety parameters obtained with SSAM for single-lane roundabouts show that roundabouts are less safe than other types of intersections [24]. This is in stark contrast to the fact that roundabouts are recognized as the safest intersection types of all possible at-grade intersection configurations.

3. Materials and Methods

The method used in the present study was based on the micro-simulation of different traffic scenarios starting from a specific urban road context characterized by a sequence of standard roundabouts, actually present in Italy. This context was then compared, at the level of performance offered, with a project configuration in which the main intersections were hypothesized as turbo-roundabouts designed according to Dutch regulations (CROW, 2008). In particular, two software programs were used which made it possible to obtain both the outputs related to the operational performance offered to users in all the traffic scenarios considered, and the results related to the performance in terms of safety.

3.1. Site Selection

The study context is a road section along the urban artery that forms the ring road of the metropolis of Catania. It is characterized by the presence of two multi-lane round-abouts (only in one of them, however, the horizontal signs separating the two lanes on the circulatory roadway are clearly visible). These two roundabouts are located near the neighborhoods of Nesima and San Nullo. They are about 1.20 km apart and both have a large diameter; the Nesima roundabout (R1) has a diameter of 70 m, while the San Nullo roundabout (R2) has a diameter of 90 m. Between the two roundabouts, almost in the middle, there is a standard intersection (N) with a secondary road called Via Sebastiano Catania. The two roundabouts were first rebuilt in a CAD environment and then redesigned by building two turbo-roundabouts from the central turbo-block based on the literature standard [6,55,56] (Figure 2 and Table 2).

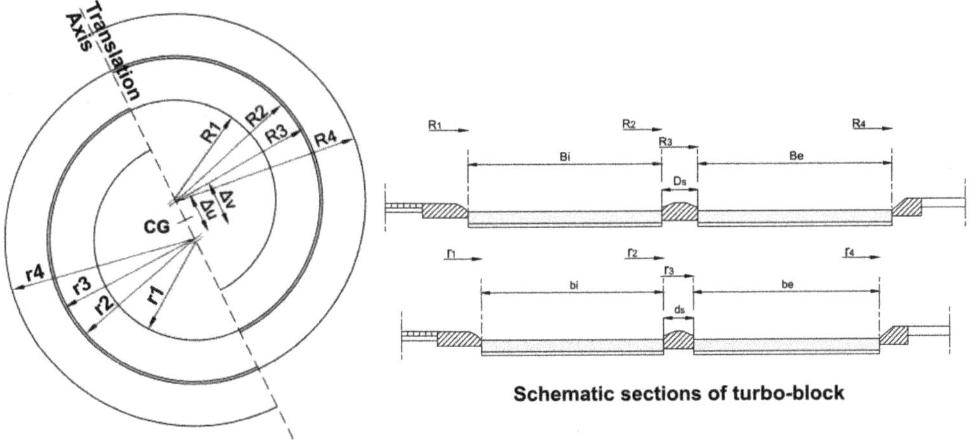

Figure 2. Design elements of turbo-roundabouts (turbo-block and translation axis).

Table 2. Design parameters of turbo-roundabouts.

	Type of Turbo-Roundabout			
Design parameter	*Mini*	*Standard*	*Medium*	*Large*
R_1	10.45 m	12.00 m	14.95 m	19.95 m
R_2	15.85 m	17.15 m	20.00 m	24.90 m
R_3	16.15 m	17.45 m	20.30 m	25.20 m
R_4	21.20 m	22.45 m	25.25 m	29.95 m
r_1	10.95 m	12.50 m	15.45 m	20.45 m
r_2	15.65 m	16.95 m	19.80 m	24.70 m
r_3	16.35 m	17.65 m	20.50 m	25.40 m
r_4	20.70 m	21.95 m	24.75 m	29.45 m
D_s	0.30 m	0.30 m	0.30 m	0.30 m
d_s	0.70 m	0.70 m	0.70 m	0.70 m
B_i	5.40 m	5.15 m	5.05 m	4.95 m
B_e	5.05 m	5.00 m	4.95 m	4.75 m
b_i	4.70 m	4.45 m	4.35 m	4.25 m
b_e	4.35 m	4.30 m	4.25 m	4.05 m
D_v	5.75 m	5.30 m	5.15 m	5.15 m
D_u	5.05 m	5.00 m	4.95 m	4.75 m

Figure 3 shows the two roundabouts along the road section under consideration and the two alternative design configurations consisting of the TR1 and TR2 turbo-roundabouts.

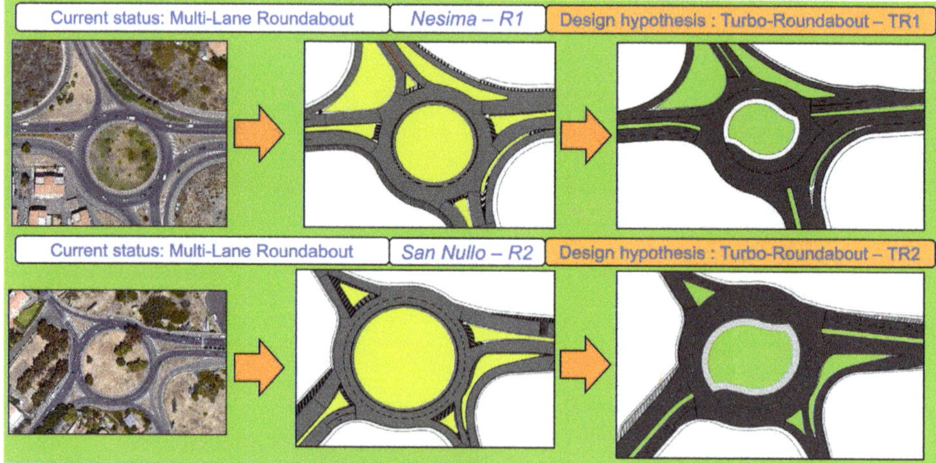

Figure 3. Multi-lane existing roundabouts (R1 and R2) and design hypothesis (turbo-roundabouts, TR1 and TR2).

3.2. Traffic Scenarios

The traffic conditions for the section in question were derived from the preparatory studies for the preparation of the Urban Traffic Plan (PUT) of the City of Catania (2014) and used as a reference for the traffic scenario referred to as "scenario n. 1". In particular, the peak period from 12:00 to 13:00 was considered, characterized by a total traffic flow of $Q_1 = 4594$ vehicles per hour. In order to compare different traffic conditions and consequently analyze the differences in operational and safety performance, three additional traffic scenarios were defined, characterized by increasing vehicle flows starting at Q_1.

In particular, a final scenario (scenario n. 4) was assumed, in which the total traffic flow was increased by 25% compared to scenario n. 1, and 2 other scenarios were defined (scenarios n. 2 and n. 3), among which the 25% increase was fairly distributed.

Finally, for the subsequent simulations, four scenarios were used with the total traffic flow (Qi) values given below:

- Scenario n. 1: Q_1 = 4594 vehicles/hour
- Scenario n. 2: Q_2 = 4977 vehicles/hour
- Scenario n. 3: Q_3 = 5359 vehicles/hour
- Scenario n. 4: Q_4 = 5742 vehicles/hour

Figure 4 shows the graph of the analyzed network part.

Figure 4. Graph of the analyzed part of the road network.

This graph also shows the 10 centroids used to assign traffic flows to the nodes (intersections) and edges (road segments) of the network.

The street sections of the main street were indicated with capital letters, the side streets with small letters.

3.3. Micro-Simulation Process

The micro-simulation was performed using two pieces of software:

(1) AIMSUN Next for the calculation of the operational performance corresponding to the traffic scenarios described in the previous paragraph;

(2) SSAM for the estimation of the safety indicators, starting from the kinematic parameters associated with all the vehicle trajectories obtained as output of the AIMSUN Next software.

The AIMSUN software and the VISSIM software are both provided by the Department of Civil Engineering and Architecture of the University of Catania. For the present study, the authors preferred the AIMSUN software because, as shown in Section 2 (see in particular Table 1), it is more reliable than VISSIM in generating the most realistic trajectories possible. Since one of the objectives of this study was to also perform the evaluations of the surrogate safety measures, it was decided to use the most appropriate micro-simulation software to create the trajectories compatible with driving on multi-lane roundabouts and turbo-roundabouts.

Moreover, regarding the use of SSAM for the estimation of safety indicators, the authors were encouraged by the good results in the literature (also in Section 2) regarding the reliability of the model in the case of turbo-roundabouts and multi-lane roundabouts.

3.3.1. AIMSUN Model

AIMSUN Next is a software package that is developed by Siemens Business. The basis of this software is a microscopic traffic simulator developed at the Department of Statistics and Operations Research of the Politècnica de Catalunya University by Barceló and Casas (2002), Spain. The structure of AIMSUN can be described in terms of the following two elements:

(1)   geometric scheme of the road network.
(2)   modelling of vehicle behavior.

The AIMSUN mesoscopic model uses a representation of the road network based on a directed graph consisting of the following three geometric elements:

- Centroids. They are the source of vehicles entering and exiting the network.
- Nodes (intersections). They are treated as node servers in the mesoscopic representation. In the node servers, vehicles are directed from one section to a turning and then to their next section. These turnings connect the lanes of the originating section to the lanes of the destination section. All vehicles are assumed to travel unimpeded, i.e., at free-flow speed, in turnings.
- Edges. They are the segments that connect the nodes. Each edge contains information about its geometry (e.g., the number of lanes, shoulder width, etc.).
- The behavioral models used in AIMSUN are the following:
- Behavioral models in edges: Car-following models and lane-changing models.
- Behavioral models in nodes (intersections): Gap acceptance and lane choice models.

The car-following model is a simplified version of the Gipps car-following model, which is used for the microscopic level and considers two components; deceleration and acceleration. The deceleration and acceleration constraints are simplified to obtain the following expressions [24,57]:

$$\begin{aligned} x(t,n) &\leq (t - R_T, n) + S_{n(max)} \cdot R_T \\ x(t,n) &\leq (t - R_T, n - 1) - E_L \end{aligned} \tag{1}$$

$$S_n(t + R_T) = -d_c \cdot R_T + \sqrt{d_c^2 \cdot R_T^2 + d_c \cdot \left[2 \cdot d - S_n(t) \cdot R_T + \frac{S_{n-1}^2(t)}{d_{c(e)}} \right]} \tag{2}$$

where:

- t = simulation time;
- n = vehicle number ordered by its arrival time on the lane;
- $x(t, n)$ = position of vehicle n at time t;
- S_n = speed of the nth vehicle;
- $S_{n(max)}$ = maximum speed of the vehicle (the minimum between the desired maximum speed of the vehicle and the maximum speed of the edge);
- E_L = effective length of the vehicle (vehicle length plus minimum distance between vehicles);
- d = distance between the vehicles;
- R_T = reaction time of the driver of the follower vehicle;
- d_c = maximum deceleration of the considered vehicle;
- $d_{c(e)}$ = estimate of the desired deceleration of the leading vehicle;
- d = distance between the vehicles.

The original Gipps car-following model was used to calculate the speed in the next simulation step. In AIMSUN, car-following and lane-change models are used to calculate the edge travel time. The number of vehicles in any segment is limited by the capacity of the edge.

The gap-acceptance model is used to model the give-way behavior. In particular, the model is used to decide which of two vehicles has priority in a conflicting movement in the intersections. It considers the travel time from both vehicles to the collision point, then determines how long it will take the vehicles to clear the intersection, and finally makes the decision.

The maximum give-way time parameters are used to determine when drivers become impatient if they cannot find a gap. If the driver has waited longer than this time, the safety distance—normally set at twice the reaction time—is reduced linearly to zero.

From AIMSUN trajectory files (.trj) are obtained. These files contain the routes and speed changes that drivers adopt in the simulated scenarios (Figure 5).

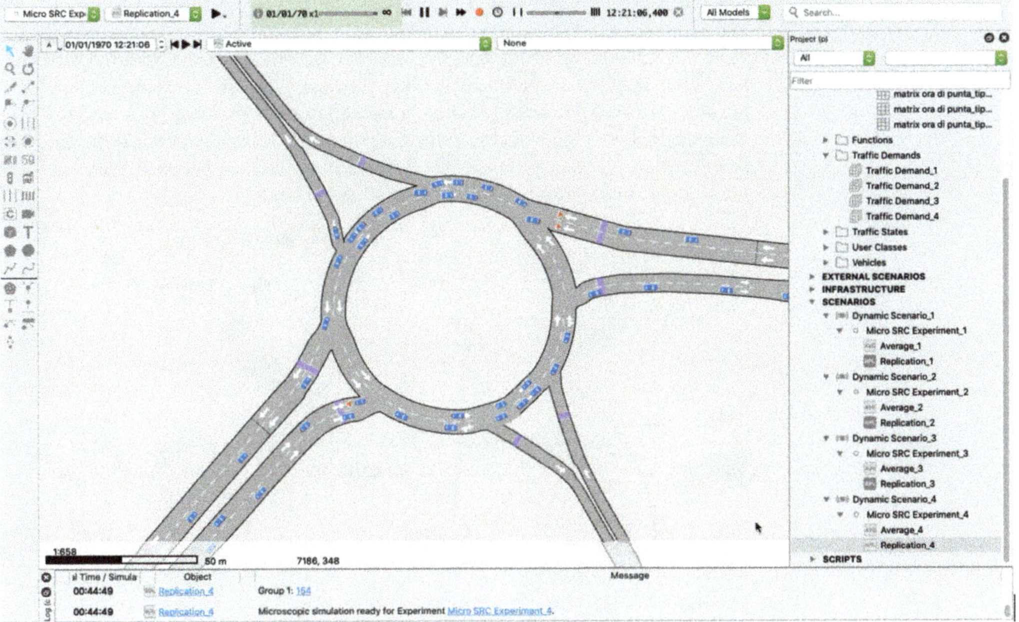

Figure 5. Example of a AIMSUN output for the simulation of one of the examined roundabouts.

Among the output data available from AIMSUN Next, the following were considered in this study:

(1) Total number of stops: A stop for a vehicle happens whenever its speed decreases below the queue entry speed and while it remains below the queue exit speed parameter. Once the vehicle speed goes above the queue exit speed parameter the vehicle is no longer considered in a queue nor stopped. A new stop will be added to the number of stops statistics when the vehicle speed goes below queue entry speed again.

(2) Delay Time: Average delay time per vehicle per kilometer (seconds/km). This is the difference between the expected travel time (the time it would take to traverse the system under ideal conditions) and the travel time. It is calculated as the average of all vehicles and then converted into time per kilometer. It does not include the time spent in a virtual queue.

3.3.2. SSAM Model

SSAM (Surrogate Safety Assessment Module) is software developed by Siemens Energy and Automation, Inc. in collaboration with the Federal Highway Administration in 2008. This software uses trajectory data generated by micro-simulators to identify potential conflicts based on the conflict definition specified by the modeler.

A trajectory file is created by the micro-simulator during the model run and contains information about the position and movement of each vehicle. The most important data in the trajectory file can be divided into the following four classes: (1) Dimension, (2) Timestep, (3) Vehicle, (4) Conflict [58–60].

- The Dimension class contains information about the spatial characteristics of the rectangular bounding box of the microsimulation environment.
- The Timestep class contains a record of the current time step since the start of the simulation. This variable allows SSAM to position the vehicles in time.
- The Vehicle class contains information about the spatial characteristics of the vehicle and the speed and acceleration values used to predict vehicle motion.
- The Conflict class contains information about the conflict angle, which is an approximate angle for a hypothetical collision between colliding vehicles based on the estimated heading of each vehicle. Depending on the values of this angle, the resulting conflict may be a rear-end collision, a lane change, or crossing movement. Specifically, the rear-end angle is used to define a potential collision when following and lane changing, and the crossing angle defines potential collisions in head-on scenarios, such as maneuvering through an intersection (Figure 6).

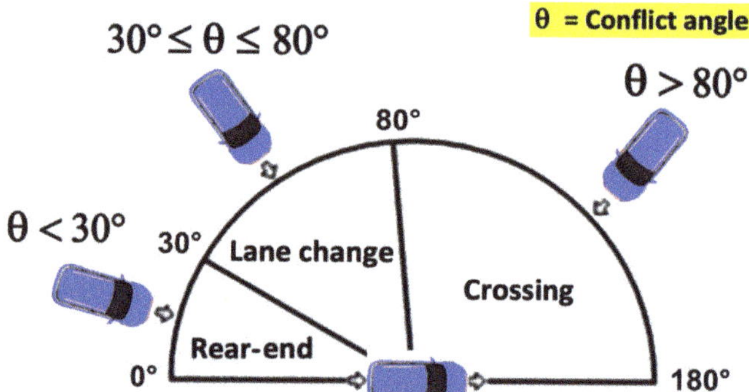

Figure 6. Representation of conflict angles according to SSAM.

Using the four classes of parameters listed above, SSAM can determine whether a vehicle's trajectory will collide with that of another vehicle and report information about this interaction. This information includes the following surrogate safety measures:

1. Time-to-collision minimum (TTC_{min})—the minimum time-to-collision value (in seconds) observed during the conflict, for two vehicles to collide if speeds and directions do not change [61]. For TTC_{min}, the study was accomplished with conflicts specified as $0.1\ s < TTC_{min} < 1.5\ s$ [62].
2. Post-Encroachment-Time (PET)—the time (in seconds) between when the first vehicle last occupied a position and the time when the second one arrived at the same position [63]. For the PET, conflicts were specified as $0.1\ s < PET < 5\ s$ [62].
3. Maximum speed (MaxS)— the maximum speed (m/s) of two vehicles involved in the conflict event [64].
4. Difference in vehicle speeds (DeltaS)—the absolute value of difference in speeds (m/s) of two conflicting vehicles [64].
5. Initial Deceleration Rate (DR)—the magnitude of the deceleration action (m/s^2) of a driver the moment he begins an evasive braking maneuver [65].

6. Maximum deceleration rate (MaxD)—the maximum deceleration (m/s^2) of the through vehicle [45].

In this study, four of the six surrogate measures that SSAM reports were used: TTC$_{min}$, PET, MaxS and DR.

In addition, other parameters were evaluated that are representative of the level of safety provided by the infrastructure configurations considered. These parameters are the total number of conflicts generated in the considered time interval and the number of conflicts divided into the three types that can be evaluated by SSAM (crossing, rear end, lane change) (Figure 7).

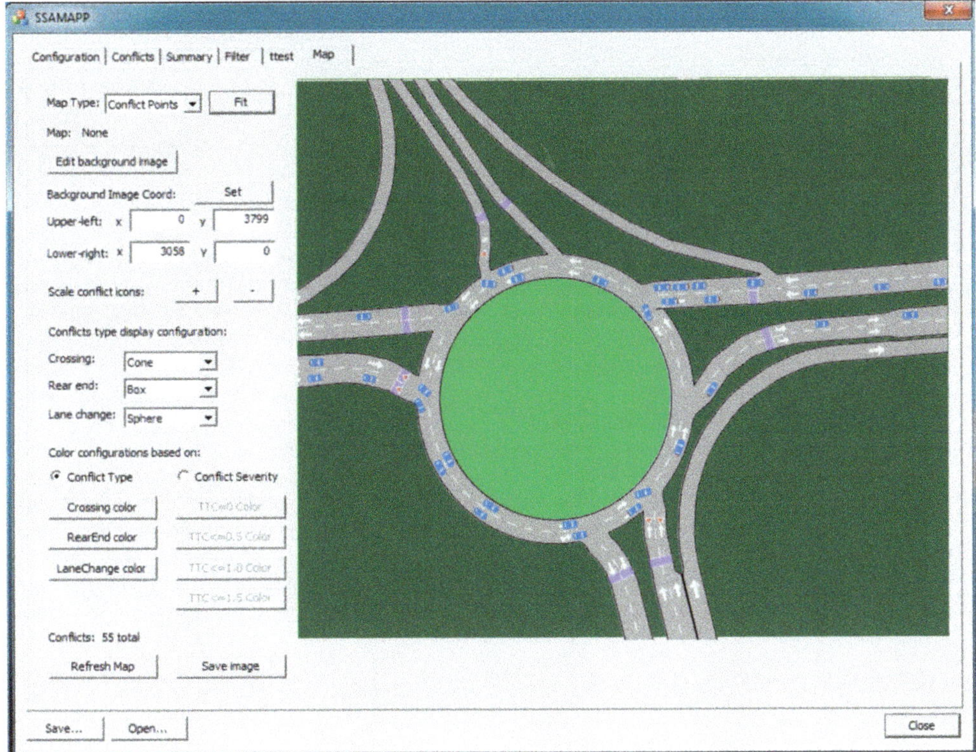

Figure 7. Example of conflicts in SSAM for one of the examined roundabouts.

4. Results and Discussion

The procedure described in the previous section made it possible to estimate the values of 10 indicators: Two performance indicators (Total number of stops and Delay Time) and eight safety indicators (TTC$_{min}$, PET, Maximum speed (MaxS), Initial Deceleration Rate (DR), Total number of conflicts, Crossing Conflicts, Rear End Conflicts and Lane Change Conflicts).

The results of the simulations are shown in Tables 3 and 4. Specifically, Table 3 shows the values of the 10 parameters obtained from the simulations for the four traffic scenarios with respect to the existing configuration, i.e., the configuration with the two multi-lane roundabouts, while Table 4 shows the results of the four simulations with respect to the project configuration that assumed the presence of turbo-roundabouts.

Table 3. Performance and safety indicators for the road configuration where standard multi-lane roundabouts are present.

	Total Number of Stops	Delay Time [s/km]	TTC_{min} [s]	PET [s]	Maximum Speed (MaxS) [m/s]	Initial Deceleration Rate (DR) [m/s^2]	Total Number of Conflicts	Crossing Conflicts	Rear End Conflicts	Lane Change Conflicts
Scenario n. 1	1113	24.75	0.93	1.85	6.44	2.84	3364	10	2911	443
Scenario n. 2	1313	35.89	0.91	1.91	5.82	2.87	4699	23	3982	694
Scenario n. 3	2819	74.34	0.88	1.84	4.67	2.51	9028	34	7783	1211
Scenario n. 4	3176	91.40	0.89	1.87	4.57	2.50	9683	45	8360	1278

Table 4. Performance and safety indicators for the road configuration where turbo-roundabouts are present.

	Total Number of Stops	Delay Time [s/km]	TTC_{min} [s]	PET [s]	Maximum Speed (MaxS) [m/s]	Initial Deceleration Rate (DR) [m/s^2]	Total Number of Conflicts	Crossing Conflicts	Rear End Conflicts	Lane Change Conflicts
Scenario n. 1	1926	64.16	1.15	2.31	4.86	2.24	2009	1	1798	112
Scenario n. 2	1999	73.44	1.23	2.49	3.98	1.96	2417	2	2310	156
Scenario n. 3	2056	107.44	1.25	2.47	3.82	1.83	2659	2	2490	167
Scenario n. 4	2368	136.92	1.24	2.43	3.74	1.73	2741	4	2567	183

For a better representation of the results obtained, it was decided to use the Kiviar diagram (or radar diagram or spider diagram). A radar chart is a 2D chart that represents multivariate data by assigning an axis to each variable and plotting the data as a polygonal shape across all axes. All axes have the same origin, and the relative position and angle of the axes are usually not informative. The equiangular spokes from the origin to the point on each axis represented by the variable are called radii. Typically, a radar chart looks like an irregular polygon, or like several irregular polygons stacked on top of each other, all with the same center point.

The Kiviar diagrams in Figures 8–11 show the percentage changes along the rays for the 10 parameters considered, calculated by comparing the values obtained with respect to the project configuration with turbo-roundabouts with the values of the configuration with multi-lane- roundabouts. In particular, the diagrams in Figures 8–11 allow the evaluation, for each scenario, of how much each of the parameters evaluated in relation to the route with turbo-roundabouts varies compared to the sequence of road sections where the main intersections are the standard roundabouts. The diagram in Figure 12, on the other hand, makes it possible to compare all the calculated indicators for all four traffic scenarios at the same time.

The following considerations can be derived from the analysis of Tables 3 and 4 and Figures 8–12.

Total number of stops: The first simulation scenario, to which the lower traffic flows correspond, shows that the multi-lane roundabout is the best design solution. This consideration also applies to the second simulation scenario. Consistent with the first two scenarios, the presence of turbo-roundabouts results in a 73% (first scenario) and 52% (second scenario) increase in the total number of stops. However, when traffic flow continues to increase, a decrease in the total number of stops is observed in the presence of turbo-roundabouts (-27% for scenario n. 3; -25% for scenario n. 4), which translates into a decrease in queuing phenomena. Thus, from the point of view of the indicator considered, the configurations in which the turbo-roundabouts have to manage significant traffic flows are more functional than those in which there are standard roundabouts.

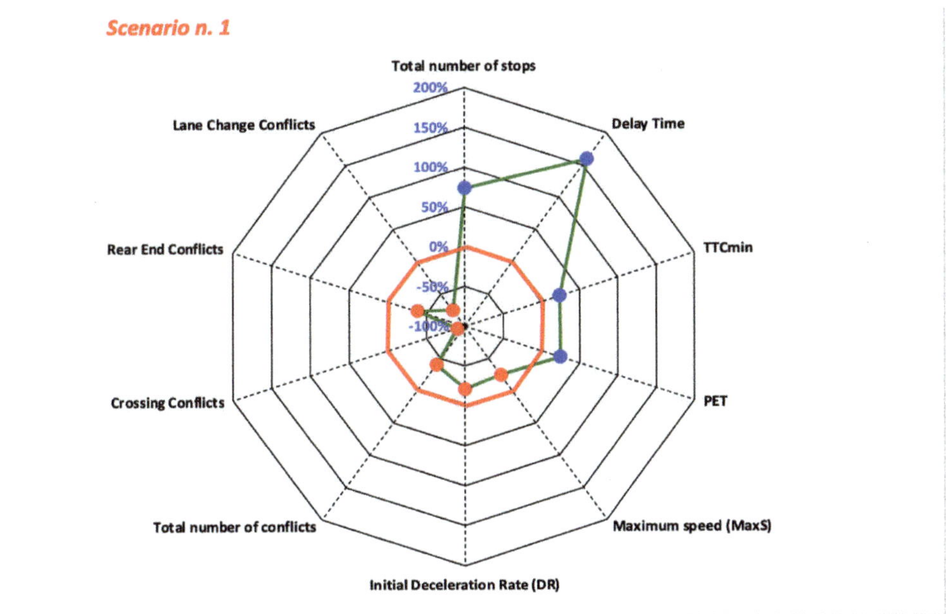

Figure 8. Percentage variation of the 10 indicators considered (Scenario n. 1).

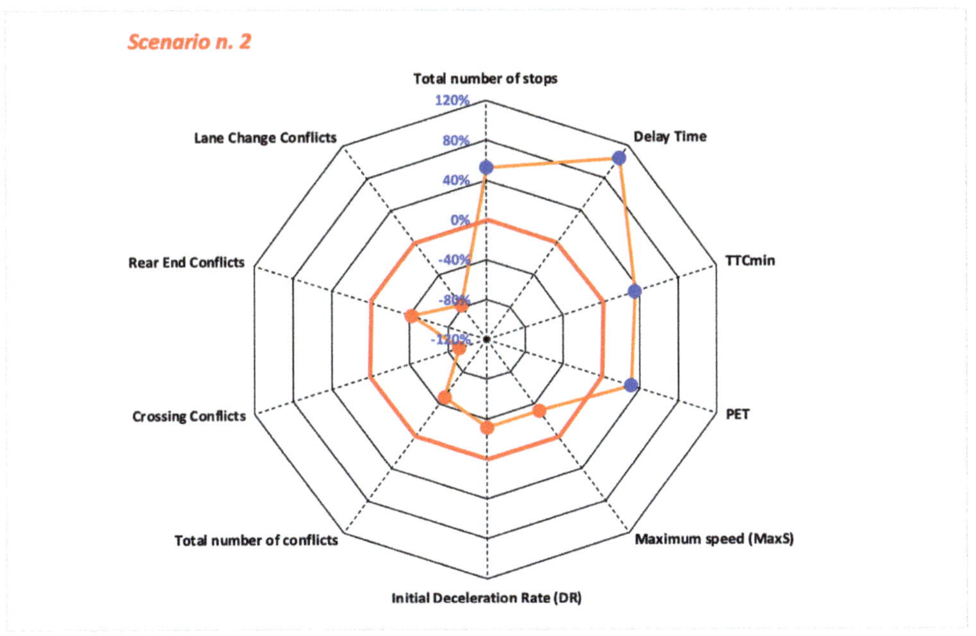

Figure 9. Percentage variation of the 10 indicators considered (Scenario n. 2).

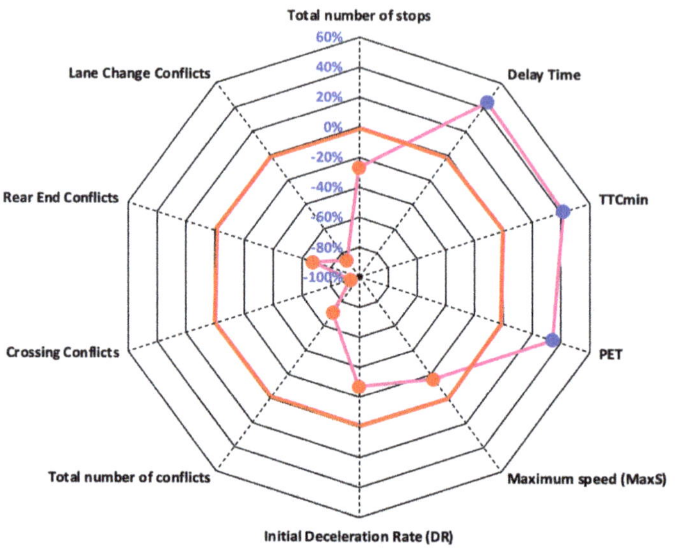

Figure 10. Percentage variation of the 10 indicators considered (Scenario n. 3).

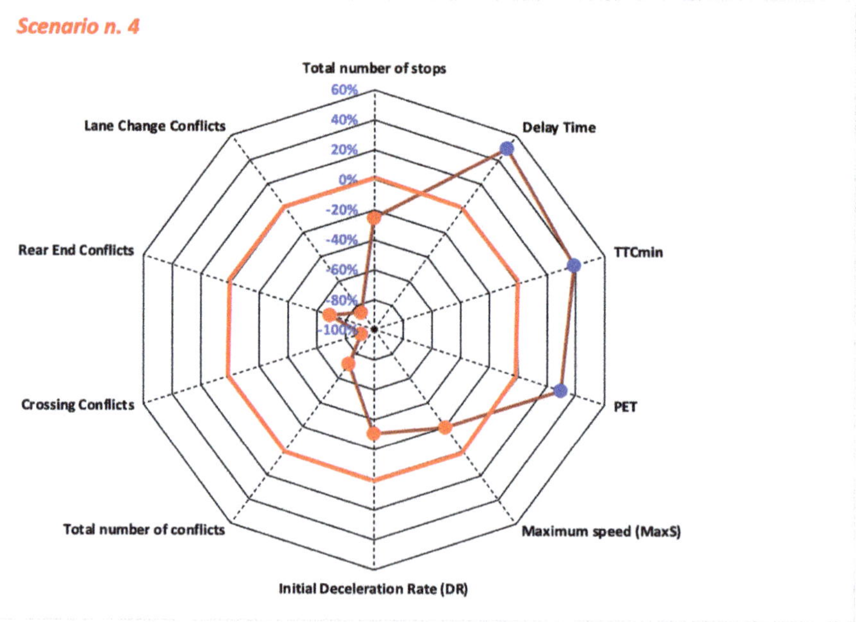

Figure 11. Percentage variation of the 10 indicators considered (Scenario n. 4).

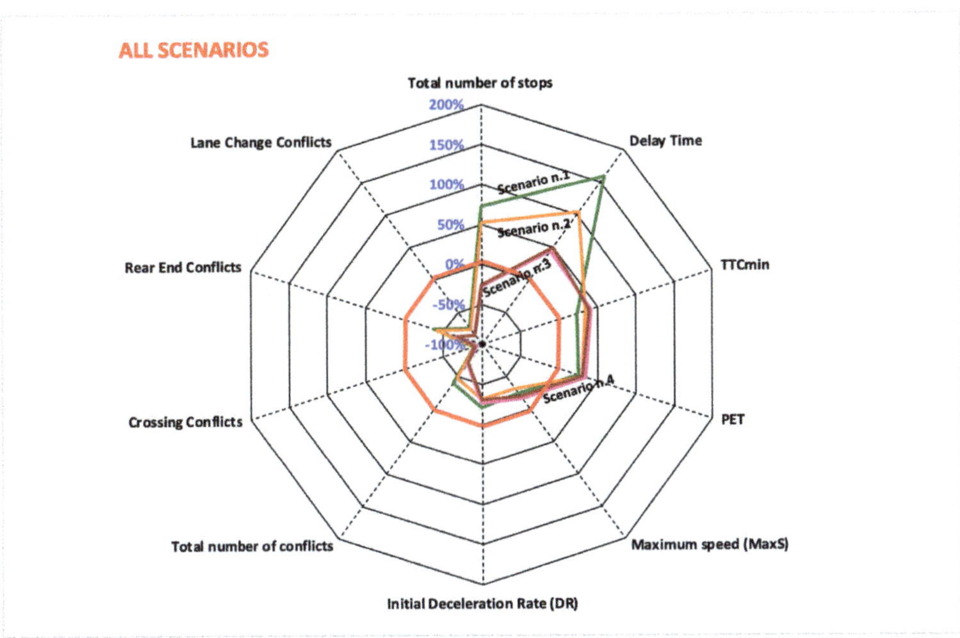

Figure 12. Percentage variation of the 10 indicators considered (All scenarios).

Delay Time: Similar to the other performance indicator analyzed in the previous point, multi-lane roundabouts are more functional than turbo-roundabouts when traffic volumes are medium to low. In fact, delays are about two and a half times higher in scenario n. 1 than in turbo-roundabouts and about twice as high in scenario n. 2. For high traffic volumes, the performance of turbo-roundabouts improves, although high delay values still occur, which in both scenario n. 3 and scenario n. 4 are about 1.5 times the values determined for standard roundabouts. The results related to the evaluation of this parameter are consistent with the uncertainty found in the literature related to operational performance of turbo-roundabouts in general. Indeed, many studies show that the values of operational performance (capacity, delays, queues) strongly depend on the geometric configurations of the turbo-roundabouts and on the distribution of vehicle flows among the different approaches [6,7,9,35–37,40,41]. Therefore, the evaluations of operational performance often do not coincide (see the analysis performed in Section 2). Regarding delays, the results of this study are in contrast with studies (such as [42,43]) that show that turbo-roundabouts, where the traffic flows are higher on the main road, and have significantly lower delays than conventional multi-lane roundabouts. In fact, the part of the network investigated in this study is characterized by a main road, the "Circonvallazione", which is significantly more loaded with traffic than the secondary approaches, yet the total delays always remain high (although they tend to decrease with increasing traffic). The delays valued in this study may be more consistent with the study showing that delays are slightly lower for turbo-roundabouts than for multi-lane roundabouts when traffic flow is balanced [31]. This could mean that when traffic flow is unbalanced, delays on turbo-roundabouts tend to increase. However, it would be necessary to perform simulations under different traffic conditions to obtain positive feedback on the validity of the above statement.

TTC$_{min}$: All traffic scenarios associated with the configuration in which the multi-lane roundabouts are present show TTC$_{min}$ values averaging 0.9 s, which is well below the threshold for this indicator (1.5 s). In contrast, TTC$_{min}$ values for turbo- roundabouts average 1.2 s. This shows not only that this indicator is 24% to 42% higher than for multi-lane roundabouts, but also that thanks to the turbo-roundabout, drivers always have more time to avoid conflicts, thus providing road users with a higher level of safety. The TTC_{min} values obtained in this study are quite comparable to the average values obtained in a study conducted in Bogotà, where the TTC_{min} value in a turbo-roundabout simulated with the SSAM software was 1.32 s [53].

PET: This indicator, similar to the previous one, also shows a higher level of safety of the turbo-roundabouts compared to the multi-lane configurations. The values of PET in the scenarios associated with turbo-roundabouts range from 2.3 sec (scenario n. 1) and 2.5 sec (scenario n. 2), are always higher than in the existing configuration with the multi-lane roundabouts and show significant percentage increases between 25% and 34%.

Maximum speed (MaxS): As traffic conditions vary, this indicator takes values from about 4.5 m/s to 6.5 m/s for simulation scenarios where multi-lane roundabouts are present. The aforementioned values, on the other hand, decrease by percentages ranging from about 16% (scenario n. 3) to 32% (scenario n. 2) when there are conflicts between vehicles on the route where turbo-roundabouts are present. Thus, the change in this indicator shows how the presence of turbo-roundabouts leads to conflicts that occur at significantly lower speeds than conflicts on a route where multi-lane roundabouts are present. So even in this case, turbo-roundabouts are safer on average than standard roundabouts.

Initial Deceleration Rate (DR): The values of this safety indicator are always higher than 2.5 m/s^2 for traffic scenarios related to the road section characterized by the presence of multi-lane roundabouts. However, for the simulations performed on the road section equipped with turbo-roundabouts, three out of four scenarios have values of DR lower than 2 m/s^2. In terms of percentage change, the comparison between the configuration with turbo-roundabouts and the configuration with multi-lane roundabouts shows an average reduction of about 28%, without much variation between scenarios. Thus, the simulations show that the specific influence of traffic conditions on the change in this parameter is not particularly evident. Therefore, it is reasonable to conclude that turbo-roundabouts induce drivers to face conflict situations at lower speeds than standard roundabouts and that they also require little deceleration to avoid a potential accident. In this way, drivers can handle the potentially hazardous conditions with greater calm and without sudden braking, which benefits safety.

Total number of conflicts: The safety benefits of a turbo-roundabout become even clearer when we consider the change in this indicator. Already in scenario n. 1, the reduction of total conflicts in the configuration with turbo-roundabouts is obvious (−40%). This difference becomes even more evident as traffic volumes increase. In fact, there is a change in conflicts of about −48% in scenario n. 2 and about 70% for scenario n. 3 and scenario n. 4. These results are quite consistent with studies showing a percentage reduction in conflict of the same magnitude as in this study. In particular, a study comparing the safety performance of a two-lane roundabout and a turbo-roundabout with traffic flows similar to those in scenarios 1 and 2 considered in this study showed a percentage reduction of 45% in the total number of conflicts in the turbo-roundabout [52]. A study comparing an existing two-lane roundabout and a turbo-roundabout simulated with VISSIM and SSAM in the city of Bogotá (Colombia) under traffic conditions similar to scenarios 3 and 4 in this study showed a 72% reduction in the total number of conflicts in the turbo-roundabout [53]. Other studies, based on conflict analysis techniques applied to nine layouts with different demand scenarios, showed 40–50 % reductions in accident rates [6,10]. Further consideration of conflicts can be made by looking at the types of conflicts in more detail. Specifically:

➢ *Crossing conflicts:* These types of conflicts almost cancel out in the turbo-roundabout configuration compared to the multi-lane roundabout configuration. It should be noted, however, that the percentage of these conflicts, as could logically be expected in a context where the two main intersections are roundabouts, is already low in each of the scenarios considered. Therefore, it is not considered appropriate to highlight this result.

> *Rear End Conflicts:* Although this type of conflict generally has the least severe consequences, it is the most common in all of the traffic scenarios considered. In scenarios n. 3 and n. 4, which refer to the road configuration characterized by the presence of multi-lane roundabouts, the number of conflicts even exceeds 8000 on average. The presence of turbo-roundabouts drastically reduces these conflicts, especially in configurations with high traffic volumes: −38% (scenario n. 1), −42% (scenario n. 2), −68% (scenario n. 3), −69% (scenario n. 4).

> *Lane change Conflicts:* The two multi-lane roundabouts have two lanes on the circulatory roadway. It was logical to expect that the simulations would yield a large number of lane change conflicts; nearly 450 in scenario n. 1, about 700 in scenario n. 2, and over 1200 in both scenario n. 3 and scenario n. 4. The design of turbo-roundabouts would result in a very significant reduction in these conflicts. This is confirmed by the following reductions identified in the simulations: −75% (scenario n. 1), −78% (scenario n. 2), −86% (scenario n. 3) and −86% (scenario n. 4).

Finally, it is interesting to note, as can be clearly seen in Figure 9, that scenarios n. 3 and n. 4 are very similar in terms of the variation of all the parameters considered. This means that from a certain traffic volume, which affects the studied part of the road network and, consequently, the roundabouts present there, the conditions of safety and functional operation stabilize and do not vary significantly.

5. Conclusions

Today, turbo-roundabouts are an almost exclusively European reality; of the 600 or so turbo-roundabouts in the world, not even a dozen are in operation on non-European continents. The undisputed homeland of turbo-roundabouts is Holland, where there are more than 370 turbo-roundabouts to date. There are countries, such as France, where there are no turbo-roundabouts, and others, such as Spain and the UK, where some turbo-roundabouts are in the trial phase. In Italy there are no turbo-roundabouts, and at the moment they are not even experimental, although the advantages of this type of intersection are now obvious. It is true that the benefits in terms of improved operational performance are still controversial.

Although this study refers to a particular case study where the part of the simulated road network is characterized by a main road with much higher traffic volumes than the minor roads, it confirms that multi-lane roundabouts on the circulatory roadway, although not allowed by current legislation, are more efficient than turbo-roundabouts for low to medium traffic volumes. On the other hand, the operational performance of turbo-roundabouts improves significantly at high traffic volumes; under these conditions, they actually help reduce the total number of stops compared to standard roundabouts, and average delays, while remaining higher than those of multi-lane roundabouts, tend to be comparable to those of multi-lane roundabouts (it is likely that average delays are noticeably reduced for turbo-roundabouts with more uniform vehicle flows on the various approaches). In any case, under high traffic conditions, turbo-roundabouts result in reduced congestion and improved fluidity of traffic (the so-called green wave) compared to conventional multi-lane roundabouts.

In terms of safety performance, however, this work confirms what other researchers have already found; variations in key safety indicators testify to significant advantages of turbo-roundabouts compared with multi-lane roundabouts. Rear-end and lane change conflicts are also significantly lower compared to standard roundabouts (in turbo-roundabouts, the inner and outer lanes do not intersect).

One of the next goals of this research group is to simulate additional case studies where existing roundabouts and/or existing standard intersections are replaced with turbo-roundabouts. In this way, we will try to overcome the limitations of the current research, which currently lie mainly in the specificity of the case study used, especially in terms of traffic conditions that are unbalanced. New study scenarios characterized by different configurations of turbo-roundabouts, both in terms of geometric design and distribution of vehicle flows on the different approaches, would certainly make it possible to create the conditions for generalizing the research results, which is not possible at this stage, even if it is believed to have taken an important first step towards this goal.

However, it is believed that the case study treated, typical of an Italian urban reality, is in any case important to pursue an important objective; that is, to induce the legislator to evaluate the possibility of designing turbo-roundabouts in Italy; if it is true that multi-lane roundabouts seem to be better than turbo-roundabouts from the point of view of operational performance, it is also true that multi-lane roundabouts are prohibited and, in any case, less safe than single-lane roundabouts. Moreover, from the point of view of safety, turbo-roundabouts have undeniable advantages over multi-lane solutions. In Italy, the revision of the Decree of 19 April 2006, which regulates the geometric design of intersections, has been discussed for several years. The authors believe that today it is feasible to consider turbo-roundabouts as another design solution that can be used alongside modern roundabouts. In this context, the Italian legislator could take advantage of consolidated project standards; in particular, the Dutch one (CROW).

Author Contributions: Conceptualization, N.D. and S.L.; methodology, N.D.; software, N.D. and S.L.; validation, S.L.; formal analysis, N.D. and S.L.; investigation, N.D. and S.L.; resources, N.D. and S.L.; data curation, S.L.; writing—original draft preparation, N.D. and S.L.; writing—review and editing, N.D. and S.L.; visualization, N.D.; supervision, S.L. All authors have read and agreed to the published version of the manuscript.

Funding: This research received no external funding.

Institutional Review Board Statement: Not applicable to this study because effects on humans are essentially nonexistent.

Informed Consent Statement: Not applicable to this study because no subjects were involved in the research trials.

Data Availability Statement: No new data has been created.

Conflicts of Interest: The authors declare no conflict of interest.

References

1. Rella Riccardi, M.; Augeri, M.G.; Galante, F.; Mauriello, F.; Nicolosi, V.; Montella, A. Safety Index for evaluation of urban roundabouts. *Acc. Anal. Prev.* **2022**, *158*, 106858. [CrossRef] [PubMed]
2. Leonardi, S.; Distefano, N.; Pulvirenti, G. Italians' public opinion on road roundabouts: A web based survey. *Transp. Res. Proc.* **2020**, *45*, 293–300. [CrossRef]
3. Distefano, N.; Leonardi, S.; Consoli, F. Drivers' Preferences for Road Roundabouts: A Study based on Stated Preference Survey in Italy. *KSCE J. Civ. Eng.* **2019**, *23*, 4864–4874. [CrossRef]
4. Distefano, N.; Leonardi, S. Experimental investigation of the effect of roundabouts on noise emission level from motor vehicles. *Noise Control Eng. J.* **2019**, *67*, 282–294. [CrossRef]
5. Ignaccolo, M.; Zampino, S.; Maternini, G.; Tiboni, M.; Leonardi, S.; Inturri, G.; Le Pira, M.; Cocuzza, E.; Distefano, N.; Giuffrida, N.; et al. How to redesign urbanized arterial roads? The case of Italian small cities. *Transp. Res. Proc.* **2022**, *60*, 196–203. [CrossRef]
6. Tollazzi, T. *Alternative Types of Roundabouts: An Informational Guide*; Springer International Publishing: New York, NY, USA, 2015.
7. Elhassy, Z.; Abou-Senna, H.; Radwan, E. Performance evaluation of basic turbo roundabouts as an alternative to conventional double-lane roundabouts. *Transp. Res. Rec.* **2021**, *2675*, 180–193. [CrossRef]
8. Balado, J.; Díaz-Vilarino, L.; Arias, P.; Novo, A. A safety analysis of roundabouts and turbo roundabouts based on Petri nets. *Traffic Inj. Prev.* **2019**, *20*, 400–405. [CrossRef]
9. Ciampa, D.; Diomedi, M.; Giglio, F.; Olita, S.; Petruccelli, U.; Restaino, C. Effectiveness of unconventional roundabouts in the design of suburban intersections. *Eur. Transp. /Trasp. Eur.* **2020**, *80*, 1–16. [CrossRef]
10. Mauro, R.; Cattani, M.; Guerrieri, M. Evaluation of the Safety Performance of Turbo Roundabouts by Means of a Potential Accident Rate Mode. *Balt. J. Road Bridge Eng.* **2015**, *10*, 28–38. [CrossRef]
11. Petru, J.; Krivda, V. An Analysis of Turbo Roundabouts from the Perspective of Sustainability of Road Transportation. *Sustainability* **2021**, *13*, 2119. [CrossRef]
12. Pitlova, E.; Kocianova, A. Case Study: Capacity Characteristics Comparison of Single-lane Roundabout and Turbo-roundabouts. *Procedia Eng.* **2017**, *192*, 701–706. [CrossRef]
13. Šarić, A.; Lovrić, I. Multi-Lane Roundabout Capacity Evaluation. *Front. Built Environ.* **2017**, *3*, 42. [CrossRef]

14. Someswara, R.B.; Kadali, B.R. Review Of Traffic Safety Evaluation At T-intersections Using Surrogate Safety Measures In Developing Countries Context. *IATSS Res.* **2022**, *3*, 307–321. [CrossRef]
15. Sobhani, A.; Young, W.; Sarvi, M. A simulation based approach to assess the safety performance of road locations. *Transp. Res. Part C Emerg. Technol.* **2013**, *32*, 144–158. [CrossRef]
16. Young, W.; Sobhani, A.; Lenne, M.G.; Sarvi, M. Simulation of safety: A review of the state of the art in road safety simulation modelling. *Accid. Anal. Prev.* **2014**, *66*, 89–103. [CrossRef]
17. Killi, D.V.; Vedagiri, P. Proactive evaluation of traffic safety at an unsignalized intersection using Microsimulation. *J. Traffic Logist. Eng.* **2014**, *2*, 140–145. [CrossRef]
18. Essa, M.; Sayed, T. Simulated traffic conflicts: Do they accurately represent field- measured conflicts? *Transp. Res. Rec.* **2015**, *2514*, 48–57. [CrossRef]
19. Mahmud, S.M.S.; Ferreira, L.; Hoque, M.S.; Tavassoli, A. Micro-simulation modelling for traffic safety: A review and potential application to heterogeneous traffic environment. *IATSS Res.* **2019**, *43*, 27–36. [CrossRef]
20. Ulak, M.B.; Ozguven, E.E.; Moses, R.; Sando, T.; Boot, W.; Abdel Razig, Y.; Sobanjo, J.O. Assessment of traffic performance measures and safety based on driver age and experience: A microsimulation based analysis for an unsignalized T-intersection. *J. Traffic Transp. Eng.* **2019**, *6*, 455–469. [CrossRef]
21. Guo, Y.; Sayed, T.; Zheng, L.; Essa, M. An extreme value theory based approach for calibration of microsimulation models for safety analysis. *Simul. Model. Pract. Theory.* **2021**, *106*, 102172. [CrossRef]
22. Torday, A.; Baumann, D.; Dumont, D. Indicator for microsimulation-based safety evaluation. In Proceedings of the 3rd STRC Swiss Transport Research, Monte, Verita, 19–21 March 2003.
23. Caliendo, D.; Guida, M. Microsimulation approach for predicting crashes at unsignalized intersections using traffic conflicts. *J. Transp. Eng.* **2012**, *138*, 1453–1467. [CrossRef]
24. Astarita, V.; Festa, D.C.; Giofrè, V.P.; Guido, G. Surrogate safety measures from traffic simulation models a comparison of different models for intersection safety evaluation. *Transp. Res. Proc.* **2019**, *37*, 219–226. [CrossRef]
25. Ozbay, K.; Yang, H.; Bartin, B.; Mudigonda, S. Derivation and validation of new simulation-based surrogate safety measure. *Transp. Res. Rec.* **2008**, *2083*, 105–113. [CrossRef]
26. Ambros, J.; Turek, R.; Paukrt, J. Road safety evaluation using traffic conflicts: Pilot comparison of micro-simulation and observation. In Proceedings of the International Conference on Traffic and Transport Engineering, Belgrade, Serbia, 27–28 November 2014.
27. Guido, G.; Astarita, V.; Giofré, V.; Vitale, A. Safety performance measures: A comparison between microsimulation and observational data. *Procedia Soc. Behav. Sci.* **2011**, *20*, 217–225. [CrossRef]
28. Astarita, V.; Guido, G.; Vitale, A.; Giofré, V. A new microsimulation model for the evaluation of traffic safety performances. *Eur. Transp. /Trasp. Eur.* **2012**, *51*, 1–16.
29. Astarita, V.; Giofré, V.; Guido, G.; Vitale, A. Calibration of a new microsimulation package for the evaluation of traffic safety performances. *Procedia Soc. Behav. Sci.* **2012**, *54*, 1019–1026. [CrossRef]
30. Silva, A.; Vasconcelos, L.; Santos, S. Moving from Conventional Roundabouts to Turbo-Roundabouts. *Procedia Soc. Behav. Sci.* **2014**, *111*, 137–146. [CrossRef]
31. Mauro, R.; Branco, F. Comparative Analysis of Compact Multilane Roundabouts and Turbo-Roundabouts. *J. Transp. Eng.* **2010**, *136*, 316–322. [CrossRef]
32. Bulla, L.; Castro, W. Analysis and Comparison between Two-Lane Roundabouts and Turbo Roundabouts Based on a Road Safety Audit Methodology and Microsimulation: A Case Study in Urban Area. In Proceedings of the 3rd International Conference on Road Safety and Simulation, Indianapolis, IN, USA, 14–16 September 2011; Available online: http://onlinepubs.trb.org/onlinepubs/conferences/2011/RSS/2/Bulla,L.pdf (accessed on 14 April 2022).
33. Silva, A.; Santos, S.; Gaspar, M. Turbo-Roundabout Use and Design. In Proceedings of the CITTA 6th Annual Conference on Planning Research, Responsive Transports for Smart Mobility, Coimbra, Portugal, 17 May 2013.
34. Engelsman, J.; Uken, M. Turbo Roundabouts as an Alternative to Two Lane Roundabouts. In Proceedings of the 26th Southern African Transport Conference (SATC 2007), Pretoria, South Africa, 9–12 July 2007; pp. 581–589. Available online: http://www.lags.corep.it/doc/turbo-roundabout.pdf (accessed on 21 April 2022).
35. Giuffrè, O.; Guerrieri, M.; Granà, A. Evaluating Capacity and Efficiency of Turbo-Roundabouts. In Proceedings of the 88th Annual Meeting of the Transportation Research Board, Washington, DC, USA, 11–15 January 2009.
36. Vasconcelos, L.; Silva, A.; Seco, A. Capacity of Normal and Turbo-Roundabouts: Comparative Analysis. *Proc. Inst. Civ. Eng. Transp.* **2014**, *167*, 88–99. [CrossRef]
37. Vasconcelos, L.; Silva, A.B.; Seco, Á.M.; Fernandes, P.; Coelho, M.C. Turboroundabouts: Multicriterion Assessment of Intersection Capacity, Safety, and Emissions. *Transp. Res. Rec. J. Transp. Res. Board.* **2014**, *2402*, 28–37. [CrossRef]
38. Tollazzi, T.; Mauro, R.; Guerrieri, M.; Rençelj, M. Comparative analysis of four new alternative types of roundabouts: Turbo, flower, target and four-flyover roundabout. *Period. Polytech. Civ. Eng.* **2016**, *60*, 51–60. [CrossRef]
39. Silva, A.B.; Mariano, P.; Silva, J.P. Performance assessment of turbo-roundabouts in corridors. *Transp. Res. Procedia* **2015**, *10*, 124–133. [CrossRef]

40. Kwakwa, O.; Adams, C. Assessment of Turbo and Multilane Roundabout Alternatives to Improve Capacity and Delay at a Single Lane Roundabout Using Microsimulation Model Vissim: A Case Study in Ghana. *Am. J. Civ. Eng. Archit.* **2016**, *4*, 106–116. [CrossRef]

41. Gallelli, V.; Iuele, T.; Vaiana, R. Conversion of a Semi-two Lanes Roundabout into a Turbo-Roundabout: A Performance Comparison. *Procedia Comput. Sci.* **2016**, *83*, 393–400. [CrossRef]

42. Giuffrè, O.; Granà, A.; Marino, S. Comparing performances of turbo-roundabouts and double-lane roundabouts. *Mod. Appl. Sci.* **2012**, *6*, 70–79. [CrossRef]

43. Giuffrè, O.; Granà, A.; Marino, S. Turbo-Roundabouts vs Roundabouts Performance Level. *Procedia Soc. Behav. Sci.* **2012**, *53*, 590–600. [CrossRef]

44. Fernandes, P.; Salamati, K.; Rouphail, N.M.; Coelho, M.C. The effect of a roundabout corridor's design on selecting the optimal crosswalk location: A multi-objective impact analysis. *Int. J. Sustain. Transp.* **2017**, *11*, 206–220. [CrossRef]

45. Gettmann, D.; Pu, L.; Sayed, T.; Shelby, S. *Surrogate Safety Assessment Model and Validation: Final Report*; United States. Federal Highway Administration. Office of Safety Research and Development: McLean, VA, USA, 2008. Available online: http://www.fhwa.dot.gov/publications/research/safety/08051/08051.pdf (accessed on 28 April 2022).

46. Huang, S.; Guo, L.; Yang, Y.; Casas, I.; Sadek, A.W. Dynamic demand estimation and microscopic traffic simulation of a university campus transportation network. *Transp. Plan. Technol.* **2012**, *35*, 449–467. [CrossRef]

47. So, J.; Hoffmann, S.; Lee, J.; Busch, F.; Choi, K. A prediction accuracy-practicality tradeoff analysis of the state-of-the-art safety performance assessment methods. *Transp. Res. Proc.* **2016**, *15*, 794–805. [CrossRef]

48. Al-Ghandour, M.; Schroeder, B.; Williams, B.; Rasdorf, W. Conflict models for single-lane roundabout slip lanes from microsimulation. *Transp. Res. Rec.* **2011**, *2236*, 92–101. [CrossRef]

49. Chai, C.; Wong, Y.D. Micro-simulation of vehicle conflicts involving right-turn vehicles at signalized intersections based on cellular automata. *Accid. Anal. Prev.* **2014**, *63*, 94–103. [CrossRef] [PubMed]

50. Dijkstra, A.; Marchesini, P.; Bijleveld, F.; Kars, V.; Drolenga, H.; van Maarseveen, M. Do calculated conflicts in micro-simulation model predict number of crashes? *Transp. Res. Rec.* **2010**, *2147*, 105–112. [CrossRef]

51. Vasconcelos, L.; Neto, L.; Seco, A.; Silva, A. Validation of the surrogate safety assessment model for assessment of intersection safety. *Transp. Res. Rec.* **2014**, *2432*, 1–9. [CrossRef]

52. Vasconcelos, L.; Silva, A.; Seco, A. Safety analysis of turbo-roundabouts using the SSAM technique. In Proceedings of the CITTA 6th Annual Conference on Planning Research, Coimbra, Portugal, 17 May 2013.

53. Bulla-Cruz, L.; Lyons, L.; Darghan, E. Complete-Linkage Clustering Analysis of Surrogate Measures for Road Safety Assessment in Roundabouts. *Rev. Colomb. Estad.* **2021**, *44*, 91–121. [CrossRef]

54. Chen, Y.; Persaud, B.; Lyon, C. Effect of Speed on Roundabout Safety Performance: Implications for Use of Speed as Surrogate Measure. In Proceedings of the Transportation Research Board 90th Annual Meeting, TRID, Washington, DC, USA, 23–27 January 2011.

55. Džambas, T.; Ahac, S.; Dragčević, V. Design of turbo roundabouts based on the rules of vehicle movement geometry. *J. Transp. Eng.* **2016**, *142*, 05016004. [CrossRef]

56. CROW. *Turborotondes*; CROW Publication no.257. Ede; CROW: Dutch, The Netherlands, 2008.

57. Ciuffo, B.; Casas, J.; Montanino, M.; Perarnau, J.; Punzo, V. Gaussian process metamodels for sensitivity analysis of traffic simulation models: Case study of AIMSUN mesoscopic model. *Transp. Res. Rec.* **2013**, *2390*, 87–98. [CrossRef]

58. Virdi, N.; Grzybowska, H.; Waller, S.T.; Dixit, V. A safety assessment of mixed fleets with connected and autonomous vehicles using the surrogate safety assessment module. *Accid. Anal. Prev.* **2019**, *131*, 95–111. [CrossRef]

59. Pu, L.; Joshi, R. Surrogate Safety Assessment Model (SSAM)—Software User Manual. Federal Highway Administration, Virginia, 2008. Available online: https://www.fhwa.dot.gov/publications/research/safety/08050/08050.pdf (accessed on 4 March 2022).

60. Guo, Y.; Essa, M.; Sayed, T.; Haque, M.M.; Washington, S. A Comparison between Simulated and Field-Measured Conflicts for Safety Assessment of Signalized Intersections in Australia. *Transp. Res. Part C Emerg. Technol.* **2019**, *101*, 96–110. [CrossRef]

61. Hayward, J.C. Near-miss determination through use of a scale of danger. *Highw. Res. Rec.* **1972**, *1*, 24–34. Available online: http://onlinepubs.trb.org/Onlinepubs/hrr/1972/384/384-004.pdf (accessed on 4 February 2022).

62. Pulvirenti, G.; De Ceunynck, T.; Daniels, S.; Distefano, N.; Leonardi, S. Safety of bicyclists in roundabouts with mixed traffic: Video analyses of behavioural and surrogate safety indicators. *Transp. Res. Part F. Traffic Psychol. Behav.* **2021**, *76*, 72–91. [CrossRef]

63. Laureshyn, A. In Search of the Severity Dimension of Traffic Events: Extended Delta-v as a Traffic Conflict Indicators. *Accid. Anal. Prev.* **2017**, *98*, 46–56. [CrossRef] [PubMed]

64. Gettman, D.; Head, L. Surrogate safety measures from traffic simulation models. *Transp. Res. Record* **2003**, *1840*, 104–115. [CrossRef]

65. Johnsson, C.; Laureshyn, A.; De Ceunynck, T. In search of surrogate safety indicators for vulnerable road users: A review of surrogate safety indicators. *Transp. Rev.* **2018**, *38*, 765–785. [CrossRef]

 *sustainability*

Article

A Methodological Framework to Assess Road Infrastructure Safety and Performance Efficiency in the Transition toward Cooperative Driving

Maria Luisa Tumminello [1], Elżbieta Macioszek [2], Anna Granà [1,*] and Tullio Giuffrè [3]

1   Department of Engineering, University of Palermo, Viale delle Scienze ed 8, 90128 Palermo, Italy;
    marialuisa.tumminello01@unipa.it
2   Department of Transport Systems, Traffic Engineering and Logistics, Faculty of Transport and Aviation
    Engineering, Silesian University of Technology, Krasińskiego 8 Street, 40-019 Katowice, Poland;
    elzbieta.macioszek@polsl.pl
3   Faculty of Engineering and Architecture, University of Enna Kore, Viale della Cooperazione,
    94100 Enna, Italy; tullio.giuffre@unikore.it
*   Correspondence: anna.grana@unipa.it

Abstract: There is increasing interest in connected and automated vehicles (CAVs), since their implementation will transform the nature of transportation and promote social and economic change. Transition toward cooperative driving still requires the understanding of some key questions to assess the performances of CAVs and human-driven vehicles on roundabouts and to properly balance road safety and traffic efficiency requirements. In this view, this paper proposes a simulation-based methodological framework aiming to assess the presence of increasing proportions of CAVs on roundabouts operating at a high-capacity utilization level. A roundabout was identified in Palermo City, Italy, and built in Aimsun (version 20) to describe the stepwise methodology. The CAV-based curves of capacity by entry mechanism were developed and then used as target capacities. To calibrate the model parameters, the capacity curves were compared with the capacity data simulated by Aimsun. The impact on the safety and performance efficiency of a lane dedicated to CAVs was also examined using surrogate measures of safety. The paper ends with highlighting a general improvement with CAVs on roundabouts, and with providing some insights to assess the advantages of the automated and connected driving technologies in transitioning to smarter mobility.

Keywords: roundabout; microsimulation; surrogate safety measures; road safety; connected and automated vehicles; traffic operations

check for updates

Citation: Tumminello, M.L.;
Macioszek, E.; Granà, A.; Giuffrè, T.
A Methodological Framework to
Assess Road Infrastructure Safety
and Performance Efficiency in the
Transition toward Cooperative
Driving. *Sustainability* **2023**, *15*, 9345.
https://doi.org/10.3390/su15129345

Academic Editors: Salvatore
Leonardi and Natalia Distefano

Received: 20 April 2023
Revised: 21 May 2023
Accepted: 5 June 2023
Published: 9 June 2023

1. Introduction

The technological breakthroughs in transportation systems are an integral part of the transformations of cities [1]. To get increasingly smart cities, intelligent transportation Systems and big data applications have been increasingly targeted over time to design road traffic services truly geared toward increased road users' safety, congestion reduction, energy saving, and driving comfort. Furthermore, the potential of these technologies is that vehicles, system users, and road infrastructures can be integrated to properly improve mobility [2]. The technological development of transportation has also been directed toward the vehicle's automation and connectivity between the vehicles and road infrastructures [3]. There are six levels of vehicular autonomy from fully manual to fully autonomous driving which describe the human–machine shared interaction on roads [4]. Connectivity forges an inspiring environment for vehicle-to-infrastructure (V2I), vehicle-to-pedestrian (V2P), vehicle-to-vehicle (V2V), vehicle-to-device (V2D), and vehicle-to-network (V2N) communications [5]. Connected and automated vehicles (CAVs) are expected to perform the driving tasks by using the cooperative adaptive cruise control (CACC) system based on vehicle-to-vehicle and infrastructure-to-vehicle communications, which facilitates the exchange of information between vehicles and enhances connectivity [5–7]. In this regard, Wang et al. [7] proposed a method to estimate the motion state of preceding vehicles and surrounding vehicles to improve the safety control of intelligent connected vehicles; simulation and real vehicle tests confirmed that the prediction approach properly balanced the communication among vehicles and their performances. Despite the advantages of cooperative driving, however, there are still many and not easy challenges to tackle [8]. In this view, a novel practical application concerned the development of a machine-to-machine (M2M)-based cooperative driving protocol, specifically focused on V2I communication and tested on a real-world merging crossroad [9]. According to the results, there is the potential to expand the application on different types of roads to make autonomous driving experience safer.

Transition toward cooperative driving systems still requires understanding of the key issues involved in adapting the geometry of road infrastructures to the kinematics of CAVs in order to achieve the proper balance between road safety and traffic efficiency [10]. Although cooperative driving is expected to improve operating performances of road infrastructures, it is not yet clear how connectivity may affect the car's ability to move through intersections and roundabouts in which users experience curvilinear trajectories while entering, crossing the intersection area, and exiting [11]. To this day, CAVs are not yet available to all users, and all levels of automation are not yet available on the market, where their entry is expected to happen gradually also in relation to the physiological rate of replacement of old cars with new ones [5].

In a responsible society perspective, which directs the road design choices toward the use of materials and constructive ways sustainable, road engineers should know, in advance, the response to the expected performance requirements of a given road infrastructure, where heterogenous fleets made of human-driven vehicles (HDVs) and CAVs are simultaneously mixed in traffic [6,12,13]. In this regard, microscopic traffic simulation models are already configured as valuable tools to model operational performances at a single node or corridor level and to assess the expected benefits of increasingly high market penetration rates of CAVs [14,15]. Although roundabouts are renowned among the road infrastructures for their potential in making road traffic safer and in reducing delays, fuel consumption, and construction and maintenance costs [16], microsimulation can be of great interest to evaluate to what extent the curvilinear design of roundabouts affects traffic patterns and operations in presence of high levels of automation and connectivity [16–18].

To fill the above gaps, the paper aims to propose a novel simulation-based methodological framework aiming to assess the interactions between CAVs and HDVs on roundabout systems operating at a high-capacity utilization rate, as well as their impacts on road safety and performance efficiency. In this regard, the adjustment factors for roundabout systems provided by [17] were applied to develop the CAV-based capacity curves in mixed traffic situation. It should be also noted that the beforementioned factors were developed by using microsimulation which considered all the high-reliability elements necessary to fully implement cooperation among vehicles [17].

The framework in this paper consists of the following main sequential steps: (1) conceptualizing the roundabout system; (2) identifying the mechanisms that regulate the entry of vehicles on the circulatory roadway where circulating traffic moves around the central island; (3) traffic data collection, processing, and mining to determine the capacity target values for every entry mechanism and CAV penetration rate; (4) building the roundabout network model and simulation from free-flowing traffic to capacity for calibration purposes; (5) assessing the impact of CAVs on traffic throughput; (6) assessing the CAVs impact on safety and performance efficiency for roundabouts with a dedicated lane to CAVs compared to the traffic situation with CAVs and HDVs sharing the lanes. With a view to providing an overview of the six sequential steps of the proposed methodological framework, a roundabout was identified in the road network of Palermo City, Italy, and then built in Aimsun Next (Aimsun from now on) [18] to model operations at capacity.

Due to the current lack of high levels of automation and connectivity in traffic, the CAV-based capacity curves corrected using the adjustment factors proposed by [17] for different market penetration rates of CAVs were employed as an alternative source of target capacity data [19]. To calibrate the model parameters, the CAV-based capacity curves were then compared with the data simulated by Aimsun. The safety and performance efficiency of roundabout dedicated lanes for CAVs were also compared to the mixed traffic situation with HDVs and CAVs. For the safety assessment, surrogate measures were used by combining Aimsun with the Surrogate Safety Assessment Model (SSAM) [20].

Results highlighted general improvements in roundabout performance with CAVs in traffic compared to the scenario made only by human driven vehicles; moreover, they provided some insights to assess the advantages of the CAVs to consider in the cities' successful strategies aimed at improving mobility. The proposed stepwise approach can assist transportation engineers and decision makers in assessing the level-of-service of road infrastructures at the design or implementation phase, in the transition toward cooperative driving.

This research includes the following scientific and public contributions:

Scientific: It identifies some parameters of cooperative driving on a roundabout model that meets the geometry and traffic characteristics of a real-life counterpart in order to examine the effects of the changes in design and driving behavior from a safety and efficiency perspective in the transition to growing proportions of CAVs in traffic.

Public: It highlights a general performance improvement with CAVs on roundabouts compared to the base case with HDVs only, and it provides some insights to assess the expected safety and operational advantages of connected and automated driving to tackle the future challenges in mobility.

The organization of the paper includes a brief overview of the related research on the topic in Section 2; Section 3 presents the proposed framework, includes the materials and methods applied to a roundabout case study, and describes the reasons behind the assumptions and choices made; Sections 4 and 5 present and discuss the results, respectively; the final section presents the conclusions of the research and some future developments.

2. Related Research

The risks of crash events and significant losses in the economic and social dimension have always been associated with increased road mobility. Since crash-related data show that driver behavior is one of the main factors influencing road crashes, analysts are currently questioning the entry of CAVs into full service in the near future and their impact on road safety and performance efficiency [5,21]. The participation of CAVs in road traffic will minimize or remove human factors during driving, since certain functions will be carried out with limited human participation or automatically [21]. However, fully autonomous driving will transfer the driving tasks to a computer system; as a consequence, the classifications provided by the literature denote that full automation does not yet exist, but currently requires specific human support [4].

The introduction of CAVs in traffic will change traffic conditions and transform safety standards for road design, maintenance and infrastructure administration, and traffic modeling and assessment tools for road management. In this view, Rahman and Abdel-Aty [22] evaluated vehicle platooning on expressways by employing surrogate measures of safety; the results provided useful information for different market penetration rates of connected and automated vehicles in traffic. In turn, Rahman et al. [23] used VISSIM [24] to simulate the implementation of V2V communication technologies and to assess the effectiveness of CAVs in adverse visibility conditions on a US interstate; specifically, they simulated both connected vehicles without platooning and connected vehicles with platooning, and then employed surrogate measures of safety to assess the reduction in the crash risk. Their results showed improved safety in fog conditions and improvements in average speed as the market penetration rates of CAVs increased. Another study used microsimulation to model a road network and to evaluate the contraflow evacuation operation in CAV environment [25]. The study found improvements in the system performance for contraflow operations with CAVs in the evacuation traffic; the results showed reduced delay and travel times for the evacuation route, as well as increased speeds with at least 30% of CAVs in traffic.

In the transition toward cooperative driving, there is a need to examine the main performance efficiency and safety issues with CAVs and HDVs simultaneously moving in road traffic (see, e.g., [26]). In this regard, the literature reports some studies that have examined how to improve the gap acceptance at roundabout entries [27,28]. Another key aspect is to take into account the kinematic and dynamic needs of CAVs on curved trajectories which can make the interpretation of the intentions of other vehicles difficult to anticipate [29,30].

Most of the studies in this field of investigation, however, predominantly focused on trajectory control optimization of CAVs to improve mobility in mixed traffic situations, and speed optimization to minimize the total delay time [31,32]. In this regard, Wu et al. [32] proposed a method to control vehicle cooperation and to generate collision-free trajectories for CAVs at isolated roundabouts. The results showed improved throughput and average speeds, as well as reduced total travel times. In turn, Jalil et al. [33] conceptualized a novel holistic coordination system for CAVs at T-shaped roundabouts in order to optimize the traffic states of each approaching vehicle and to minimize the total delay time; the authors highlighted improvements in the average speed, traffic density, idling, and fuel consumption of vehicles around the roundabout where speed optimization ensured smooth crossing.

There is the further issue concerning the CAV ability to receive information from the other vehicles, roadway infrastructure, and traffic control centers in order to anticipate the driving actions of the preceding vehicles. Thus, a control system for the aware-situation connected driving should take into account the specific features of path planning and navigation at roundabouts that include entries, merging, turning maneuvers, lane changing, and exits [11,34]. In this regard, another study emulated a driverless vehicle in a roundabout and tested a control lateral system in a 3D simulator [35]. Major conclusions concerned the need to better optimize the automatic geometry recognition in terms of entry radii and deviation from the reference circle; the authors found defects especially when the vehicle used the greatest possible ability to maneuver. Another study simulated CAVs negotiating a roundabout which is used as a controller to implement vehicle-to-infrastructure communication [29]. The authors tested different combinations of geometry and traffic patterns in order to assess the operational performance of CAVs; however, they stressed the need for further study to generalize the conclusions.

Referring to safety effects of CAVs on roundabouts, it is not yet known to what extent people will be ready to accept smaller gaps when negotiating a roundabout. In this regard, a study investigated the impacts of autonomous vehicles on roundabout safety using VISSIM [36]. The results showed a 32% reduction in total conflicts by modeling CAVs with defensive behavior on roundabouts, while the simulation of assertive behavior worsened their performance efficiency. Similar results were provided by [37] that revealed increased conflicts with automated vehicles in traffic.

Traffic microsimulation can be a useful approach to examine the performance response of a given road infrastructure on corridor or network in the transitioning to a fully CAV fleet; they can be useful tools both to evaluate changes in road safety and traffic operations with CAVs and HVDs coexisting in traffic, and to develop novel tools to assess and to manage road safety (see, e.g., [38]). In this view, the adjustment factors for CAVs on roundabouts provided by the Highway Capacity Manual 7th Edition were the starting point of the analysis we made [17].

3. Materials and Methods

This section presents the microsimulation-based methodological framework designed for evaluating road infrastructure safety and performance efficiency with CAVs in traffic. With a view to assessing whether cooperative driving is compatible with roundabout navigation, the structure of the stepwise approach included subsequent steps for identifying a roundabout case study and the mechanisms of entry capacity, collecting traffic data and developing the CAV-based capacity curves, simulating traffic scenarios from free-flowing traffic to capacity for calibration purposes, and assessing the CAV impact on traffic performances. On the basis of the above steps, the safety and performance efficiency of a roundabout with a CAV dedicated lane has been evaluated compared to the mixed traffic situation with HDVs and CAVs in traffic. Figure 1 shows an overview of the proposed procedure applied to roundabouts.

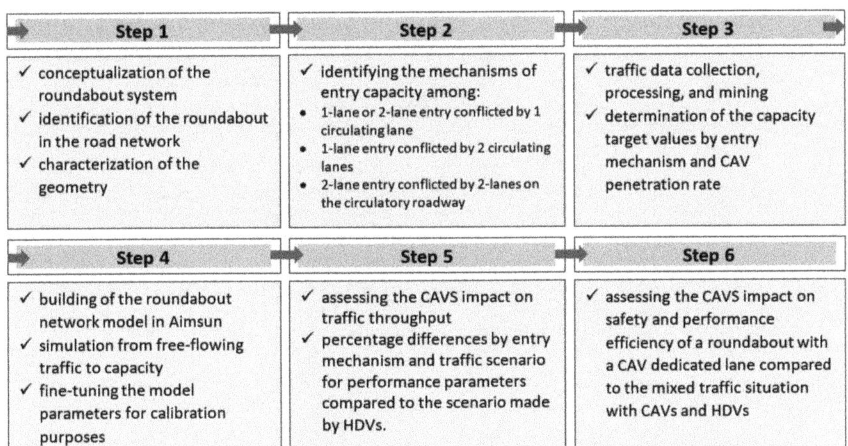

Figure 1. Overview of the use of the methodological procedure applied to roundabouts.

3.1. Step 1: Conceptualization of the Roundabout System, Identification of the Roundabout in the Road Network, and Characterization of Its Geometry

There was a need to identify a case study in the road network of Palermo City, Italy, in order to conceptualize a roundabout system and to feed the network model in Aimsun (see Section 3.4). A two-lane large diameter roundabout was selected to describe the proposed methodology. The roundabout case study is installed in the suburban area that connects the city center to the seaside town of Mondello at the intersection of F. Besta and L. Enaudi neighborhood streets in the east–west direction and G. Lanza di Scalea arterial road in the north–south direction (see Figure 2a for a view of the roundabout). The geometry of the roundabout is consistent with Italian standards for road intersections [39]; the roundabout has an inscribed circle radius of 35.50 m comprehending the non-traversable central island and 8.00 m wide circulatory roadway with two lanes, 4.00 m wide entry and exit lanes, and entry approaches with deflection angles greater than $43°$. The roundabout size allowed designing an alternative configuration with a lane dedicated to CAVs in order to assess its safety and performance efficiency compared to the roundabout where the CAVs and HDVs share the entry and circulating lanes.

3.2. Step 2: Identification of the Mechanisms of Entry Capacity

Step 2 consisted of the identification of the mechanisms of entry capacity for each traffic flow entering the roundabout conflicted by the circulating traffic on the circulatory roadway. The roundabout driving rules apply to entering vehicles that must give way to circulating vehicles proceeding counterclockwise around the central island; furthermore, vehicles preselect the appropriate entry lane and enter the roundabout when a gap wide enough occurs between subsequent vehicles in the circulatory roadway [16]. In other words, the priority rules establish how to negotiate every point of potential conflict in which the entering vehicles merge with the circulating vehicles before reaching their desired exit. Since roundabouts work with yield conditions, the combination of the number of entry and circulating lanes affects the negotiation among interacting traffic streams and, thus, the gap acceptance behavior on which entry capacity depends [16,28]. Moreover, the driving behavior depends on the type of vehicle given that shorter gaps may occur only where the leader vehicle and the follower vehicle are both CAVs [5]. The roundabout case study is devoid of any raised lane divisors on the entry approaches and the circulatory roadway. On the basis of observations in the field at entries, two mechanisms of entry capacity were identified; thus, it was assumed that both the left lane and the right lane at entries were conflicted by two circulating lanes (i.e., the outer lane and inner lane of the circulatory roadway). The real-life roundabout was the starting point to build the theoretical layout that met the geometric and operating characteristics of the case study. For each mechanism of entry capacity (i.e., the left lane and the right lane at entries), a fleet made only by human driven vehicles was used to model the target curves of capacity in the base situation; in turn, the target curves with CAVs were built using market penetration rates equal to 20%, 40%, 60%, 80%, and 100% CAVs, and corrected using the factors proposed by [17]. The identified mechanisms of entry capacity were then used in the subsequent simulation with Aimsun to define the right-of-way among conflicting movements and to simulate the gap acceptance behavior (see Section 3.4).

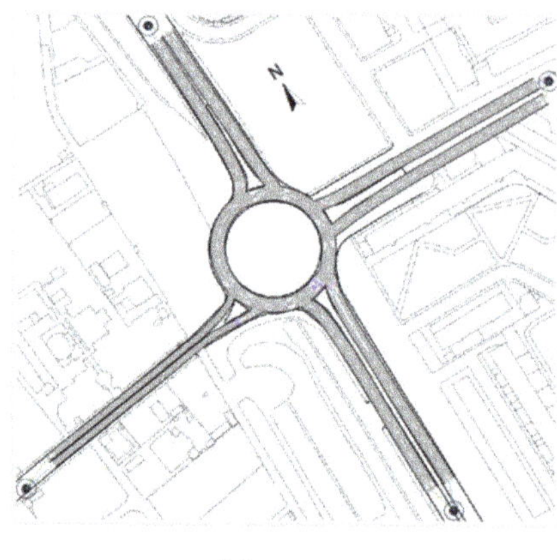

(a)   (b)

Figure 2. The roundabout case study: (**a**) the south entry view (latitude 38.177443 and longitude 13.309095 in decimal degrees); (**b**) the roundabout network model built in Aimsun and visualization of the centroids used as the origin and destination of the trips in the case study.

3.3. Step 3: Traffic Data Collection, Processing and Mining, and Determination of the Capacity Target Values for Every Entry Mechanism and CAV Penetration Rate

In Step 3, traffic surveys were carried out during 5 min time intervals in the peaks from 7:00 to 8:30 a.m. and from 6:30 to 8:00 p.m. on Tuesday to Thursday in November 2022. Survey data considered an entry traffic flow of 3422 vehicles per hour; 11% of trucks were collected in the field. Since the roundabout is installed in the suburban area, the pedestrian and bicycle traffic resulted irrelevant. There were balanced traffic flows from all the legs.

The general equation used to determine the CAV-based capacity curves reflecting the presence of different proportions of CAVs is as follows:

$$C_{e,CAVs} = f_a \cdot a \cdot e^{-f_b \cdot b \cdot Q_c}, \qquad (1)$$

where $C_{e,CAVs}$ is the CAV-based capacity curve by entry lane, adjusted also for heavy vehicles (pc/h), a and b define the intercept and slope parameters, respectively, and f_a and f_b are the planning-level adjustment factors of the parameters a and b, respectively. The CAV-based capacity curves were corrected with the adjustment factors given by [17] for different market penetration rates of CAVs and adapted to the case study in Figure 2b; in the absence of high levels of automation and connectivity in traffic, they were employed as an alternative source of target capacity values [19]. Figure 3 shows the surface functions for each interaction mechanism at entries for the roundabout in Figure 2b. A fleet made only by human driven vehicles was considered to develop the capacity function for the base traffic situation; the curves for CAVs were developed using percentages of CAVs (0%, 20%, 40%, 60%, 80%, and 100% with increments of 20%).

It is well known that roundabout capacity depends on the yielding process at entries, the distribution of gaps in the circulating traffic streams, the driver's decision to enter whether enough space occurs to complete the entry maneuvers, and the follow-up times required by drivers in a queue [16]. Moreover, the CAV behavior specifies the gap-acceptance process at entries. A CAV activates the CACC system where the conflicting vehicle is a CAV in order to have information on the kinematics (i.e., position and speed) of the conflicting CAV. Furthermore, a CAV may activate the adaptive cruise control (ACC) to enter whether a human driven vehicle is the conflicting vehicle [17]. In this view, CAVs are expected to provide greater increases in capacity because they have the potential to accept smaller gaps safely than HDVs [17,26]. The lane change maneuvers and the differences in driving skills among vehicles may make the negotiation of two-lane roundabouts more complicated than the single-lane counterparts [16]. According to [17], the surface functions in Figure 3 show improved capacity as the percentages of CAVs increase; thus, more vehicles can accept smaller gaps safely. From this, it would follow that the benefits of cooperative driving may be reached more easily.

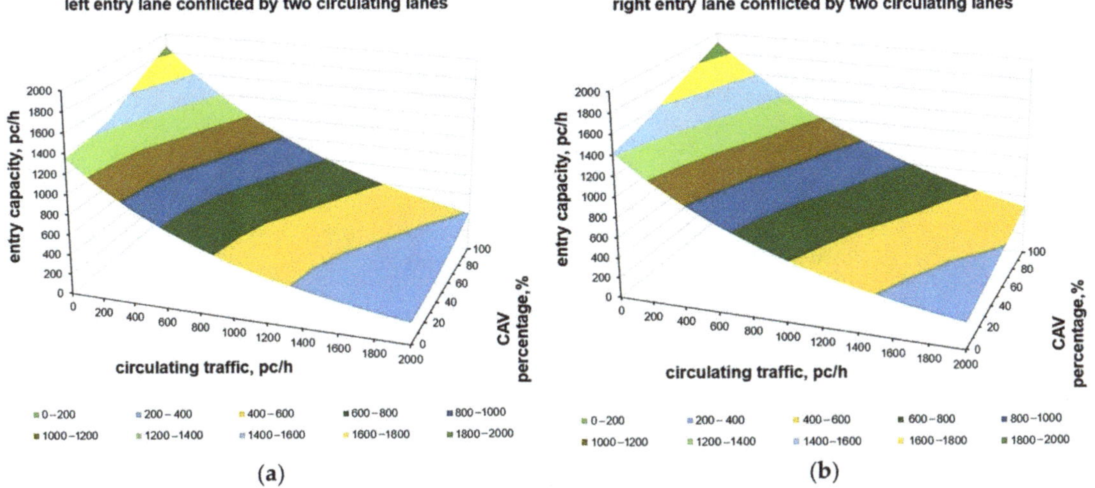

Figure 3. Surface functions of entry capacity under different proportion of CAVs in traffic for the case study in Figure 2b: (**a**) left lane; (**b**) right lane.

3.4. Step 4: Building the Roundabout Network Model and Simulation from Free-Flowing Traffic to Capacity for Calibration Purposes

In Step 4, the roundabout network model was built in Aimsun (see Figure 2b) to simulate operating conditions from free-flowing traffic to capacity for the two mechanisms of entry capacity (i.e., the right entry lane or the left entry lane) identified in Section 3.2. Different market penetration rates of CAVs were set to simulate the traffic scenarios (i.e., from base to 5); each mixed traffic situation included a percentage x of CAVs varying from 0% to 100% and the corresponding $(1 - x)$ percentage of HDVs; percentage increases of 20% were applied. In each traffic scenario, we also assumed that 100% of CAVs were equipped with the CACC system and only 30% of HDVs were equipped with the adaptive cruise control (ACC) system. However, there was the awareness that ACC systems provide minimum gap times that are similar to, or longer than, the gaps used by human drivers, thus contributing to decreased entry capacity when in use [17,18]. Moreover, the required vehicle-to-vehicle communication abilities were considered to be in operation at a high degree of system reliability. Before simulations ran, the coded roundabout network model and traffic demand data were reviewed using the matrices derived from traffic counts. Ten simulations, each of them lasting 60 min, were performed and included the time to load traffic and to reach steady state for initializing the roundabout network model. This phase was necessary for avoiding potential coding errors before calibration. The results confirmed the capability of Aimsun to reproduce the traffic observed in throughout 5 min sampling intervals in peak hours; the GEH index was found to be less than five (about more than 95% of the cases) in each simulation [19]. There was a need to split demand data into two O/D matrices, each with x percentage of CAVs and $(1 - x)$ percentage of HDVs according to the abovementioned scenarios. To simulate saturated traffic conditions and the reaching of capacity, nine subsequent O/D matrices were derived and assigned to the subject entry lane so that it reached the saturated condition, while the circulating traffic was increasing (i.e., 0 to 1800 veh/h with an increase of 200). The simulation from free-flowing traffic to capacity ran so that the number of vehicles reasonably expected to enter the roundabout reflected the entry capacity values as returned by detectors on the roundabout network model. For calibration purposes, the model parameters were manually fine-tuned to improve the match between the simulated data with the target capacity curves for each entry mechanism and each proportion of CAVs. A global calibration was first performed, followed by link-tuning of the network model [19]. According to [19,40], few model parameters were identified, individually examining them and adjusting them in Aimsun in order to obtain simulation output close to the CAV-based curves (see the results in Table 1).

Calibration of HDVs included the speed limit acceptance and the time gap. The speed limit acceptance is the degree of acceptance of the speed limit by the drivers [18]. Typically, if the speed limit acceptance is higher than one, the maximum speed on a road link can be higher than the speed limit, otherwise, when setting the speed limit acceptance lower than one, the maximum speed can be lower than the speed limit. Furthermore, the time gap measured from the rear to the front bumper of subsequent vehicles is calibrated to override the time headway between the front bumpers of subsequent vehicles. The default value of this car-following parameter of 0 s means that the time headway can be used instead of the gap, whereas other values can cause wider headways and can affect the follower vehicle's deceleration in relation to the leader vehicle's kinematics [18].

The calibration also regarded the reaction time that drivers use to adapt their speed to the speed variation of the next vehicle (see also [41]). A higher capacity may result with lower reaction time since the driver can be able to find and to accept smaller gaps safely before entering the roundabout. It should be noted that this car-following parameter of Aimsun can be set to the same value for all vehicles and is equal to the timestep of Aimsun. However, the reaction times of CAVs are shorter than HDVs; thus, the reaction time value under mixed traffic was set equal to a weighted average of the values of each vehicle class in relation to their percentage by scenario.

According to the sensitivity analysis and manual calibration, the maximum acceleration that a vehicle can reach in any circumstance, the safety margin factor, and the sensitivity factor were fine-tuned only for CAVs. Higher values of the maximum acceleration compared to the default ones returned improved vehicle performances [42]. The safety margin factor, in turn, explains whether the vehicles can negotiate an intersection; values higher than the default ones correspond to larger headways that are usually performed by cautious drivers, while values lower than the default ones correspond to a more assertive driving behavior. The fine-tuned values of the safety margin in Table 1 resulted consistent with what Aimsun recommends to reflect the effect of a given geometry on the maneuvers to be performed [18]. At last, the sensitivity factor concerns the follower behavior when the deceleration of the leader should be estimated. According to [18], a sensitivity factor below one matches assertive behavior, while a sensitivity factor greater than one matches cautious behavior during driving. Further parameters, such as the longitudinal and lateral spacing between two vehicles, tested in the simulations, turned out to be quite insignificant since all the simulated vehicles were similarly sized.

Table 1. Results of the fine-tuning parameter process by entry mechanism.

Entry Mechanism	Parameters	Default Values	Fine-Tuned Values					
			Penetration Rate of CAVs (%)					
			0	20	40	60	80	100
Left entry lane	Speed acceptance [1]	1.10	0.97	0.97	0.97	0.97	0.97	0.97
	Time gap [1]	0.00	1.33	1.33	1.33	1.33	1.33	1.33
	Reaction time [2]	0.80	0.95	0.89	0.84	0.78	0.73	0.67
	Max acceleration [3] $[m/s^2]$	3.00	4.00	4.00	4.00	4.00	4.00	4.00
	Safety margin factor [3]	1.00	0.50	0.50	0.50	0.50	0.50	0.50
	Sensitivity factor [3]	1.00	0.50	0.50	0.50	0.50	0.50	0.50
	GEH	58.33 [4]	90.63	100	100	97.22	94.44	91.67
Right entry lane	Speed acceptance [1]	1.10	0.95	0.95	0.95	0.95	0.95	0.95
	Time gap [1]	0.00	1.00	1.00	1.00	1.00	1.00	1.00
	Reaction time [2]	0.80	0.94	0.78	0.76	0.74	0.72	0.70
	Max acceleration [3] $[m/s^2]$	3.00	3.50	3.50	3.50	3.50	3.50	3.50
	Safety margin factor [3]	1.00	0.40	0.40	0.40	0.40	0.40	0.40
	Sensitivity factor [3]	1.00	0.50	0.50	0.50	0.50	0.50	0.50
	GEH	60.40 [4]	100	100	100	100	100	100

[1] The model parameter was calibrated only for HDVs; [2] the reaction time under mixed traffic was given by a weighted average of the reaction time value for each vehicle class, where the weights were the percentage of CAVs and HDVs by scenario; [3] the model parameter was calibrated only for CAVs; [4] the GEH index was employed as a criterion for accepting the model.

It should be noted that the vehicle cooperation to create a gap was activated in order to allow lane changes on the two-lane circulatory roadway and entries; this parameter ranges from level 0.00 to 1.00 (where 1.00 means high aggressiveness); the value of 0.50 was set to implement the speed limit of 50 km/h on the case study. According to [26], other model parameters were excluded, not proving further benefits in the calibration process. At last, the fine-tuned parameters in Table 1 supported the aim to provide a realistic tradeoff among the different attributes of the cooperative driving on roundabouts in order to avoid too large (or short) headways which could cause an unlikely reduction (or increase) of entry capacity. Moreover, the table shows the GEH results by varying the CAV penetration rates in traffic. According to [19], the fine-tuned model could be accepted since the deviation between the CAV-based capacity curves used as target values and simulated data were smaller than 5 in (at least) 85% of the cases. The two-sample *t*-test tested the null hypothesis, or the equality of the means of two groups of CAV-based curves and simulated capacity data for each entry mechanism; it also ensured that there was no statistical difference between them (at the significance level $\alpha = 0.05$) and, therefore, was likely due to chance. The F-test statistic was also calculated to test the equality of the sample variances. As an example, Table 2 shows the results for the right lane. There is evidence to conclude that both the t-value and the F-value should be larger than their respective critical values to reject the null hypothesis at the significance level of 0.05. The table also shows the values of the root-mean-square normalized error used to quantify the overall error of the microsimulator, and the mean percentage error used to explain under- or over-prediction in Aimsun [19]. In turn, Figure 4 shows the comparison of the CAV-based capacity curves with the simulated data for the left and right entry lanes in the mixed traffic with 60% CAVs and 40% HDVs; one can observe the decrease in the entry capacity as the circulating flow increased. The same figure shows the scattergram analysis for the mechanism of right-lane entry capacity; similar results were also returned for the left entry lane but are not reported here for reasons of synthesis.

Table 2. Statistics for right entry lane capacity target values and simulated data.

Entry Capacity (pc/h)	CAV Penetration Rate (%)					
	0	20	40	60	80	100
μ_1 [1] (s.e.) [2]	832.55 (65.62)	869.22 (67.20)	988.88 (66.88)	1055.11 (68.43)	1089.33 (71.09)	1149.00 (75.51)
μ_2 [1] (s.e.) [2]	787.13 (56.07)	847.1 (57.63)	922.10 (60.95)	1020.18 (63.49)	1084.92 (66.95)	1161.48 (68.62)
95% c.i. [3]	(−126.7; 217.6)	(−154.4; 198.7)	(−113.7; 247.3)	(−151.3; 221.1)	(−190.4; 199.2)	(−216.0; 191.0)
t-value [4]	0.53	0.25	0.73	0.37	0.05	−0.12
t-critical value [5]	1.995	1.995	1.994	1.666	1.994	1.995
$p(\alpha)$-value [6]	0.60	0.80	0.50	0.71	0.96	0.90
F-value [7]	1.37	1.36	1.20	1.16	1.13	1.21
F-critical value [8]	1.757	1.757	1.757	1.757	1.757	1.757
F-prob [9]	0.36	0.35	0.59	0.66	0.72	0.57
$RMSNE = \dfrac{\sqrt{N \cdot \sum_{n=1}^{N}(Y_n^{sim}-Y_n^{obs})^2}}{\sum_{n=1}^{N} Y_n^{obs}}$	0.09	0.07	0.08	0.05	0.033	0.047
$MPE = \dfrac{1}{N} \cdot \sum_{n=1}^{N}\left(\dfrac{Y_n^{sim}-Y_n^{obs}}{Y_n^{obs}}\right)$	0.034	0.002	0.07	0.03	−0.002	−0.023

[1] μ_1 and μ_2 are the mean values of the two groups here compared; [2] s.e. is the standard error; [3] c.i. is the confidence interval; [4] t-value is the *t*-test statistic; [5] t-critical value is the critical value of the distribution; [6] $p(\alpha)$ is the probability of obtaining test statistics values equal to or greater than the target ones (in absolute value) at significance level α of 0.05; [7] F-value is the F-test statistic; [8] F-critical value is the value found in the F-distribution; [9] F-prob is the probability that the samples have equal variances; RMSNE denotes the normalized root-mean-square error and MPE denotes the mean percentage error, where N is the number of observations, while Y_n^{obs} and Y_n^{sim} are the target capacity values and the corresponding simulated values.

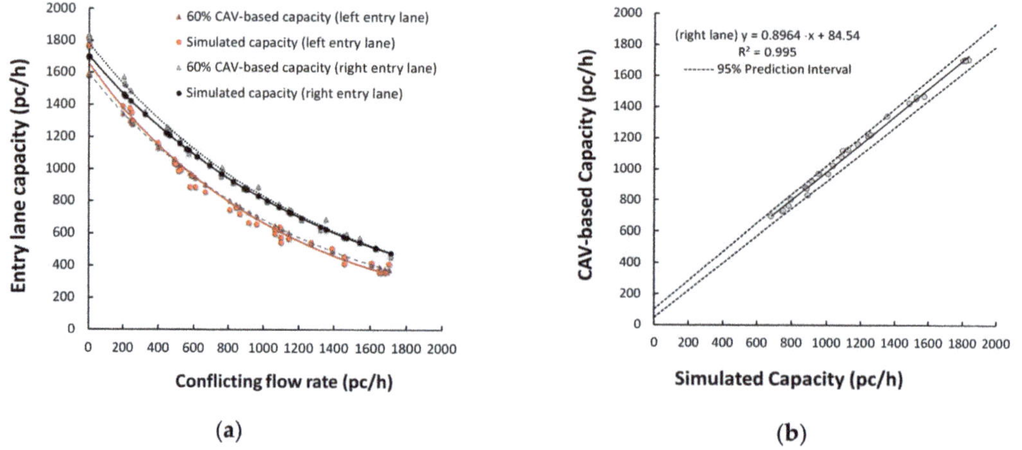

Figure 4. The roundabout model built in Aimsun: (**a**) the comparisons between the CAV-based and simulated capacities for the scenario with 60% CAVs and 40% HVDs; (**b**) scattergram analysis for the mixed traffic with 60% CAVs and 40% HDVs (right lane).

3.5. Step 5: Assessing the CAV Impact on Traffic Throughput

To investigate the performance efficiency of the large diameter roundabout modeled in Aimsun in the transitioning toward a fully CAV fleet, the CAV impact on traffic throughput was assessed compared to the base scenario made of 100% HDVs. Since the expectation of implementing CAVs in traffic is to achieve a larger throughput by creating incentives to operate the entry mechanisms among vehicles at high levels of utilization, operations at capacity were simulated (see previous section).

It should be noted that utilization can be expressed by the throughput (i.e., the vehicles entering the roundabout) compared to capacity (i.e., the maximum flow of vehicles processed by lane) [16]. Furthermore, the delay time values (s/km) were also computed since they represent the time lost by all the vehicles that the roundabout system can process compared with free-flowing traffic without the intersection installed along their path.

Thus, to explain the entry mechanisms of vehicles coming from the entry lanes, reference is made to the southbound entry in Figure 2b. On the basis of the fine-tuning process of the model parameters, the percentage differences for the entry capacity values (pc/h) and delay time values (s/km) were calculated compared to the scenario made only by HDVs as the CAV penetration rates increased (see Figure 5).

There were similar results for both lanes compared to the 100% HDVs case, since high penetration rates of CAVs improved their ability to accept smaller gaps, thus improving entry capacity and reducing delay times. Given that the assumptions related to CAVs were based on simulation and could not be calibrated to real operating conditions on the roundabout under examination, we recommend to the reader that the results presented in Figure 5 can be taken as a projection of future situations with CAVs widespread on the road network. In order to assess benefits of cooperative driving (see next step), another roundabout layout was designed by dedicating a lane to CAVs: CAVs coming from the right lane entered the roundabout, used the outer lane of the ring, and exited the right lane, thus moving on a lane reserved for them. To ensure the above behavior, the CAV dedicated lane was physically separated from the inner lane for mixed traffic with interruptions that corresponded to the entries and exits. As in the roundabout where CAVs and HDVs shared the lanes, different market penetration rates of CAVs were simulated for the situation with a lane dedicated to CAVs.

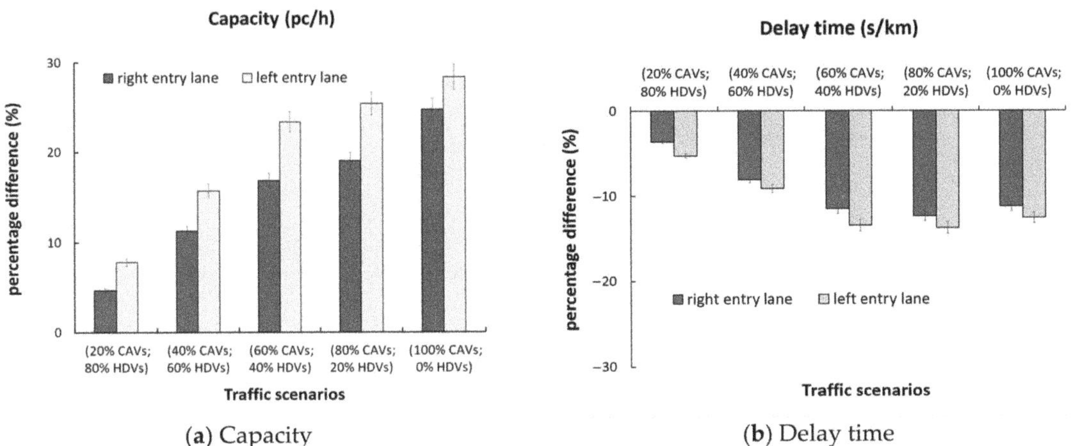

Figure 5. Traffic scenarios vs. the case made only by human driven vehicles: (**a**) capacity for left and right entry lanes; (**b**) delay time for left and right entry lanes.

3.6. Step 6: Assessing the CAV Impact on Traffic Throughput and Safety for the Roundabout with a CAV Dedicated Lane Compared to the Mixed Traffic Situation

Step 6 consisted of assessing the impact of the cooperative driving on the safety and performance efficiency of the roundabout with a lane dedicated to CAVs compared to the mixed traffic situation, as already investigated for turbo roundabouts in previous research [43]. It should be noted that, for the roundabout where a dedicated lane was designed, the parameters fine-tuned for the right lane were applied to the right-dedicated-lane, whereas the model parameters fine-tuned for the left lane were set for the left lane with HDVs only (see Table 1). The mean values of the parameters fine-tuned for HDVs and CAVs in Table 1 were selected for the roundabout layout with the shared lanes. Balanced flow patterns were assigned as described in Section 3.4. To analyze the safety performance, the Surrogate Safety Assessment Model (SSAM) [20] was combined with Aimsun.

It is well known that the surrogate measures of safety explain the safety performances of the road facilities also using vehicle trajectories provided by traffic microsimulators [20,44,45]. Thus, the SSAM reads the trajectory files generated by Aimsun and, through surrogate measures of safety (e.g., time to collision or post encroachment time), can assess the probability of occurrence of a conflict. According to the SSAM logic, all conflict events (i.e., conflicting vehicle pairs) are listed step by step, but including all the conflicts from the previous step. Ten trajectory files for each layout (i.e., the roundabout with shared lanes and the roundabout with a dedicated lane for CAVs) were extracted from Aimsun and processed by the SSAM; the number of conflicts was then drawn from the SSAM for each roundabout. In line with [46], the filters were set to process each conflicting event and to provide output data as much as possible independent from the micro-simulators. Consistent with [45], a filter was applied to consider conflicts within 30 m of the entries as happened at the roundabout in order to avoid recording of conflicts far from the line of entry.

According to a sensitive analysis, the parameters with the greatest influence on the potential conflicts between the vehicular trajectories included the time-to-collision (TTC) and the post-encroachment time (PET) [47]. In this regard, smaller values of the TTC and PET are more likely to cause a conflict, while a TTC equal to zero represents a collision; however, the TTC should be shorter than the PET [20]. The maximum threshold of the TTC was set at the value of 1.5 s, or equal to the default value of the TTC, since other threshold values reduced from the value of 1.5 s provided less overlap for the vehicle pair in the projection timeline and returned a new maximum threshold of the TTC [46]. It should be noted that the SSAM updates the TTC values of each pair of vehicles as long as the projection timeline is without overlaps. However, a crash occurs when the projection reaches zero but the vehicles overlap. The conflict can be considered after the TTC value exceeds the threshold value again [20].

In turn, the threshold value of the PET, or the time gap between one vehicle leaving and another vehicle entering the conflict area, was set to 2.50 s, whereas the default value is 5.00 s [20]. A PET is associated with a timestep by conflict; once a conflict has ended, the final PET value can be recorded, but the TTC value can be less than its threshold value. The minimum values of TTC and PET were set to 0.10 s since zero values were processing errors to be deleted [47]. The SSAM also recorded the maximum speed of the vehicles throughout the conflict; the trajectory files returned similar values of the surrogate safety measure, however, consistent with the urban speed limit of 50 km/h. The angle (i.e., the conflict angle) of hypothetical collision between the conflicting vehicles returned values ranging from $0°$ (indicating a direct rear approach of the second vehicle) to around $-135°$ (indicating an approach of the second vehicle from the left). The conflict type parameter allowed classifying the conflicts: a rear-end conflict happened when the absolute value of the conflict angle was less than $30°$; a crossing conflict happened when the absolute value of the conflict angle was wider than $85°$; otherwise, a lane-changing conflict occurred. It should be said that a rear-end conflict involves two vehicles in the same lane at the same time, while lane changing involves two vehicles which have changed lane. In the cases in which the vehicles enter or exit the roundabout during a conflict event, the SSAM logic considered a rear-end or lane-changing conflict according to the conflict angle values and the underlying roundabout configuration with a lane dedicated to CAVs or shared lanes. Other surrogate safety measures concerning driving behavior remained at the default values to avoid unrealistic maneuvers.

Figure 6 shows the percentage difference in travel time (s/km) compared to the scenario with HDVs only, while Figure 7 shows the number of conflicts on the roundabout where CAVs and HDVs share the same lanes or CAVs used a dedicated lane. Figure 8 shows the number of conflicts by type. It should be noted that the number of total conflicts and conflicts by type were in both cases the average number of conflicts recorded by 10 trajectory files elaborated by the SSAM.

It should be noted that the analysis carried out in this study necessarily reflects the assumptions underlying the conceptualization of the roundabout network model operating as an isolated node of the road network. In this regard, how accurately the simulated traffic conflicts are consistent with measurable on-the-field conflicts relates to matters outside the objectives of this research activity; among other things, this is still an open field of research (see, e.g., [48]).

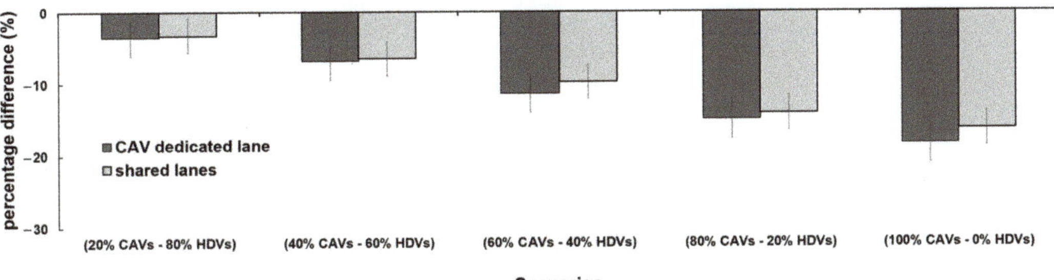

Figure 6. Travel time percentage differences in the roundabout layout with a dedicated lane for CAVs compared to the roundabout layout with shared lanes.

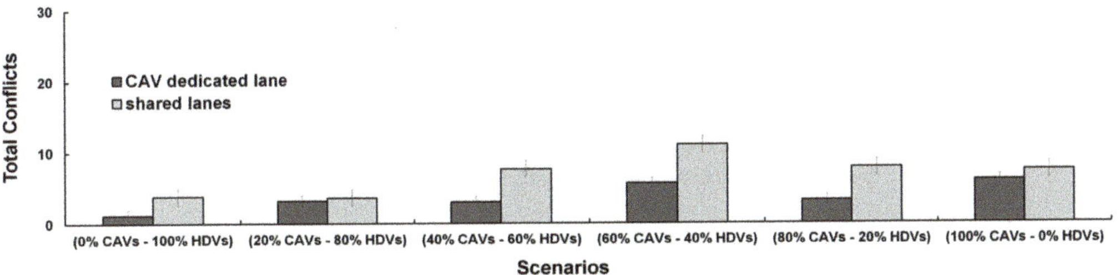

Figure 7. Total conflicts in the roundabout with a lane dedicated to CAVs compared to the layout with shared lanes.

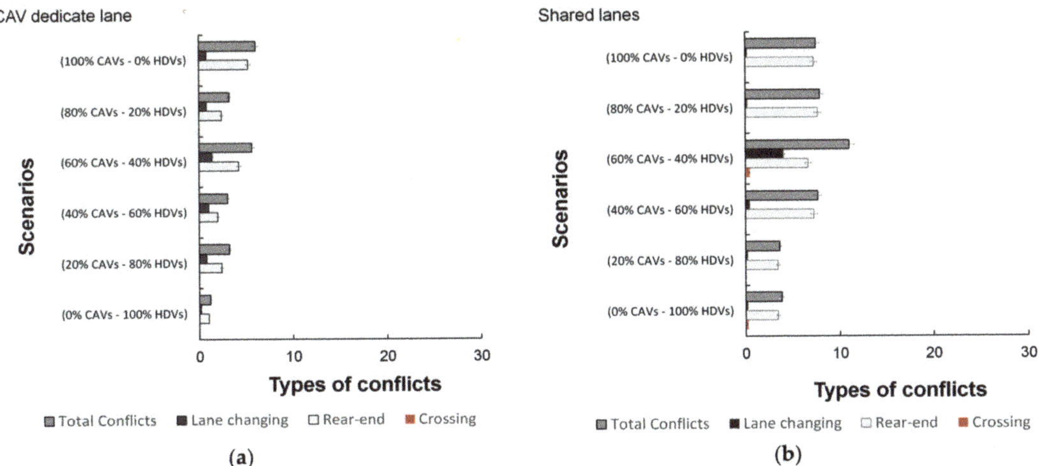

Figure 8. Types of conflicts at the examined roundabouts: (**a**) case with a CAV dedicated lane; (**b**) case with shared lanes.

4. Results

Operational and safety performances improved by dedicating an exclusive lane to CAVs compared to the roundabout where vehicles share the lanes and lane-changing may occur (see Figures 6 and 7). The lane dedicated to CAVs separates the two classes of vehicles here considered, thus ensuring platooning, and effectively drives the travel time and number of total conflicts down [34,49]. Figure 6 shows a slightly higher reduction in the travel times (i.e., the total possible routes of vehicles recorded by the control detectors) on the roundabout with a lane dedicated to CAVs than the roundabout with the shared lanes. The benefit of the lane dedicated to CAVs is evident in the scenario with 100% CAVs, where the percentage reduction in travel time is equal to about 18% compared to the case of all human-driven vehicles; the percentage reduction in travel time is equal to about 16% where CAVs and HDVs share the lanes compared to the case of all human-driven vehicles.

Concerning the total conflicts, the percentage reduction is evident where the dedicated lane is designed than the layout where the entry and circulating lanes are shared by the two classes of vehicles (see Figure 7). Despite the decreasing trend, the expected safety benefits of the cooperative driving than traditional driving systems tend to decrease in the scenario with 100% CAVs. It should be noted that the assumptions of assertive driving behavior used in the simulation runs (see Table 1) may have reduced the safety margin among the vehicles, thus affecting the negotiation among CAVs at roundabouts. According to [34], high traffic volumes made of 100% CAVs may compromise the safety benefits since shorter headways between subsequent vehicles contribute to increasing rear-end conflicts (see Figure 8). In this regard, the SSAM returned a percentage of about 85% rear-end conflicts where a lane dedicated to CAVs is designed compared to the percentage of 95% where the shared lanes are operating. In this regard, the SSAM considers the entry and exit maneuvers as lane-changing events, whose number may depend on the behavioral assumptions of the underlying simulation model and the assumed threshold values. Although the SSAM filters were set in line with [46], where two microsimulators were used to avoid the setting of software-dependent parameters, a comparison with other microsimulators may be necessary to test further aspects of CAV driving on roundabouts. Together with the hypotheses of assertive behavior used to simulate CAVs, the lane separation may also force the trajectories of entry or exit performed by them and then increase the percentage of lane-changing conflicts returned by the SSAM (see Figure 8). This type of conflict is, however, expected in the mixed traffic situation, and it is more evident where the two classes of vehicles are almost balanced (see the scenario with 60% CAVs and 40% HDVs in Figure 8b). Thus, the different distribution by type of conflict can be also due to the different way to select the driving lane up to the desired exit in the roundabout layouts here examined. At last, although the results point to the promising effects of growing penetration rates of CAVs, it should be noted that they show only a projection of what might happen if the cooperative driving systems were fully widespread on the road network.

5. Discussion

The novel opportunities of cooperative driving can further increase road safety and efficiency since vehicles can move in a coordinated manner also with harmonizing effect on mixed traffic situations at different levels and scales of complexity [5,50]. Connectivity will help road users to better anticipate upcoming events (e.g., the horizontal and vertical alignment of the road routes, the traffic states, and neighboring vehicle movements), future events, and situations that may be perceived as potentially hazardous during driving; in turn, automation will allow vehicles to adjust their movements more precisely to anticipate imminent events. However, the implementation of CAVs in real traffic where they have to interact with nearby human-driven vehicles is still in development and far from the full experimental validation [51]. While many studies have shown a myriad of prominent applications on vehicle design requirements, V2V or V2I communications, jam dynamics, CAV platooning, and reduced infrastructure footprint on highways (e.g., [7,9,52–54]), only a few studies have been undertaken to determine how CAVs can negotiate intersections and roundabouts, a fact which also clearly influences land-use and urban planning (e.g., [30,55,56]).

Assumptions have to be made to model the gap-acceptance behavior of CAVs entering a roundabout; the types of conflicting vehicles, together with the entry mechanism and the way in which the priority is negotiated, determine the critical and follow-up headways on which the capacity depends [5,17]. Among the factors that affects the CAV ability to improve capacity, a higher proportion of CAVs can realize the competitive advantages of connectivity.

The simulation of free-flowing traffic to capacity allowed exploring the entire range of operations for the entry mechanisms here examined (i.e., the right lane and the left lane). In this regard, the CACC feature enabled by V2V communication allowed CAVs to accept shorter gaps safely than traditional vehicles. Thus, Figure 4a shows the decrease in the entry capacity as the circulating flow increased. The comparison of the CAV-based capacity curves with the simulated data in the same figure for the scenario with 60% of CAVs and 40% of HDVs confirmed the ability of Aimsun to capture variations in driving behavior for the vehicle classes within the dataset, and to return simulated data of capacity fitting well with the corresponding target values (see Table 1). According to [34], the accuracy of the results depended mainly on the accuracy of the calibration. The versatility of Aimsun was also confirmed by the statistical tests (see Table 2 and Figure 4b).

The throughput of the roundabout improved with the presence of enough CAVs in the network (see also [57]). The CACC systems when in use allowed the simulated vehicles to accept shorter time gaps than human drivers and to increase the roadway capacity (see Figure 5a). Slightly higher percentage differences for the left lane than the right lane depended on a greater readiness to enter required to vehicles coming from the left lane; they mainly needed to turn left or travel through the roundabout and, once entered, had to move forward more quickly than vehicles coming from the right lane. It can be observed that, in the highest CAV penetration rate, the percentage increase in starting capacity was 25% for the right lane; an additional percentage increase of 3% occurred for the left lane compared to the case of human driven vehicles. Similar results were also shown in the literature regarding the impact of automated driving on roundabout capacity [58] and mixed traffic situations [59]. In turn, the delay time values tended to significantly reduce up to the scenario made of 60% CAVs and 40% HDVs with percentage differences of about 11% for the right lane and 13% for the left lane; the percentage reductions tended to stabilize for higher penetration rates of CAVs (see Figure 5b).

The size of selected roundabout allowed designing a lane dedicated to CAVs, while the suburban character of its context of installation made the traffic situation easily managed by CAVs. Aimsun was employed to calibrate the model parameters and then coupled with the SSAM to perform the safety analysis.

The results show that the travel times decreased in the roundabout designs under examination as the CAV penetration rate increased. In the condition of low penetration rate of CAVs (i.e., 20% CAVs and 80% HDVs), the percentage reduction in travel times was about 3% compared to the case of all-human driven vehicles in both roundabout layouts (see Figure 6); the benefits of the dedicated lane were evident at high penetration rates of CAVs (i.e., from 60% to 100% CAVs). By way of example for the mixed traffic situation with 60% CAVs and 40% HDVs, Figure 6 shows percentage differences of 11.40% and 9.80% in the travel times on the roundabout with a lane dedicated to CAVs and the roundabout with shared lanes, respectively. There was a percentage reduction of 18.40% in travel times in the scenario 100% CAVs compared to the case made only by human drivers for the roundabout with a dedicated lane because of better driving performance of a fully CAV fleet than the case of all-human driven vehicles; the percentage reduction was about 16% in the roundabout without a lane dedicated to CAVs when the comparison was performed at the edges of the range (i.e., 100% CAVs vs. 100% HDVs).

In turn, Figures 7 and 8 show the efforts to assess the safety performance of the roundabouts under examination. According to [43,44,60,61], the configuration with a dedicated lane separated the CAVs and HDVs, thus relegating the lane changing mainly to the approach areas to perform entry and exit maneuvers, but depending on the availability of acceptable gaps where vehicles could advance side by side.

The dedicated lane for CAVs halved the number of total conflicts compared to the initial counterpart with shared lanes; the redesign of the large roundabout into a layout with a CAV dedicated lane returned a percentage reduction of total conflicts on average just over 50% (see scenarios from 40% to 80% CAVs) compared to the shared traffic situation (see Figure 7). The percentage difference decreased in the scenario with a fully CAV fleet due to the concomitant increase in rear-end conflicts; there were percentages of rear-end conflicts of about 85% and 95% in the design option with the CAV dedicated lane and in the counterpart without it, respectively (see Figure 8). This was also confirmed by the simulations of the vehicles turning left (or traveling through the roundabout) which tended to preselect the left lane before entering, and the vehicles turning right which tended to preselect the right lane to enter regardless of the presence of the raised lane divisors, thus modifying the assumed conflict patterns [17].

In conclusion, the methodological framework to assess road infrastructure safety and performance efficiency in the transition to cooperative driving yielded some fruitful results. In this view, the paper presents the research efforts aimed at better understanding the performance of CAVs where the curvilinear design may lead to misinterpretation of driving intentions or simply complicate the negotiation of the system.

However, it may be useful to address some further issues.

(1) This research was primarily focused on the comparison in terms of safety and operational performances at the single road entity level. Moreover, the roundabout network model simulated in Aimsun meets the geometry and traffic characteristics of a real-life large roundabout chosen as case study where vehicular traffic flows were balanced along the major and minor directions of driving. Future developments should also be conducted at road corridor or network level, varying not only the roundabout geometry but also the traffic demand matrices to investigate the effects of different geometric shapes (i.e., outer diameter size, number of entry, exit and circulating lanes, etc.), spacing, and traffic patterns on the performance efficiency and speed management on roundabouts.

(2) Despite the observed effects on traffic safety and efficiency due to the design of dedicated lanes (with mandatory or optional use) to separate CAVs from HDVs on roundabouts, nothing can be said about conflicts attributable only to CAVs in mixed traffic situations and their severity. Thus, future research actions should be directed toward addressing the methodological limitations in the analysis of shared situations where CAVs and HDVs interact, in order to better incorporate the abovementioned conflict characteristics into decision support tools.

(3) The research results were given in the terms of an evaluation framework of the model's practical application and simulation verification, but they showed only a projection of what might happen if CAVs were fully widespread on the road network. This perspective also highlights the need for efficient methods to assess the potential of CAVs and to enhance their throughput through an intelligent road management in view of future mobility strategies. It is appropriate to deepen issues on smart roundabout design to make the road network in operation suitable for the progressive transition toward the full implementation of CAV technologies. There is also a need to hypothesize how a control area performs in order to implement communications among CAVs with the road infrastructure manager system (see [62] for the turbo roundabout case).

6. Conclusions

There is increasing interest in CAVs, since their full implementation will transform road transportation and promote social and economic change. Transition toward cooperative driving systems still requires understanding of the key issues involved in adapting the geometry of road infrastructures to the kinematics of CAVs in order to achieve the proper balance between road safety and traffic efficiency. Despite the clear benefits of CAVs, several limitations will need to be addressed before their widespread implementation becomes possible. There is a need to manage HDVs and CAVs mixed in traffic, especially at roundabouts where the curvilinear feature of geometric design may complicate the mutual interpretation of the driving intentions among vehicles.

On the basis of the above, this paper proposed a simulation-based methodological framework to assess the impacts of CAVs on operational and safety performances at roundabouts. The size of the selected roundabout allowed installing a lane dedicated to CAVs, while the suburban character of its context made the traffic situation easily managed by CAVs. Microscopic simulation from free-flowing traffic to capacity on the roundabout designed in Aimsun, consistent with the geometry and traffic characteristics of the real counterpart, allowed identifying some behavioral parameters of cooperative driving suitable to simulate road situations with growing proportions of CAVs. Aimsun was used to calibrate the model parameters and was coupled with the SSAM to perform the safety performance analysis. Despite the effects on traffic throughput and safety with CAVs, the advantage of a dedicated lane is to separate HDVs from CAVs, thus reducing human error and potential conflicts among vehicles.

The benefit of the lane dedicated to CAVs was evident in the traffic made only by CAVs where the percentage reduction in travel time was equal to about 18% compared to the traffic made only by human drivers; the percentage reduction in travel time was equal to about 16% where CAVs and HDVs share the lanes compared to the case with 100% human-driven vehicles. Concerning the total conflicts, the percentage reduction was more evident where the dedicated lane was designed than the situation where the entry and circulating lanes were shared by CAVs and HDVs. However, the expected safety benefits of cooperative driving compared to traditional driving systems tended to decrease in the scenario made only by CAVs. The assumptions of assertive driving behavior used in simulation may have reduced the safety margin among CAVs, thus affecting their negotiation and contributing to increasing rear-end conflicts. However, nothing can be said about conflicts due only to CAVs in mixed traffic situations and conflict severity based on the analysis tools available to date.

Thus, further research actions should be directed toward addressing the methodological limitations in the analysis with CAVs to better incorporate the conflict characteristics into decision support tools. Moreover, many more intersections and traffic conditions should be studied to have a more comprehensive and detailed vision of CAVs at corridor or road network levels and to assess the potential benefits of these technologies in the face of growing demands for smart mobility. Since the roads of the future need high levels of adaptation, automation, and resilience, guidance to select the intersection types suitable to maintain stable operation against natural disruptions or manmade events should also be provided. Thus, which intersection geometry or type of control mode (i.e., stop signs, roundabouts, or traffic signals) may affect the efficiency, safety and resilience at intersections and roundabouts before and after a disruption will be a research question to answer in future developments of the research. Lastly, future developments should also include a comprehensive sustainability assessment of the energy and emission impacts of the full lifecycle of CAVs from an operational perspective, to better understand the broader direct and indirect effects at the mobility system level.

Author Contributions: Conceptualization, M.L.T., E.M., A.G. and T.G.; methodology, M.L.T., E.M., A.G. and T.G.; software, M.L.T., E.M., A.G. and T.G.; validation, M.L.T., E.M., A.G. and T.G.; formal analysis, M.L.T., E.M., A.G. and T.G.; investigation, M.L.T., E.M., A.G. and T.G.; resources, M.L.T., E.M., A.G. and T.G.; data curation, M.L.T., E.M., A.G. and T.G.; writing—original draft preparation, M.L.T., E.M., A.G. and T.G.; writing—review and editing, M.L.T., E.M., A.G. and T.G.; visualization, M.L.T., E.M., A.G. and T.G.; supervision, E.M., A.G. and T.G.; project administration, M.L.T., E.M., A.G. and T.G.; funding acquisition, M.L.T., E.M., A.G. and T.G. All authors have read and agreed to the published version of the manuscript.

Funding: This research received no external funding.

Institutional Review Board Statement: Not applicable.

Informed Consent Statement: Not applicable.

Data Availability Statement: Data can be provided upon kind request to the corresponding authors.

Acknowledgments: The authors would like to thank the reviewers for their comments which improved the paper.

Conflicts of Interest: The authors declare no conflict of interest.

References

1. Javed, A.R.; Shahzad, F.; Rehman, S.U.; Zikria, Y.B.; Razzak, I.; Zunera, J.; Xu, G. Future smart cities: Requirements, emerging technologies, applications, challenges, and future aspects. *Cities* **2022**, *129*, 103794. [CrossRef]
2. Lv, Z.; Shang, W. Impact of Intelligent transportation systems on energy conservation and emission reduction of transport systems: A comprehensive review. *Green Technol. Sustain.* **2023**, *1*, 100002. [CrossRef]
3. Rana, M.M.; Hossain, K. Connected and Autonomous Vehicles and Infrastructures: A Literature Review. *Int. J. Pavement Res. Technol.* **2021**, *14*, 1–21. [CrossRef]
4. SAE International. *Taxonomy and Definitions for Terms Related to Driving Automation Systems for On-Road Motor Vehicles*; SAE International: Warrendale, PA, USA, 2018.
5. Brost, M.; Deniz, Ö.; Österle, I.; Ulrich, C.; Senzeybek, M.; Hahn, R.; Schmid, S. Energy Consumption of Connected and Automated Vehicles. In *Electric, Hybrid, and Fuel Cell Vehicles, Encyclopedia of Sustainability Science and Technology*, 2nd ed.; Elgowainy, A., Ed.; Springer: New York, NY, USA, 2021; pp. 216–224.
6. Karbasi, A.H.; Mehrabani, B.B.; Cools, M.; Sgambi, L.; Saffarzadeh, M. Comparison of Speed-Density Models in the Age of Connected and Automated Vehicles. *Transp. Res. Record* **2022**, *2677*, 849–865. [CrossRef]
7. Wang, Y.; Yan, Y.; Shen, T.; Bai, S.; Hu, J.; Xu, L.; Yin, G. An Event-Triggered Scheme for State Estimation of Preceding Vehicles Under Connected Vehicle Environment. *IEEE Trans. Intell. Veh.* **2023**, *8*, 583–593. [CrossRef]
8. Duggal, A.S.; Singh, R.; Gehlot, A.; Gupta, L.R.; Akram, S.V.; Prakash, C.; Singh, S.; Kumar, R. Infrastructure, mobility and safety 4.0: Modernization in road transportation. *Technol. Soc.* **2021**, *67*, 101791. [CrossRef]
9. Jang, J.; Baek, J.; Lim, K.; Ro, Y.; Yoon, S.; Jang, S. A Study on V2I based Cooperative Autonomous Driving. In Proceedings of the 2023 International Conference on Electronics, Information, and Communication (ICEIC), Singapore, 5–8 February 2023; pp. 1–3. [CrossRef]
10. Liu, Y.; Tight, M.; Sun, Q.; Kang, R. A systematic review: Road infrastructure requirement for Connected and Autonomous Vehicles (CAVs). *J. Phys. Conf. Ser.* **2019**, *1187*, 042073. [CrossRef]
11. Martin-Gasulla, M.; Elefteriadou, L. Traffic management with autonomous and connected vehicles at single-lane roundabouts. *Transp. Res. Part C Emerg. Technol.* **2021**, *125*, 102964. [CrossRef]
12. Guo, Q.; Ban, X.B.; Abdul Aziz, H.M. Mixed traffic flow of human driven vehicles and automated vehicles on dynamic transportation networks. *Transp. Res. Part C Emerg. Technol.* **2021**, *128*, 103159. [CrossRef]
13. Gong, B.; Wang, F.; Lin, C.; Wu, D. Modeling HDV and CAV Mixed Traffic Flow on a Foggy Two-Lane Highway with Cellular Automata and Game Theory Model. *Sustainability* **2022**, *14*, 5899. [CrossRef]
14. Ivanjko, E.; Kušic, K.; Greguric, M. Simulational analysis of two controllers for variable speed limit control. *Proc. Inst. Civ. Eng.—Transp.* **2022**, *175*, 413–425. [CrossRef]
15. Raju, N.; Farah, H. Evolution of Traffic Microsimulation and Its Use for Modeling Connected and Automated Vehicles. *J. Adv. Transp.* **2021**, *2021*, 2444363. [CrossRef]
16. National Academies of Sciences, Engineering, and Medicine. *Roundabouts: An Informational Guide*, 2nd ed.; The National Academies Press: Washington, DC, USA, 2010. [CrossRef]
17. National Academies of Sciences, Engineering, and Medicine. *Highway Capacity Manual 7th Edition: A Guide for Multimodal Mobility Analysis*; The National Academies Press: Washington, DC, USA, 2022. [CrossRef]
18. *Aimsun Next*, Version 20. Dynamic Simulator User Manual. TSS-Transport Simulation Systems: Barcelona, Spain, 2020.
19. Barceló, J. *Fundamentals of Traffic Simulation*; Springer: London, UK, 2010.
20. Gettman, D.; Pu, L.; Sayed, T.; Shelby, S.G. *Surrogate Safety Assessment Model and Validation: Final Report*; Georgetown Pike (US) Report FHWA HRT 08–051; Federal Highway Administration: Washington, DC, USA, 2008.
21. Michałowska, M.; Ogłoziński, M. Autonomous Vehicles and Road Safety. In *Smart Solutions in Today's Transport: 17th International Conference on Transport Systems Telematics, TST 2017, Katowice–Ustroń, Poland, 5–8 April 2017, Selected Papers 17*; Springer International Publishing: Cham, Switzerland, 2017; Volume 715, pp. 191–202. [CrossRef]
22. Rahman, M.S.; Abdel-Aty, M. Longitudinal safety evaluation of connected vehicles' platooning on expressways. *Accid. Anal. Prev.* **2018**, *117*, 381–391. [CrossRef] [PubMed]
23. Rahman, M.S.; Abdel-Aty, M.; Wang, L.; Lee, J. Understanding the Highway Safety Benefits of Different Approaches of Connected Vehicles in Reduced Visibility Conditions. *Transp. Res. Rec. J. Transp. Res. Board* **2018**, *2672*, 91–101. [CrossRef]

24. *VISSIM*, Version 5.30. User Manual PTV Planug Transport Verkehr AG. Manufacture: Karlsruhe, Germany, 2011.
25. Ekram, A.; Rahman, M.S. Effects of Connected and Autonomous Vehicles on Contraflow Operations for Emergency Evacuation: A Microsimulation Study. In Proceedings of the Transportation Research Board 97th Annual Meeting, Washington, DC, USA, 7–11 January 2018.
26. Tumminello, M.L.; Macioszek, E.; Granà, A.; Giuffrè, T. Simulation-Based Analysis of "What-If" Scenarios with Connected and Automated Vehicles Navigating Roundabouts. *Sensors* **2022**, *22*, 6670. [CrossRef] [PubMed]
27. Qiao, J.; Zhang, D.; de Jonge, D. Virtual Roundabout Protocol for Autonomous Vehicles. In *Advances in Artificial Intelligence. AI 2018*; Mitrovic, T., Xue, B., Li, X., Eds.; Lecture Notes in Computer Science; Springer: Cham, Switzerland, 2018; Volume 11320, pp. 773–782.
28. Mauro, R. *Calculation of Roundabouts. Capacity, Waiting Phenomena and Reliability*; Springer: Berlin/Heidelberg, Germany, 2010. [CrossRef]
29. Martin-Gasulla, M.; Elefteriadou, L. Single-Lane Roundabout Manager under Fully Automated Vehicle Environment. *Transp. Res. Rec. J. Transp. Res. Board* **2019**, *2673*, 439–449. [CrossRef]
30. Wu, Y.; Zhu, F. Junction management for connected and automated vehicles: Intersection or roundabout? *Sustainability* **2021**, *13*, 9482. [CrossRef]
31. Mohebifard, R.; Hajbabaie, A. Trajectory control in roundabouts with a mixed fleet of automated and human-driven vehicles. *Comput. -Aided Civ. Infrastruct. Eng.* **2022**, *37*, 1959–1977. [CrossRef]
32. Wu, R.; Jia, H.; Yang, L.; Miao, H.; Lin, Y.; Zhang, Y. A distributed trajectory control strategy for the connected automated vehicle in an isolated roundabout. *IET Intell. Transp. Syst.* **2022**, *16*, 232–251. [CrossRef]
33. Jalil, K.; Xia, Y.; Zahid, M.N.; Manzoor, T. A Speed Optimization Strategy for Smooth Merging of Connected and Automated Vehicles at T-Shape Roundabout. *IEEE Access* **2022**, *10*, 76953–76965. [CrossRef]
34. Anagnostopoulos, A.; Kehagia, F. CAVs and roundabouts: Research on traffic impacts and design elements. *Transp. Res. Procedia* **2020**, *49*, 83–94. [CrossRef]
35. Pérez, J.; Milanés, V.; de Pedro, T.; Vlacic, L. Autonomous driving manoeuvres in urban road traffic environment: A study on roundabouts. *IFAC Proc.* **2011**, *44*, 13795–13800. [CrossRef]
36. Morando, M.; Truong, L.; Vu, H. Investigating safety impacts of autonomous vehicles using traffic micro-simulation. In Proceedings of the Australasian Transport Research Forum, Auckland, New Zealand, 27–29 November 2017.
37. Tibljas, D.; Giuffre, T.; Surdonja, S.; Trubia, S. Introduction of Autonomous Vehicles: Roundabouts Design and Safety Performance Evaluation. *Sustainability* **2018**, *10*, 1060. [CrossRef]
38. Gora, P.; Katrakazas, C.; Drabicki, A.; Islam, F.; Ostaszewski, P. Microscopic traffic simulation models for connected and automated vehicles (CAVs)—State-of-the-art. *Procedia Comput. Sci.* **2020**, *170*, 474–481. [CrossRef]
39. Italian Ministry of Infrastructure and Transport. *Functional and Geometric Standards for Road Intersections*; Italian Ministry of Infrastructure and Transport: Rome, Italy, 2006.
40. Hollander, Y.; Liu, R. The principles of calibrating traffic microsimulation models. *Transportation* **2008**, *35*, 347–362. [CrossRef]
41. Giuffrè, O.; Granà, A.; Tumminello, M.L.; Sferlazza, A. Capacity-based calculation of passenger car equivalents using traffic simulation at double-lane roundabouts. *Simul. Model. Pract. Theory* **2018**, *81*, 11–30. [CrossRef]
42. Gipps, P.G. A Behavioural Car-Following Model for Computer Simulation. *Transp. Res. Part B* **1981**, *15*, 105–111. [CrossRef]
43. Giuffrè, T.; Granà, A.; Trubia, S. Safety evaluation of turbo-roundabouts with and without internal traffic separations considering autonomous vehicles operation. *Sustainability* **2021**, *13*, 8810. [CrossRef]
44. Han, I.; Pham, D.T. Safety analysis of roundabouts and avoidance of conflicts for intersection-advanced driver assistance systems. *Cogent Eng.* **2022**, *9*, 2112813. [CrossRef]
45. Giuffrè, T.; Trubia, S.; Canale, A.; Persaud, B. Using microsimulation to evaluate safety and operational implications of newer roundabout layouts for European road networks. *Sustainability* **2017**, *9*, 2084. [CrossRef]
46. Giuffrè, O.; Granà, A.; Tumminello, M.L.; Giuffrè, T.; Trubia, S. Surrogate Measures of Safety at Roundabouts in AIMSUN and VISSIM Environment. In *Roundabouts as Safe and Modern Solutions in Transport Networks and Systems*; Macioszek, E., Akçelik, R., Sierpiński, G., Eds.; TSTP 2018. Lecture Notes in Networks and Systems; Springer: Cham, Switzerland, 2019; Volume 52. [CrossRef]
47. Saleem, T.; Persaud, B.; Shalaby, A.; Ariza, A. Can Microsimulation be used to Estimate Intersection Safety? *Transp. Res. Rec.* **2018**, *2432*, 142–148. [CrossRef]
48. Bulla-Cruz, L.A.; Laureshyn, A.; Lyons, L. Event-based road safety assessment: A novel approach towards risk microsimulation in roundabouts. *Measurement* **2020**, *165*, 108192. [CrossRef]
49. Lo Cigno, R.; Segata, M. Cooperative driving: A comprehensive perspective, the role of communications, and its potential development. *Comput. Commun.* **2022**, *193*, 82–93. [CrossRef]
50. Taiebat, M.; Brown, A.L.; Safford, H.R.; Qu, S.; Xu, M. A Review on Energy, Environmental, and Sustainability Implications of Connected and Automated Vehicles. *Environ. Sci. Technol.* **2018**, *52*, 11449–11465. [CrossRef] [PubMed]
51. Ge, J.I.; Avedisov, S.S.; He, C.R.; Qin, W.B.; Sadeghpour, M.; Orosz, G. Experimental validation of connected automated vehicle design among human-driven vehicles. *Transp. Res. Part C Emerg. Technol.* **2018**, *91*, 335–352. [CrossRef]
52. Stange, V.; Kühn, M.; Vollrath, M. Manual drivers' experience and driving behaviour in repeated interactions with automated Level 3 vehicles in mixed traffic on the highway. *Transp. Res. Part F Traffic Psychol. Behav.* **2022**, *87*, 426–443. [CrossRef]

53. Kim, B.; Heaslip, K.P. Identifying suitable car-following models to simulate automated vehicles on highways. *Int. J. Transp. Sci. Technol.* **2023**, *12*, 652–664. [CrossRef]
54. Xie, Y.; Gartner, N.H.; Chowdhury, M. Editors' notes: Special issue on connected and autonomous vehicles. *Transp. Res. Part C: Emerg. Technol.* **2017**, *78*, 34–36. [CrossRef]
55. Vitale, F.; Roncoli, C. Distributed Formation Control for Managing CAV Overtaking and Intersection Maneuvers. *IFAC-PapersOnLine* **2022**, *55*, 198–203. [CrossRef]
56. Long, K.; Gao, Z.; Jiang, Z.; Ma, C.; Hu, J.; Yang, X. Optimization based trajectory planner for multilane roundabouts with connected automation. *J. Intell. Transp. Syst.* **2022**, *27*, 411–422. [CrossRef]
57. Suh, W.; Kim, J.I.; Kim, H.; Ko, J.; Lee, Y.-J. Mathematical Analysis for Roundabout Capacity. *Math. Probl. Eng.* **2018**, *2018*, 1–8. [CrossRef]
58. Boualam, O.; Borsos, A.; Koren, C.; Nagy, V. Impact of autonomous vehicles on roundabout capacity. *Sustainability* **2022**, *14*, 2203. [CrossRef]
59. Ard, T.; Austin Dollar, R.; Vahidi, A.; Zhang, Y.; Karbowski, D. Microsimulation of energy and flow effects from optimal automated driving in mixed traffic. *Transp. Res. Part C Emerg. Technol.* **2020**, *120*, 102806. [CrossRef]
60. Mauro, R.; Cattani, M.; Guerrieri, M. Evaluation of the safety performance of turbo roundabouts by means of a potential accident rate model. *Balt. J. Road Bridge Eng.* **2015**, *10*, 28–38. [CrossRef]
61. Tollazzi, T.; Rencelj, M. Comparative Analyse of The Two New Alternative Types of Roundabouts—Turbo and Flower Roundabout. *Balt. J. Road Bridge Eng.* **2014**, *9*, 164–170. [CrossRef]
62. Guerrieri, M. A theoretical model for evaluating the impact of Connected and Autonomous Vehicles on the operational performance of turbo roundabouts. *Int. J. Transp. Sci. Technol.* **2023**. [CrossRef]

Article

Assessing the Future Streetscape of Rimini Harbor Docks with Virtual Reality

Rachid Belaroussi [1], Margherita Pazzini [2,*], Israa Issa [2], Corinne Dionisio [1], Claudio Lantieri [2], Elena Díaz González [3], Valeria Vignali [2] and Sonia Adelé [1]

1   COSYS-GRETTIA, University Gustave Eiffel, F-77447 Marne-la-Vallée, France
2   Department of Civil, Chemical, Environmental and Materials Engineering (DICAM), University of Bologna, 40126 Bologna, Italy
3   Higher School of Engineering and Technology, Universidad de La Laguna, 38071 San Cristóbal de La Laguna, Spain
*   Correspondence: margherita.pazzini2@unibo.it

Abstract: The human factor plays an important role in the successful design of infrastructure to support sustainable mobility. By engaging users early in the design process, information can be obtained before physical environments are built, making designed spaces more attractive and safer for users. This study presents the collected data of a virtual reality (VR) application in which user perception has been evaluated within an urban redevelopment context. The area under consideration is the Canal of the Port of Rimini (Italy), a degraded area not connected to the city center. The redevelopment of degraded urban areas is the first step towards achieving the sustainability aims set out in the Sustainable Development Goals. Prior to this work, evaluation methods were developed in the decision-making process, considering different social, economic, and environmental aspects in order to obtain a priority scale of interventions for urban regeneration. Architectural solutions were proposed to represent targeted and specific interventions that are designed precisely for the context to which they are dedicated in order to make the Canal Port area a continuum with its urban context and to improve its perception by tourists and inhabitants. To assess these proposed infrastructure modifications, two models of VR were created, one relevant to the current condition and one representing the future condition after redevelopment of the area. Virtual visits to the Canal of the Port of Rimini were created under two scenarios, namely, the current situation and the future situation after redevelopment of the infrastructure. Then, human participants were involved through two different questionnaires. The first allowed participants validate the VR model created by comparing it with the real context, while the second served to evaluate the perceptions of users by comparing the two VR models of the canal before and after the intervention. The results of this empirical research highlight the benefits of engaging users early in the design process and improving the user experience before implementing renovation of the infrastructure.

Keywords: urban redevelopment; human-centered design; virtual reality; streetscape; built environment

Citation: Belaroussi, R.; Pazzini, M.; Issa, I.; Dionisio, C.; Lantieri, C.; González, E.D.; Vignali, V.; Adelé, S. Assessing the Future Streetscape of Rimini Harbor Docks with Virtual Reality. *Sustainability* **2023**, *15*, 5547. https://doi.org/10.3390/su15065547

Academic Editors: Salvatore Leonardi and Natalia Distefano

Received: 23 February 2023
Revised: 14 March 2023
Accepted: 16 March 2023
Published: 21 March 2023

1. Introduction

1.1. Creating Resilient Urbanism with Streetscape Design

Our analysis began with on-site inspections to assess the urban and territorial system of the Canal of the Port of Rimini in Italy, illustrated in Figure 1, to identify deficiencies in terms of services, connections, availability of spaces, cycle-pedestrian paths, and carriageable roads [1]. This analysis led us to identify needs to support the functionality of the port as a whole, on the basis of which various alternative project scenario were proposed for the urban redevelopment of the Canal Port.

After proposition of a regeneration project of the canal by architects, described in [2], a before–after comparison between the design scenarios was necessary in order to evaluate the improvement of the future layout of the canal. We decided to do this using virtual reality; the present paper describes the scene construction and methodology used to visually compare the baseline scenario and the project situation. Using TwinMotion software, a model was created in virtual reality representing the existing infrastructure and current situation of the Port of Rimini. A second VR model was created, the post-intervention scenario, to analyze mobility and infrastructure in the study area by evaluating opportunities and risks in the new port area envisioned as part of the project.

The research previously begun in [1,2] suggests a method of supporting and justifying project proposals in the complex case of regeneration of port areas. The aim is to show how important sustainable mobility is within a deep urban redevelopment of a historical context, such as the Canal Port of Rimini. The reconnection of cycle-pedestrian paths, the redevelopment of the quays, and the creation of urban spaces for tourists and citizens are possible solutions to improve quality of life in degraded and underutilized urban areas.

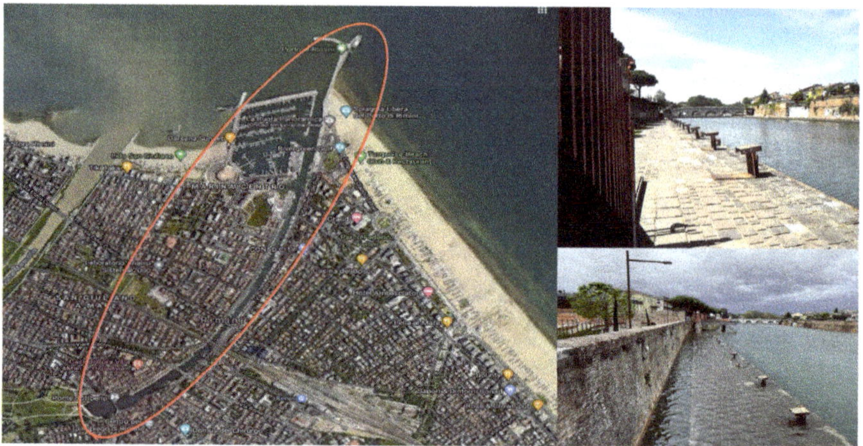

Figure 1. Canal of the Port of Rimini (in red) and pictures of its current quays. Imagery taken from Google Earth.

The design phase began with the identification of the height to which to lift the docks in order to solve the problem of frequent flooding due to tides and adverse weather conditions. Following an in-depth hydraulic study of the area, raising of the quays was justified and verified. Access to platforms and public spaces was designed to identify new functions for the benefit of the community. As a result of raising the docks, the cycling and pedestrian paths along the two banks of the Canal Port were revised accordingly. The new cycle and pedestrian infrastructure can improve public health and make cities more active and environmentally friendly. Recent studies have shown that the regeneration of urban public spaces is closely related to the presence of safe and connected cycling and pedestrian paths. The proposed solutions are currently being defined and refined, and could receive funding from the Municipality of Rimini to be implemented; however, they need to be validated. This paper proposes a methodology for prospective validation.

The solutions proposed in this contribution represent targeted and specific interventions that are designed precisely for the context to which they are dedicated in order to make the Canal Port area a continuum with its urban context and to improve its perception by tourists and inhabitants. Although the proposed solution is tailor-made for this specific case, the developed approach is based on a strong scientific basis of urban regeneration projects founded on multi-criteria analysis and sets of indicators. The applied strategy can be replicated in any other similar case requiring an urban regeneration intervention. The benefits of this urban regeneration project include:

- Improved aesthetic quality of urban spaces

- Improved environmental quality of urban spaces
- Reduction of pollutant emissions through the increase in green and permeable areas
- Increased user flow (residents and tourists) in areas that are currently poorly frequented
- Increased social well-being of the regenerated area.

All these aspects could potentially result in higher economic productivity in the area. Better urban quality may lead to an increase in the real estate value of the area and implementation of new economic activities. As a future development of the research, it is proposed to deepen an economic feasibility study of the interventions. This should be convenient for the municipality, as it does not involve actions of deep urban transformation and demolition, only simple local interventions of renovation of public spaces. The proposed urban regeneration project focuses exclusively on the redevelopment of bicycle and pedestrian routes, and aims to represent a good example of how soft mobility plays a fundamental role in urban regeneration.

1.2. Virtual Scene for Streetscape Assessment

New urban planning and design tools making the use of public space more efficient are now being sought in all cities in order to build healthy and liveable urban environments and encourage the development of infrastructure [3]. Good urban planning should consider several aspects, such as the physical environment (location, climate, resources, etc.), the social environment (planning the right areas to promote socialization among people), and the economic environment supporting business [4]. The same global and integrated vision applies to urban regeneration, which aims to establish satisfactory government conditions for areas subject to transformation [5,6] and to make lasting economic, social, physical, and environmental changes by reducing problems towards a more sustainable city [7]. Urban regeneration is a new way of rethinking the use of space by combining and harmonizing economic, social, physical, and environmental issues in the same context [8]. Urban regeneration aims to redevelop abandoned areas and transform them into new attractive centers [9,10], encouraging inclusive growth of urban spaces. By focusing on social progress rather than on progress for its own sake, urban regeneration fosters the transition from an individualistic model to a more participatory one involving the development of collective thinking.

Urban planning and regeneration, along with model design, have always taken a very long time due to the many requirements involved in proposing ideas, looking for investments, presenting and adapting projects, etc. It is often months or even years before a project is approved and work can begin. Moreover, many projects are rejected because investors cannot obtain financing. In addition, the different needs and expectations of stakeholders and different perceptions of the project are sometimes not recognized or respected, and this may give rise to conflicts [11,12]. Shared urban regeneration planning is an opportunity to understand how stakeholders perceive urban heritage and how they can contribute [13]. Moreover, the participation and empowerment of stakeholders in the processes of planning and territorial regeneration is essential to overcome disparities of power and commitment of the different parties and possible lack of communication [14]. For this reason, stakeholders should be involved from the beginning to identify the criticalities of an area, rather than asking them about proposals for urban redevelopment after the project has been approved [15]. However, although it is important, few citizens participate in urban planning [16], and when they do it is unfortunately only possible to do so passively. Du et al. (2019) [16] asserted that this can lead to the exclusion of many people, resulting regret around urban projects when they are carried out. Thanks to technology, eParticipation and mParticipation have been introduced, and mobile web devices can be useful communication channels between institutions and citizens to facilitate the participation of a digital and network-connected society.

To encourage active participation in public planning, the right tools for understanding the scope of innovation should be identified. Citizens, investors, and stakeholders sometimes reject projects because of a lack of visualization and intuitive observation of how the resulting project may appear in reality [17].

An intuitive, engaging, and user-friendly approach such as VR is capable of offering a 3D vision of the project, and can be used to connect stakeholders and groups from different strata of society by bridging the information gap. Researchers in completely different fields, such as education [18,19], entertainment [20], health [21], and marketing [22] agree that the use of virtual reality improves learning experiences, promotes cooperation [23], and can enhance creativity and commitment [24].

3D models have been used in urban planning through CAD (Computer Aided Design) software for about thirty years. From the initial execution of 2D projects with maps presented on fixed screen or drawn or printed on paper, CAD has moved to support 3D virtualization [25] thanks to data access and the internet. The resulting maps are dynamic and highly interactive tools modifiable by a human by interacting with a computer, thereby becoming a real human processing system. Depending on the context, maps are likely to become increasingly interactive and intelligent in the near future, and may even be able to imitate the human brain [26]. In their paper, Jamei et al. [27] considered VR as the next step in 3D visualization.

Today, Computer-Aided Design technologies are among the most advanced tools for urban planning and modelling processes [28]. In this context, Virtual Reality (VR) is becoming common in urban planning and design [29]. By applying virtual reality technology, a project can be displayed on a computer, allowing architects and engineers to intuitively model, visualize, and observe the entire project in a realistic three-dimensional environment [30,31]. Adjustments can be made to the project to make it similar to the desired reality. Real objects and people can be placed and moved on a surface, and thanks to image recognition, these can be digitally recorded [32]. Depending on the hardware and software configurations available, the applications of VR in urban planning can be very different. Further studies on VR for architectural representation are recommended by [33], who highlighted its importance as an effective tool in urban planning.

The World Economic Forum has recognized that Virtual Reality in the public sector can help support citizen engagement, strengthen resource management and maintenance, improve public safety and emergency services, and aid in public health, sustainability, transport, urban mobility, common heritage, and tourism. In addition, virtual reality is a useful tool to demonstrate the efficiency of infrastructure initiatives by involving stakeholders in decision-making processes that directly or indirectly affect city life.

Two extremes of Virtual Reality can be recognized depending on the level of interaction with the artificial environment, namely, immersive or non-immersive. As the definition suggests, immersive technology allows users to feel fully immersed in virtual reality, blurring the boundary between real and virtual worlds [34,35]. Through a head-mounted display (HMD), the user is transported into a three-dimensional virtual world (3D), providing a truly realistic and more visceral experience than other models [36]. On the contrary, in non-immersive virtual reality the project is displayed on the screen of a computer, television, or mobile phone; no special devices are needed, and the user is not surrounded by a virtual environment [37,38]. The difference in use between the two technologies is strongly affected by the context and the type of user they are directed at. A study of age-related differences was conducted in 2019 by Plechatá et al. [39]. The performance of the elderly was much higher when using immersive reality, while for the young it was unchanged. However, in both groups immersive reality caused stress and fatigue.

The purpose of this study is to verify the validity of an urban regeneration project using non-immersive virtual reality as an assessment tool. Two different scenarios have been created, one representing the current condition and one the post-project reality. Through an online questionnaire, the perception of the relationship between actual reality (Google Earth) and virtual reality was evaluated. Then, the two virtual scenarios were compared to evaluate citizen and stakeholder perceptions of the quality of the urban spaces before and after redevelopment.

2. Context and Architectural Data

2.1. Analysis of the Rimini Area

Rimini is a sea town in northern Italy located along the Adriatic coast. Internationally famous for its long sand beach and the sea, tourism in Rimini is linked to the hotel industry for fairs, congresses, and events. The Canal of the Port of Rimini is a rather extended structure situated near the historical centre; unfortunately, it is in a state of degradation and impoverishment, and is not adequately exploited and used. In this sense, the project carried out for the Canal of the Port of Rimini has been directed to the requalification of an entire area that, while it is considered particularly degraded, has excellent potential for use.

The Canal of the Port of Rimini is located on the original mouth of the river Marecchia, and consists of docks on two sides extending on two piers. The channel is 2.2 km long, 46 m wide at the entrance and 40 m along its length up to "Parco XXV Aprile", and divides the historic centre of Rimini from San Giuliano a Mare, an important district north of the city. All the various activities related to sea life are located on the left of the port: shipyards, mechanical workshops, the fish wholesale market, and nautical shops. Historical and cultural attractions such as "Porta Galliana" and the lighthouse, a symbol of the navy, are located on the right side. If well-connected, these could increase the charm of a place that, despite its nature of urban mobility space and the points of interest present along its course, is currently anything but a pole of attraction.

After an analysis of the urban context of the Canal, a study on the microclimate of the zone was considered, including "Parco XXV Aprile" and the seafront called "Parco del Mare", two recently requalified areas near the canal. The study showed that the average values of temperature and humidity are within the standard values of temperate climates and the prevailing directions of the winds are east and northwest, i.e., the wind is from the sea. During the day the heat island is along the coast, while at night it is in inner city. In addition, in certain periods lower temperature values are accompanied by heavy above-average rainfall, and in certain cases higher values of the average annual temperature are linked to drier years.

A comprehensive analysis of all aspects of the study area allowed us to highlight the main problems to be addressed and solved. First of all, poor planning of transport and related infrastructure is evident. The port space is used inadequately and inefficiently, and the infrastructure in the area is degraded. In addition, the links with the main nodes of sustainable mobility, such as the railway station, have serious shortcomings and do not represent a real alternative to the use of private cars. Second, the entire area of the quays is subject to frequent flooding due to a design error dating back to 1977. This discourages people from passing through. In 1980, several water drainage channels were built; today, however, they are old and filled with stagnant water and algae, especially in summer. As a result, the whole area is full of mosquitoes that prevent tourists from exploring the place. Finally, access to the quays is possible only through stairs, which greatly reduces the usability of the area and represents a serious architectural barrier.

Based on this analysis, for the redevelopment of this area any successful project must first facilitate access to the Canal by promoting sustainable transport systems, in particular, by moving from car mobility to soft mobility. In addition, the urban space of the quays and existing road infrastructure has to be harmonized and made more functional. From this perspective, the Canal of the Port of Rimini represents a useful example for evaluating a synergistic approach to urban regeneration that considers the different aspects of urban planning, infrastructure, social cohesion, and sustainability. Priority actions were identified as the reconnection of cycle and pedestrian paths with the urban transportation infrastructure and the redevelopment and raising the docks to avoid flooding due to tides and adverse sea weather conditions. In order to allow boats to pass under bridges, in particular the "Ponte della Resistenza", which is the lowest and therefore the most critical to cross, docks were raised to a height between 1.30 m a.s.l. and 1.50 m a.s.l. depending on the section of the port concerned. While this does not definitively solve the problem of flooding, it lengthens the return time with which flood events occur.

2.2. Virtual Scene Construction

Virtual reality involves an artificial three-dimensional environment that can be created with or without the use of special equipment. Virtual reality technology-based tools in city planning can be categorized into three main categories: viewing or visualization, modeling, and simulating. First, virtual reality technology as a modeling tool can be used in the process of creating 3D models of a city plan. This means that architects and engineers can use virtual reality to create a 3D model of a city [30]. Second, virtual reality as a visualization tool facilitates 3D visualization of built scenes to correct the design and development of urban cities, thereby enhancing the artistry of urban architecture and the aesthetics of urban design [40]. This can help city planners to visualize how new buildings and public spaces will look and feel and to identify potential problems before they happen. It can be used as a supplementary presentation to imagery and animations used to present 3D models for non-technical team members such as stakeholders. For instance, it can be utilized to show the public how a city will look after a major renovation or redevelopment project. Third, simulations with virtual reality technology in city planning projects can be used to predict various scenarios. This type of simulation can help in evaluating and analyzing the outcome of the various scenarios during the planning process, and can be used to make decisions based on them. Public space, for instance, is one aspect of city planning that is often used in simulations, in order to predict how specific features in a city plan will affect the public space of the area. Thus, VR is an effective way to model, visualize, and simulate the environment, and planners can use it to analyze and modify the city virtually before initiatives are properly implemented in the real world [41,42].

It is important to employ the right VR software. TwinMotion is one of the most popular and powerful virtual reality tools. It is an advanced architecture software with many essential and useful features. An overview of the city can be assessed from any angle, with options including lighting, shadowing, and a considerable amount of 3D assets. The content of the city can resolve tiny detail, including buildings, bridges, harbors, structures, road systems, green areas, facilities such as playgrounds and parks, etc., with a very realistic level of detail.

This part of the study involved the use of quantitative methods to support the participatory urban planning processes. Thus, VR technology as an effective tool in the participatory urban planning processes was applied to the case study of the Rimini Port Canal. The aim of using this tool was to depict the proposed redevelopment and re-connection of pedestrian and cycle paths. TwinMotion was used to create a design for a sidewalk that can help users navigate through the area efficiently. In addition, it was used to model different cycling and street infrastructure options. For this purpose, a 3D VR model of the Rimini Canal was created and was used to create two scenarios: one representing the current situation before urban regeneration, and the other representing the future after urban regeneration. The final phase included creating an online survey to obtain feedback from the study participants.

The design proposal for the project focused on the area between the Ponte della Resistenza Bridge and the Tiberius Bridge, as illustrated in Figure 2. A technique was applied to enable the visualization of Rimini's topography and streets in combination with the morphology of the ground in order to better understand the area. Following careful investigation, it appeared that the InfraWorks software would be the most suitable alternative.

Figure 2. Study area: VR overview and virtual scene with its urban context.

It is possible for professionals to model, evaluate, and visualize infrastructure design concepts in the context of a construction site using InfraWorks conceptual design software. Because of this particularity, it was possible to extrapolate the affected area of Rimini from the surrounding environment using the "Model Builder" command.

With this command, the user selects the region to be modeled or imports it as a polygon. The computer processes the request and generates the required model within a few minutes. The program mostly displays data from OpenStreetMap; in detail, the urban context includes:

- Highways and railroads: the model's road and rail elements are created using the OpenStreetMap highway and rail databases.
- Buildings: the OpenStreetMap collection includes information about buildings.
- Images: Microsoft® Bing Maps satellite imagery is used to cover the model's topography.
- Elevation: global terrain data are provided in Digital Elevation Models (DEM) of 10 and 30 m in resolution depending on the region of interest and its geographic location. Terrain data for the US and its territories are based on 10 m USGS DEM data from the National Elevation Dataset (NED). Between $-60°$ and $+60°$ latitudes, 30 m DEM SRTMGL1 data are utilized. For latitudes between $+60$ and $+83$ degrees, DEM ASTER GDEM v2 data with a resolution of 30 meters are utilized.
- Water: data on water bodies are derived from the OpenStreetMap dataset.

The model created in InfraWorks was exported as an .FBX file and imported into TwinMotion after being rendered in the software. The workflow of the construction of the virtual scene is illustrated in Figure 3.

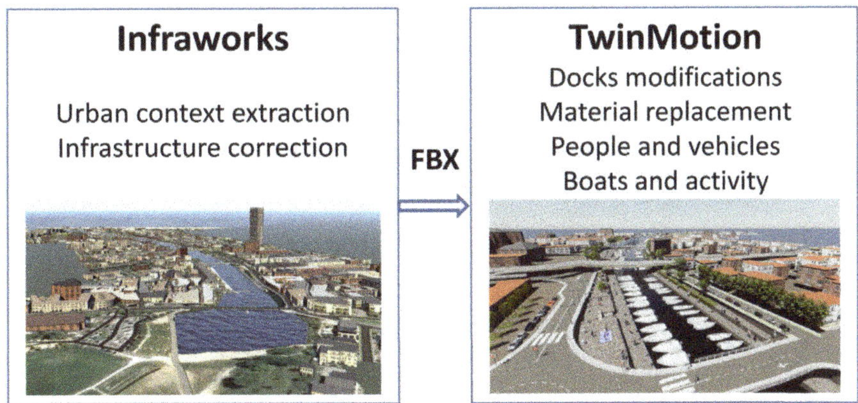

Figure 3. Immersive visit design: workflow diagram from GIS urban context extracted with Infra-Works to virtual reality refined scene with TwinMotion.

InfraWorks allowed us to extract the urban context of the project, that is, the infrastructure and buildings with texture. Nonetheless, it was a coarse model, and structural corrections had to be made manually for the elevation of bridges, roads, and roundabouts.

In addition, the docks and roads are displayed at the same altitude as the water, which does not match reality. The representation of the docks was particularly coarse, and had to be completely redone for this study. We used TwinMotion archviz software for local refinement of the 3D models.

The docks were redrawn to match the reality of the current layout of the port of Rimini based on DWG drawings of the area. Their elevation was corrected to ensure that the levels between the roads and the docks and between the docks and the water conformed to reality. The materials of the street were replaced as well in order to be more photorealistic, and we added people and vehicles to the virtual scenes. These characters in the scene only stand or sit; we chose to use a static study, as including moving people and vehicles was beyond the scope of the presented work. The same number of people were included on the docks for both the current and prospective scenarios. We added boats on the water to reflect the port environment, and activities were added following the architectural proposition described in [2].

Three videos illustrate the content of the study:

- A comparison between the real environment and virtual scene for the urban context in the reference scenario is illustrated in Video 1 www.youtube.com/watch?v=FGt6 tZ7YvXI, accessed on 20 March 2023. We used Google Earth Studio to visualize the current real environment.
- Current situation in 2022: the virtual scene of the reference situation can be seen in Video 2 www.youtube.com/watch?v=nY7cNg0jIgs, accessed on 20 March 2023.
- The urban regeneration project representing the proposed future situation can be seen in Video 3 www.youtube.com/watch?v=2olh-JRaNr4, accessed on 20 March 2023.

Figure 4 shows extracted examples of these 3D virtual scenes.

Figure 4. Before/after project comparison using virtual scenes: examples of the current version of the docks (**left**) and prospective views of the docks in the future scenario (**right**).

2.3. Assessment of the Renovation Project

2.3.1. Which Features of Streetscape Impact the Public Experience?

The term *streetscape* is applied to characterize the natural and built fabric of a street, and is outlined as the design features and visual effect of the street, especially how the paved area is structured and maintained. Sustainable streetscape design involves technology that controls flood risk and lowers carbon footprint to make sure that spaces are long-lasting and that their features are components of the wider ecosystem. This can help to produce better places to live for current and future citizens [43]. A streetscape is a public space that promotes vitality and shows a sense of belonging to a city. It can be defined as a neighborhood's character in terms of its physical infrastructure, cultural history, and interaction with other people. These factors influence the quality of life and the mental and physical wellness of its residents [44]. The design and appearance of the streetscape can help to create a visual image of a sustainable city. It is one of the most important elements in helping to attract tourists to a city [45]. Therefore, one of the most important factors for evaluating the success of the city is the design of the streetscape.

Unfortunately, there are many cities that do not have the proper standards for streetscaping, which can negatively affect the visual image of their cities [43]. In this context, our research explored the various aspects of sustainable streetscaping to determine how it can be integrated into the urban development and infrastructure framework in the Canal of Rimini area to create an attractive and safe environment.

The key performance indicators that must be designed and implemented properly to improve the area are identified as follows:

- Sociality: the ability of people to communicate is a basic and fundamental kind of social interaction.
- Walkability: facilitation of mixed-use development (i.e., residential, commercial, institutional) where people can comfortably walk to services within a reasonable distance.
- Accessibility: improving access to the docks to increase their public usage.
- Intermodality: use of multimodal transportation by making its modalities easily accessible.
- Activities: intensify life on the docks by providing more activities.
- Aesthetics: contribute to the improvement of the natural landscape and effectively interact with the modern structure of urban territory (roads, spaces, and paths) within the landscape.
- Services and functions: provide leisure services such as bars, tourism, sports facilities, and outdoor activities.
- Connectivity: ensure that individuals can move easily between and within the central city and suburbs.
- Sustainable mobility: sharing services, use of cycling and pedestrian paths, ICT services.
- Safety: make sure public places and spaces are as safe and pleasant for people as possible.

There is no shortage of research on the characteristics of spaces designed to accommodate walkers or bicycle users [46], and there are many lists of criteria and tools for checking them [47,48]. Many studies have looked at walkability; for example, [49] proposed five criteria for a walkable city: convivial, convenient, connected, clear, and comfortable. According to [50], walkability can be considered either as a characteristic of a place or as an evaluation by an individual. More broadly, walkability is related to the physical characteristics of a space, such as the width of the sidewalks, the density of traffic, the shade provided by trees, population density, and weather, as well as the characteristics of urban design that influence the perception of users from a more subjective point of view (whether it is memorable, provokes feelings, is well-defined while being connected to the rest of the city, its a human scale, and whether it is organized and clean). Walkability can include what can be achieved within a spaces, in particular the presence of shops and services [51]. Finally, walkability is related to peoples' frequenting of a place (number, type, appearance, and activities) or the presence of nature (sound, smell, and sight) [52]. Other studies that have not directly addressed walkability or cyclability have studied the "atmosphere" of an area using criteria such as safety, beauty, smell, feeling of safety (lighting, perception of spaces, level of road traffic), feeling of comfort [53], and sociability [54].

In our Rimini Port Canal case study, the objective of designing a good streetscape is to enhance social and economic interaction through improving the environmental quality, creating social well-being by establishing streets that are a good fit for public walking, and providing gathering areas where social interaction can occur regularly, while encouraging outdoor activities. Other goals include to maintain the urban fabric by enhancing the appearance of the city and to improve public health by supporting and facilitating walking and other leisure activities.

2.3.2. Participatory Sustainable Design through an Online Questionnaire

After the 3D model and immersive visits were built, a questionnaire-based approach was conducted. The aim of the questionnaire was to collect data about the quality of public spaces and social interactions based on the perceptions of stakeholders. The survey's questions were derived from the literature and references concerning the factors that influence the quality of public spaces, as well as from the results of the previous steps in this study. The subject matter included the type of life on the docks, services and functions, sociality, walkability and cyclability, aesthetics, and access to the docks.

Multiple-choice questionnaires were designed with Google Forms and sent online to various groups and individuals from the University of Bologna and University Gustave Eiffel. The participants involved in this phase were workers, researchers, and students at both Universities.

The questionnaire was grouped into four parts. Part 1 focused on the comparison between the current situation as displayed by Google Earth and as displayed by TwinMotion. Part 2 concentrated on analysis of Scenario 1 for the Rimini Canal. Part 3 was centered on analysis of Scenario 2 for the Rimini Canal. Finally, Part 4 aimed to retrieve general information about the participants. The questionnaire was completely anonymous and took less than 10 min to complete.

Part 1 aimed to validate the realism of the virtual reconstruction of the current situation of the channel of the Rimini Port Canal in 2022. Two videos were displayed simultaneously, one showing the current situation with Google Earth imagery and the other showing the virtual scene of the Rimini Port Canal as displayed by TwinMotion. Participants were asked to compare the differences or similarities between the two videos with a set of four questions: a general question, then questions about specific characteristics of the scene such as green areas, access, and bicycle spaces. Figure A1 summarizes the questions asked on this topic. In order to allow for comparison, the two videos were presented simultaneously, and used the same angle and viewing distance.

In Part 2, a two-minute video was presented showing the current situation in virtual reality before urban regeneration. Participants were asked to assess the space presented in the video based on various criteria, such as willingness to spend time there, projected use, and willingness to practice activities (see Table 1).

Table 1. Part 2 and Part 3: mean and standard deviation for each question for the current situation and for the prospective scenario.

Question	Mean		SD	
	Part 2	**Part 3**	**Part 2**	**Part 3**
In this place, I want to spend time.	4.41	5.25	1.76	1.61
When I look at this place, I can project myself into its use.	4.66	5.34	1.69	1.43
I think this is a good place to come and do sport.	4.51	5.28	1.86	1.56

Table 1. *Cont.*

Question	Mean		SD	
	Part 2	Part 3	Part 2	Part 3
In this place, I know I will always find something nice to do.	4.00	5.00	1.74	1.48
I think this place is ideal for resting or strolling.	4.56	5.06	1.81	1.56
I think this is a good place to have a drink or a bite to eat.	4.46	5.51	1.69	1.58
In this place, I want to get together with my friends or family.	4.22	5.12	1.74	1.57
In this place I think I could meet people I know or don't know.	4.38	4.93	1.70	1.62
In this place, I want to ride my bike.	3.94	4.56	2.02	1.93
I want to walk around this place.	5.06	5.43	1.62	1.44
The circulation in this space seems fluid.	4.31	5.12	1.93	1.58
I feel safe in this place.	4.88	5.46	1.57	1.31
I can easily find my way around this place.	5.54	5.60	1.23	1.17
I think this place is beautiful.	4.38	4.90	1.91	1.77
This place is an invitation to dream.	3.32	4.03	1.85	1.84
It is a pleasure for the eyes to look at this place.	3.74	4.63	1.89	1.83
The atmosphere of this place is pleasant.	4.41	4.97	1.72	1.63
I think this place has enough green space.	3.24	4.16	1.92	1.83
When I look at this place, I can quickly see how to get to the edge of the water.	5.22	5.34	1.45	1.41

In Part 3, a two-minute video was presented showing the future situation in virtual scenes after urban regeneration. Participants were asked to assess the space presented in the video based on the same criteria as in Part 2. The results are shown in Figure A4. The videos used in Parts 2 and 3 were made from the same model, with the same tour of the docks and the same number of passers-by, in order to control the effect of density on the evaluation.

For the three first parts, we used a seven-items Likert-type scale, ranging for the first part from *Totally Different* (1) to *Totally the Same* (7) and for the two next parts from *Totally Disagree* (1) to *Totally Agree* (7).

Part 4 was optional and related to general information about participants, including their age, gender, occupation, and mode of transport used. This part aimed to collect information in order to test the links between the respondents' perceptions and their individual and mobility characteristics. Figure A5 illustrates this last part of the questionnaire.

3. Experimental Results

3.1. Sample Individuals

The questionnaire was proposed online for three weeks. A set of answers from N = 68 participants was collected. The sample was roughly half male and half female, with ages between 22 to 66 years old (M = 36.52, SD = 12.99). As shown in Figure 5a, a quarter of the respondents were students and three-quarters were personnel from the two universities, either professors or researchers. As for the main transportation modes, public transportation was the main mode and was used by 40% of the respondents, as illustrated by Figure 5b; 25% generally used a private vehicle, while a third of the participants used soft transportation, which was equally split between walking and bicycling.

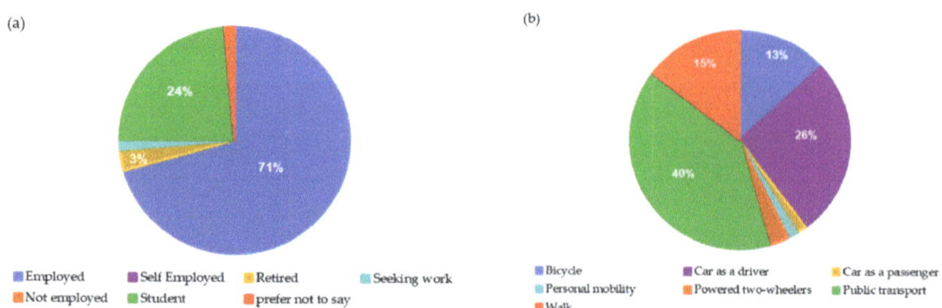

Figure 5. Characteristics of the participants: (**a**) occupation and (**b**) mode of transportation.

3.2. Realism of the Virtual Scene

We produced two visualizations of the same area of the Port of Rimini under the current scenario. As illustrated by Figure 2, Visualization 1 represents the real world images made with Google Earth Studio and Visualization 2 is its virtual counterpart tailored for this project. Four questions were included in the survey to qualify aspects related to the similarity of the represented area, green spaces, cycle spaces, and access to the docks.

It has to be noted than one limitation of this work is the resolution of the two videos presented in Visualization 1 and Visualization 2. The overall objective of the project was to provide a methodology for visual comparison between two scenes of the infrastructure before and after modification. While the resolution difference between the Google Earth Studio video and the TwinMotion video undoubtedly had an impact on the respondents' perception of the scene, it was impossible to address this as they used two different software types. However, because we are mostly interested in functional aspects of the virtual representation, we believe that ensuring a similar resolution is not an imperative element of this study.

The first part of the survey provides information about the realism of the virtual representation of the studied area. For this, we asked participants to evaluate the similarity between the real current situation and the current situation as represented by TwinMotion with respect to four aspects. Figure 6 shows the result of this first part of the study. From the point of view of the global evaluation, the participants seemed to think that the two situations were rather similar, as shown in Table 2 (M = 4.09, SD = 1.50). This allows us to consider the results obtained subsequently as valid. On the other hand, in terms of detail the respondents were more reserved on the similarity from the point of view of green spaces (M = 3.99, SD = 1.81), cycling spaces (M = 3.97, SD = 1.78), and access to platforms (M = 3.91, SD = 1.71). This is evidenced by the dispersion of the scores obtained in Figures 7–11. This obliges us to consider the results obtained for certain questions in Parts 2 and 3 with more caution.

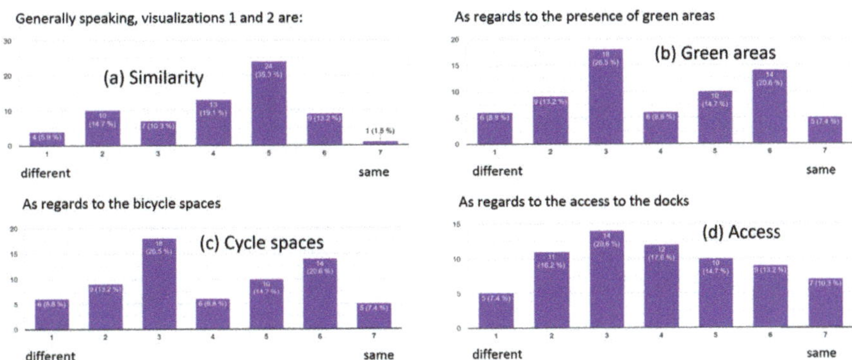

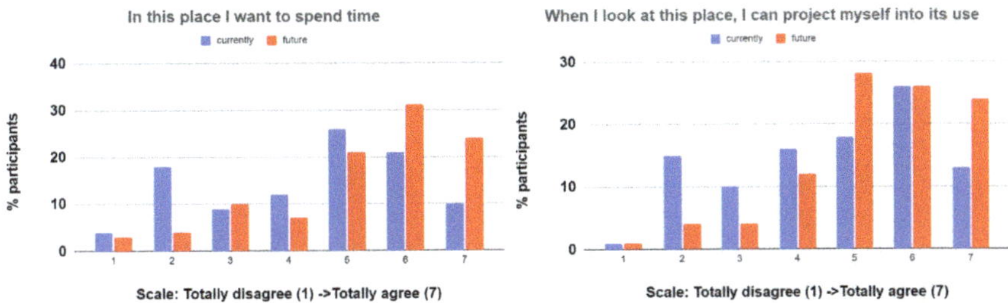

Figure 6. Part 1 answers of the 68 participants: (**a**) similarity between current situation and virtual reality; (**b**) presence of green areas; (**c**) bicycle spaces; (**d**) ways to access the docks.

Figure 7. Street life: the answers for current situation are in blue and the answers for the prospective scenario are in red.

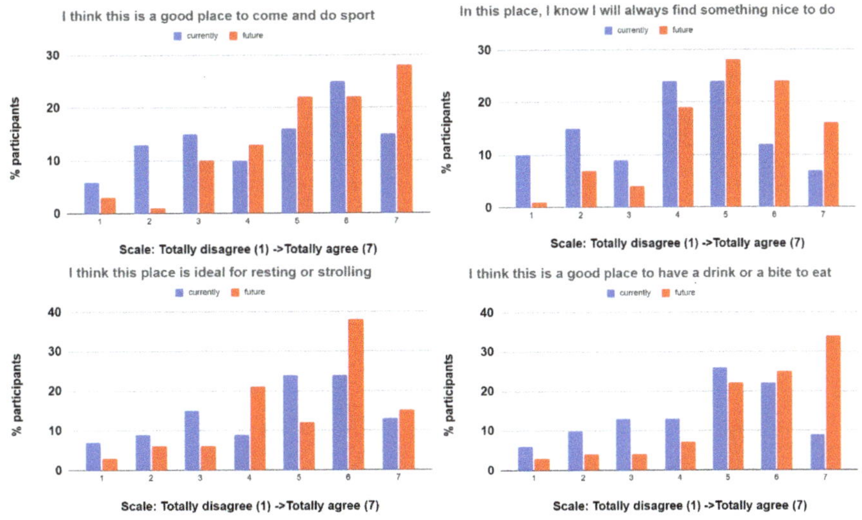

Figure 8. Services and functions: the answers for current situation are in blue and the answers for the prospective scenario are in red.

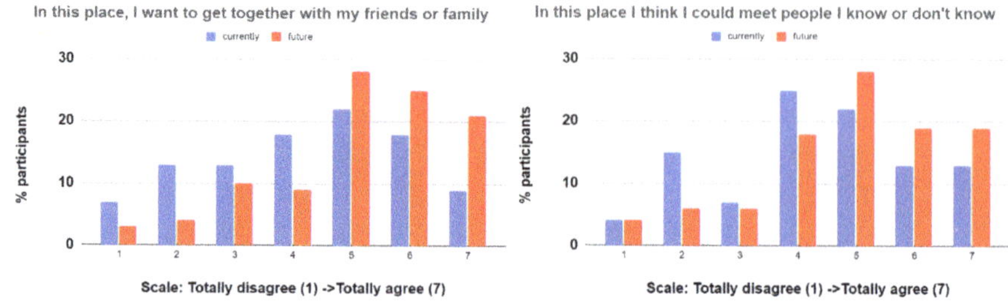

Figure 9. Sociality reflects the adequacy of a place for people to interact in.

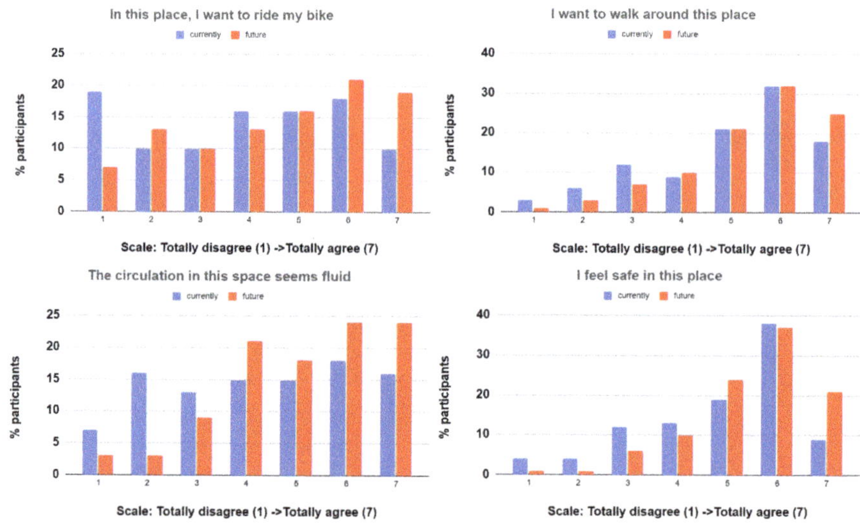

Figure 10. Characterization of walkability and cycling infrastructure.

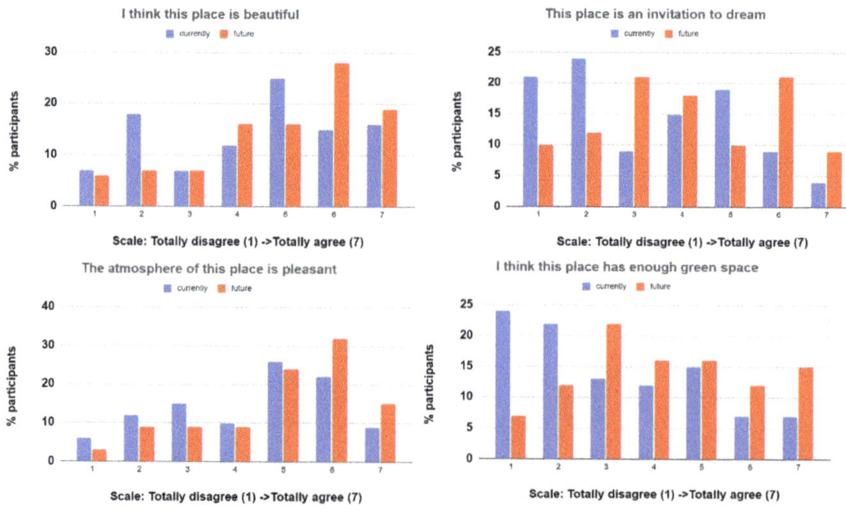

Figure 11. Aesthetics and atmosphere of the current area (blue) and renovated area (red).

Table 2. Part 1: mean and standard deviation for each question.

Question	Mean	SD
Similarity between current situation and virtual reality	4.09	1.50
Presence of green areas	3.99	1.81
Bicycle spaces	3.97	1.78
Ways to access the docks	3.91	1.71

3.3. Overview of the Comparison Before and After Renovation

In Parts 2 and 3 of the study, the perceptions of the participants about the quality of public space and usage in the project were gathered based on six indicators: life on the docks, services and functions, sociability, walkability and cyclability, aesthetics, and access to the docks.

Regarding the results obtained in Part 2 and 3 of the questionnaire, shown in Table 1, we can distinguish three main pieces of information. For several of the measured dimensions, the prospective situation does not really add value compared to the current situation. This is the case, for example, for the visibility of platform access (Mcurrent = 5.22 vs. Mfuture = 5.34) as well as for walkability (Mcurrent = 5.06 vs. Mfuture = 5.43) and ease of orientation (Mcurrent = 5.54 vs. Mfuture = 5.60). It should be noted that all these aspects are already evaluated rather favorably in the current situation, and are not those that evolve in terms of priority. On the other hand, all the other scores are higher in the future situation than in the current situation, with differences ranging from 0.52 to 1.05 points. Considering our intentions when developing the questions, it can be noted that the questions that seem to make the most difference between the current and future situation are related to the usefulness of the area (services and shops), which are questions 3 through 6. This is where we see the two largest deviations from the mean. In addition, there is an improvement in the beauty of the place (questions 14 to 18). Considering the fact that these are the elements of the place evaluated the most unfavorably for the current situation, they would seem to be priorities for implementation. In addition, it is important to carefully consider the answers to the questions relating to access to the platforms, the presence of green spaces, and cyclability.

In the following subsection, we break down the study into separate part according to the aspect surveyed. To facilitate interpretation, we clustered the answers into three groups related to the scale: for "Disagree with the statement" we regrouped the answers using a scale from (1) Totally Disagree to (3) Mildly Disagree; answers (4) are considered as "Neutral"; for "Agree with the statement", we regrouped the answers using a scale (5) Mildly Agree to (7) Totally Agree.

3.4. Life on the Docks

Regarding life on the docks, two questions were included in the survey. The results of the respondents' perception of life on the docks is shown in Figure 7. The figure compares the outcome between the two scenarios involving the current and the future situations regarding participants' responses to the questions "I want to spend time in this place" and "When I look at this place, I can project myself into its use".

Figure 7 (left) shows that in the current situation, for the statement that respondents want to spend time in this place, precisely 57% of them agree with the statement and 31% of disagree. For the future situation, 75% of the participants agreed with the statement that they wanted to spend time in this place and only 9% disagreed, indicating a much more favorable perception of spending time in this place. Thus, after making a comparison between the two scenarios, the questionnaire survey established that the majority perceived that they wanted to spend more time in the future situation, i.e., after renovation of the canal.

Figure 7 (right) reveals less differences between responses to the question "When I look at this place, I can project myself into its use". Globally, participants were satisfied projecting themselves into the use of the place in both scenarios, though the responses show more participants agreeing for the future renovated scene and more disagreeing for the current situation of Rimini Port.

3.5. Services and Functions

Regarding services and functions, four question were included in the survey. The results for the respondents' perceptions of services and functions are shown in Figure 8. The figure compares the outcomes between the two scenarios involving the current and the future situation regarding respond to the question "I think this is a good place to come and do sport", "In this place, I know I will always find something nice to do", "I think this place is ideal for resting or strolling", and "I think this is a good place to have a drink or a bite to eat".

When asked "I think this is a good place to come and do sport", in the current situation 56 % of the participants agreed or strongly agreed that the place is good for doing sport. In turn, for the future situation 77% of the participants agreed or strongly agreed with the statement that the place was a good place for sport activities.

When asked the question "In this place, I know I will always find something nice to do", in the current situation the response shows that 43% of participants agreed with the statement, 24% were neutral, and 34% did not agree, which confirms one of the results of the SWOT analysis in [1]. In the future situation, 68% of participants agreed that they would find something nice to do in this place and 19% were neutral on this aspect, revealing that most people felt they could find something nice to do in the future scenario.

When the respondents were asked to state their opinion about "I think this place is ideal for resting or strolling", 61% of them agreed with the statement in the current situation, while 65% preferred the future infrastructure, showing less difference on this aspect. This result shows that water is always inspiring for resting or strolling, whatever the other available activities.

A larger difference appears with respect to leisure activities. On the aspect of having a drink or eating in the area, 81% considered the renovated place good for this, while only 57% were satisfied with the current version of the Canal.

After conducting a comparative analysis of the two scenarios regarding their services and functions, the results of the survey revealed that most people would prefer to use the services and functions available in the future scenario.

3.6. Sociality

The essay of Tyler [54] suggested that urban engineering should refocus on the idea of sociality, that is, that the most important element of a city is its people, not its structures. What the author calls 'sociality' is the propensity of one person to interact freely with another person. People need to be able to greet known or unknown others and to have small group conversations. To build a sustainable and successful city, the city needs to ensure that it provides an environment that can facilitate conversation and sociality.

We characterize sociality with the two following statements: "In this place, I want to get together with my friends or family" and "In this place, I think I could meet people I know or don't know", as shown in Figure 9.

Placed in the future renovated area, 74% agreed with the statement that they wanted to get together with friends or family, and 66% felt that they could meet known or unknown people. On the other hand, only 49% wanted to get together in the current version of the Canal, and 48% thought they could meet people there. Clearly, the renovated area seems more auspicious for interaction than the current area, which mean that the proposed modification could solve the troubles pointed out in [1].

3.7. Walkability and Cyclability

Figure 10 shows that there is an obvious improvement in cycling infrastructure when comparing the renovated docks to the current one, with 56% of participants against 44% wishing to ride their bike in the area. Yet, this is not a high score, which shows that maybe the cycling infrastructure could be enhanced, although here there is a problem with the available space on the docks. Surprisingly enough, the walkability of the renovated docks increases agreement with the statement "I want to walk around this place" by only 7%, which show that there is room for improvement in this area as well.

The renovated docks felt safer to the respondents than the current situation, with 82% agreement against 66%; thus, the proposed modification should clearly improve the sensation of security in this public place. Moreover, the fluidity of walking trips is improved to 66% in the future scenario against 49% in the current one.

Aesthetics and Atmosphere

Figure 11 shows a clear improvement in the aesthetics and general ambiance of the renovated area compared to the current version of the docks. Around 7% more of the respondents felt that the future place is an invitation to dream and is beautiful; while this is not a particularly high difference, the conviction is higher for both statements.

The respondents considered the atmosphere to be more pleasant in the renovated place, at 71% against only 57% in the current version of the docks. This is a favorable element for the proposed modification, as the general ambiance of the current version of the docks has been deemed gloomy [2]. As for the presence of vegetation, the improvement is clear, with 60% of participants feeling that there is not enough green space in the current docks, while agreement with this statement decreased to 40% for the future scene. Nonetheless, only 43% felt that there was enough vegetation in the future scenario, which is an element that can be improved.

4. Discussion and Conclusions

The objective of this study was to assess whether it is possible to use non-immersive virtual reality as a tool to analyse the perceptions of the realization of a project by users. The aim was to demonstrate how the involvement of human participants from the earliest stages of the project is an effective means of evaluating the success or failure of a project proposal. Numerous studies have shown that the active participation of investors, stakeholders, and citizens in the design of public spaces leads to their greater involvement. In this way, any criticality not initially taken into account, or any particular requests not expressed, can be more easily identified.

The context of this work started with an analysis of the Rimini area, with the redevelopment of degraded urban areas being a first step towards achieving the sustainability aims set out in the Sustainable Development Goals. In this context, evaluation methods were developed in [1] for the decision-making process, considering different social, economic, and environmental aspects in order to obtain a priority scale of interventions for urban regeneration.

Following this work, architectural solutions were proposed in [2] that represent targeted and specific interventions designed precisely for the context to which they are dedicated in order to make the Port's canal area a continuum with its urban context and to improve its perception by tourists and inhabitants. These propositions were based on the knowledge of architectural experts. Therefore, while these two studies [1,2] provided an analysis and solutions based on socioeconomic, environmental, and architectural aspects and knowledge, their proposals need to be validated. The contribution of this paper is to provide a methodology to compare two scenarios, namely, before and after modification of the area, using virtual scenes.

To actively involve non-experts in the work to be done, intuitive tools are needed that can visualize the proposed project. The lack of such tools often creates a barrier between designers and the public, resulting in indifference and discontent around the realization of the project itself. Not understanding what is to be achieved leads to a separation of the parties involved; the public may not feel listened to and able to participate by making useful contributions.

To this end, virtual reality is a tool with a strong visual impact that allows non-specialist citizens to see and imagine the expected reality when the project is completed. This makes everyone more involved and committed, as they can understand and manage the reality around them, and even change it accordingly by expressing problems and needs. The World Economic Forum describes virtual reality use in the public sector as a means of involving citizens, optimizing resources, and improving transport with a view to environmental sustainability, thereby promoting tourism.

In this study, virtual reality has been used to analyze the case of the re-qualification of the Canal of the Port of Rimini, a famous tourist town situated in northern Italy along the Adriatic coast. Although located in the city centre, the canal port area is degraded, poorly connected to the main points of interest, and not used productively. The proposed requalification project consists of creating new access points to the canal port and raising the level of the docks to allow the construction of new cycling and pedestrian paths. This project is in accordance with recent studies that show how the redevelopment of urban public spaces is closely linked to the presence of safe and well-connected cycling and pedestrian paths, which in turn promotes environmental sustainability and the general health and livability of public spaces [1,2].

To involve stakeholders and citizens, we used the virtual reality to create two project scenarios, one representing the current condition of the Canal and the other showing the future reality after realization of the proposed project. We used an online questionnaire in several parts to obtain participant feedback. The first part was aimed at assessing the perceptions of users about the correspondence between the actual reality and virtual reality scenarios. The results after viewing a comparison video showed that the two conditions were perceived by the participants in a similar way; therefore, the analysis tool can be considered valid. More perplexities emerged from the considerations on the comparison between green spaces, cycle paths, and access to the docks that are less similar between the current reality and virtual reality scenarios. This is probably due in part to the fact that the area was shown with a top view, excluding certain details and providing only a general indication of the situation.

A number of details were highlighted in the subsequent videos and analyzed through Parts 2 and 3 of the questionnaire. While the creation of easy access and new cycling and pedestrian paths was privileged by the project, the involvement of users allowed to focus more attention on how the points of attraction of a place are of primary importance when enhancing the local infrastructure. The motivation to move is fundamental for the use of a sustainable means of transport such as the bicycle. Linked to this aspect is the assessment of the beauty of a place. After the project, users have noted that the Port's Canal was more attractive, and declared themselves more eager to frequent it.

The VR methodology used in this study proved useful in providing important results for an initial assessment of project adequacy. Future studies will consider in detail those aspects of the present study specifically related to the design of access points to the quays in order to better evaluate the perceptions of users.

Author Contributions: Conceptualization, R.B., S.A., M.P. and C.L.; methodology, S.A., C.D., M.P. and C.L.; software, R.B., I.I. and M.P.; validation, R.B., S.A. and M.P.; formal analysis, R.B., S.A. and C.D.; investigation, R.B., I.I., S.A., C.D., M.P. and E.D.G.; writing—original draft preparation, R.B., M.P., S.A., C.D. and I.I.; writing—review and editing, R.B., M.P., I.I., S.A., C.D., C.L., E.D.G. and V.V.; visualization, R.B., I.I. and M.P.; supervision, R.B., S.A., C.L. and V.V. All authors have read and agreed to the published version of the manuscript.

Funding: The research was carried out within the "FRAMESPORT" project, funded by the European Interreg Italy–Croatia project under Application ID No 10253074.

Institutional Review Board Statement: Not applicable.

Informed Consent Statement: Not applicable.

Data Availability Statement: The data that support the findings of this study are available on request from the corresponding author upon reasonable request.

Conflicts of Interest: The authors declare no conflict of interest.

Appendix A

Virtual Reality Opinion Survey

The questionnaire, which has a goal of gathering information about the project, is part of an ongoing multi-phase project centered on the urban regeneration of the city of Rimini, which is located in Italy. There are four part in this questionnaire.

During the initial phase, participants are presented with two visualizations, they are asked to identify the differences or similarities between the two.

On the next two phases, participants are presented with two different undertaking scenarios, the goal of this process is to analyse them.

In fourth part, participants are presented with general information and they are asked to

answer them.

The questionnaire is anonymous, and it only takes about 10 minutes to complete. Thanks in advance for your valuable participation!

Phase1: Analysis of the similarities between Visualization 1 and Visualization 2 in Rimini canal port.

Instructions: Please indicate your opinion about the level of similarities or differences with each of the statements below on a 7-point scale from totally different to totally the same.

<u>Two videos are displayed showing the current situation of Rimini Port Visualization 1 and Visualization 2.</u>

Note: The round arrow allows users to look back at the video if they want.

Generally speaking, visualizations 1 and 2 are : *

	1	2	3	4	5	6	7	
Totally different	○	○	○	○	○	○	○	Totally the same

As regards the presence of green areas, visualisations 1 and 2 are : *

	1	2	3	4	5	6	7	
Totally different	○	○	○	○	○	○	○	Totally the same

As regards bicycle spaces, visualizations 1 and 2 are : *

	1	2	3	4	5	6	7	
Totally different	○	○	○	○	○	○	○	Totally the same

As regards the ways to access the dock, visualizations 1 and 2 are: *

	1	2	3	4	5	6	7	
Totally different	○	○	○	○	○	○	○	Totally the same

Figure A1. Virtual reality opinion survey—Part 1.

Phase 2: Analysis of scenario 1 in Rimini Canal Port

Instruction: Please indicate your level of agreement or disagreement with each of the statements below on a 7-point scale from totally disagree to totally agree.

The video shows scenario 1 of the port of Rimini.

Note: The round arrow allows users to look back at the video if they want.

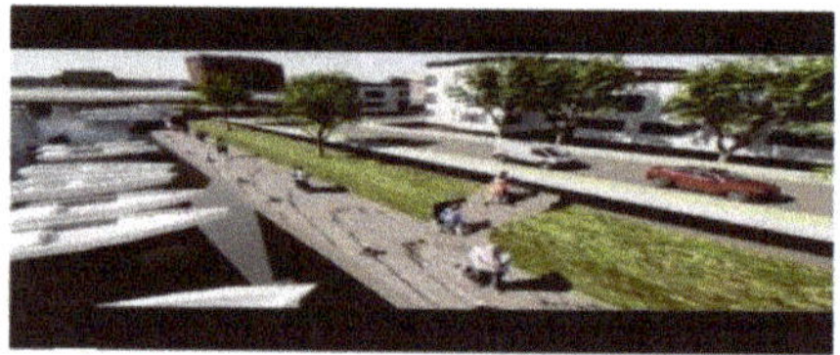

In this place, I want to spend time. *

	1	2	3	4	5	6	7	
Totally disagree	○	○	○	○	○	○	○	Totally agree

When I look at this place, I can project myself into its use. *

	1	2	3	4	5	6	7	
Totally disagree	○	○	○	○	○	○	○	Totally agree

I think this is a good place to come and do sport. *

	1	2	3	4	5	6	7	
Totally disagree	○	○	○	○	○	○	○	Totally agree

In this place, I know I will always find something nice to do. *

	1	2	3	4	5	6	7	
Totally disagree	○	○	○	○	○	○	○	Totally agree

I think this place is ideal for resting or strolling. *

	1	2	3	4	5	6	7	
Totally disagree	○	○	○	○	○	○	○	Totally agree

Figure A2. Virtual reality opinion survey—Part 2.

I think this is a good place to have a drink or a bite to eat. *

	1	2	3	4	5	6	7	
Totally disagree	○	○	○	○	○	○	○	Totally agree

In this place, I want to get together with my friends or family. *

	1	2	3	4	5	6	7	
Totally disagree	○	○	○	○	○	○	○	Totally agree

In this place I think I could meet people I know or don't know. *

	1	2	3	4	5	6	7	
Totally disagree	○	○	○	○	○	○	○	Totally agree

In this place, I want to ride my bike. *

	1	2	3	4	5	6	7	
Totally disagree	○	○	○	○	○	○	○	Totally agree

I want to walk around this place. *

	1	2	3	4	5	6	7	
Totally disagree	○	○	○	○	○	○	○	Totally agree

The circulation in this space seems fluid. *

	1	2	3	4	5	6	7	
Totally disagree	○	○	○	○	○	○	○	Totally agree

I feel safe in this place. *

	1	2	3	4	5	6	7	
Totally disagree	○	○	○	○	○	○	○	Totally agree

Figure A3. Virtual reality opinion survey—Part 2.

I can easily find my way around this place. *

	1	2	3	4	5	6	7	
Totally disagree	○	○	○	○	○	○	○	Totally agree

I think this place is beautiful. *

	1	2	3	4	5	6	7	
Totally disagree	○	○	○	○	○	○	○	Totally agree

This place is an invitation to dream. *

	1	2	3	4	5	6	7	
Totally disagree	○	○	○	○	○	○	○	Totally agree

It is a pleasure for the eyes to look at this place. *

	1	2	3	4	5	6	7	
Totally disagree	○	○	○	○	○	○	○	Totally agree

The atmosphere of this place is pleasant. *

	1	2	3	4	5	6	7	
Totally disagree	○	○	○	○	○	○	○	Totally agree

I think this place has enough green space. *

	1	2	3	4	5	6	7	
Totally disagree	○	○	○	○	○	○	○	Totally agree

When I look at this place, I can quickly see how to get to the edge of the water. *

	1	2	3	4	5	6	7	
Totally disagree	○	○	○	○	○	○	○	Totally agree

Figure A4. Virtual reality opinion survey—Part 2.

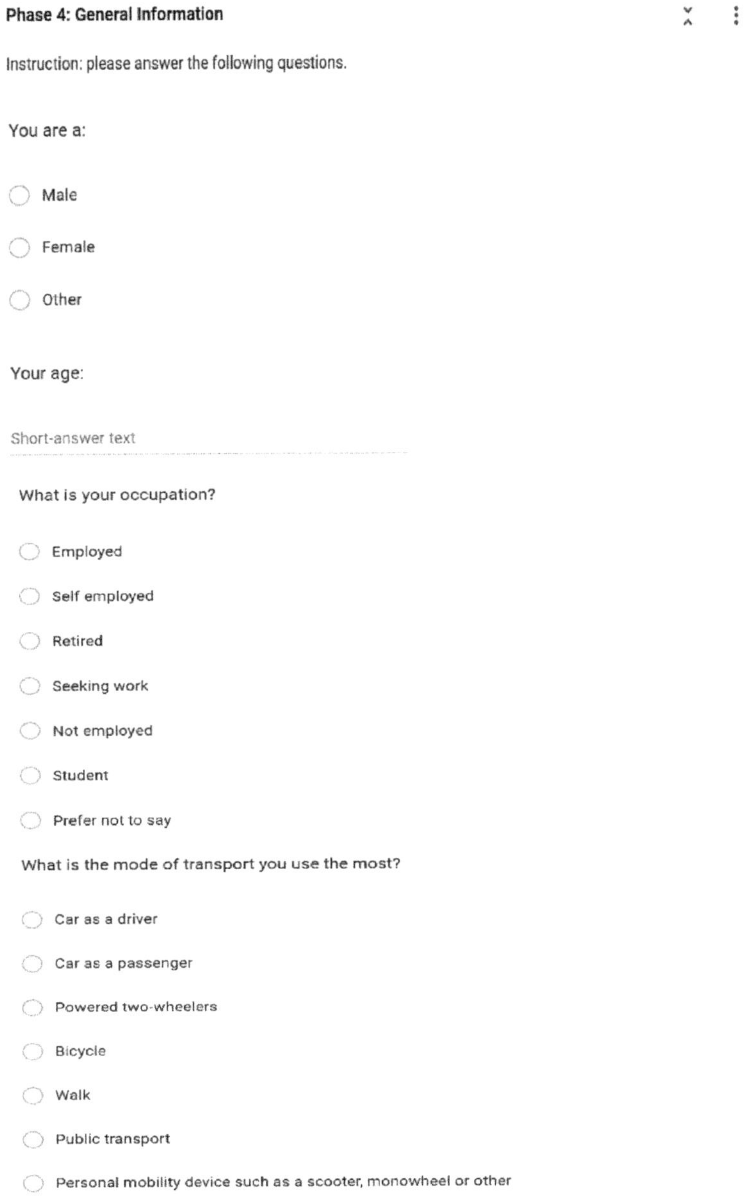

Phase 4: General Information

Instruction: please answer the following questions.

You are a:

○ Male

○ Female

○ Other

Your age:

Short-answer text

What is your occupation?

○ Employed

○ Self employed

○ Retired

○ Seeking work

○ Not employed

○ Student

○ Prefer not to say

What is the mode of transport you use the most?

○ Car as a driver

○ Car as a passenger

○ Powered two-wheelers

○ Bicycle

○ Walk

○ Public transport

○ Personal mobility device such as a scooter, monowheel or other

Figure A5. Virtual reality opinion survey—Part 4.

References

1. Pazzini, M.; Corticelli, R.; Lantieri, C.; Mazzoli, C. Multi-Criteria Analysis and Decision-Making Approach for the Urban Regeneration: The Application to the Rimini Canal Port (Italy). *Sustainability* **2023**, *15*, 772. [CrossRef]
2. Corticelli, R.; Pazzini, M.; Mazzoli, C.; Lantieri, C.; Ferrante, A.; Vignali, V. Urban Regeneration and Soft Mobility: The Case Study of the Rimini Canal Port in Italy. *Sustainability* **2022**, *14*, 14529. [CrossRef]
3. dos Santos Figueiredo, Y.D.; Prim, M.A.; Dandolini, G.A. Urban regeneration in the light of social innovation: A systematic integrative literature review. *Land Use Policy* **2022**, *113*, 105873. [CrossRef]

4. Delhoum, Y.; Belaroussi, R.; Dupin, F.; Zargayouna, M. Activity-Based Demand Modeling for a Future Urban District. *Sustainability* **2020**, *12*, 5821. [CrossRef]
5. Sessa, M.R.; Russo, A.; Sica, F. Opinion paper on green deal for the urban regeneration of industrial brownfield land in Europe. *Land Use Policy* **2022**, *119*, 106198. [CrossRef]
6. Couch, C.; Sykes, O.; Börstinghaus, W. Thirty years of urban regeneration in Britain, Germany and France: The importance of context and path dependency. *Prog. Plan.* **2011**, *75*, 1–52. [CrossRef]
7. Cucchiella, F.; Rotilio, M.; Annibaldi, V.; Berardinis, P.D.; Ludovico, D.D. A decision-making tool for transition towards efficient lighting in a context of safeguarding of cultural heritage in support of the 2030 agenda. *J. Clean. Prod.* **2021**, *317*, 128468. [CrossRef]
8. Evans, J.P.; Jones, P. Rethinking sustainable urban regeneration: Ambiguity, creativity, and the shared territory. *Environ. Plan.* **2008**, *40*, 1416–1434. [CrossRef]
9. Martinez-Fernandez, C.; Audirac, I.; Fol, S.; Cunningham-Sabot, E. *Shrinking Cities: Urban Challenges of Globalization*; Wiley: Hoboken, NJ, USA, 2012; Volume 36, pp. 213–225. [CrossRef]
10. Leccis, F. Regeneration programmes: Enforcing the right to housing or fostering gentrification? The example of Bankside in London. *Land Use Policy* **2019**, *89*, 104217. [CrossRef]
11. Liu, Y.; Dupre, K.; Jin, X.; Weaver, D. Dalian's unique planning history and its contested heritage in urban regeneration. *Plan. Perspect.* **2020**, *35*, 873–894. [CrossRef]
12. Demgenski, P. Dabaodao: The planning, development, and transformation of a Chinese (German) neighbourhood. *Plan. Perspect.* **2019**, *34*, 311–333. [CrossRef]
13. Liu, Y.; Jin, X.; Dupre, K. Engaging stakeholders in contested urban heritage planning and management. *Cities* **2022**, *122*, 103521. [CrossRef]
14. Erfani, G.; Roe, M. Institutional stakeholder participation in urban redevelopment in Tehran: An evaluation of decisions and actions. *Land Use Policy* **2020**, *91*, 104367. [CrossRef]
15. Ratanaburi, N.; Alade, T.; Saçli, F. Effects of stakeholder participation on the quality of bicycle infrastructure. A case of Rattanakosin bicycle lane, Bangkok, Thailand. *Case Stud. Transp. Policy* **2021**, *9*, 637–650. [CrossRef]
16. Du, G.; Kray, C.; Degbelo, A. Interactive Immersive Public Displays as Facilitators for Deeper Participation in Urban Planning. *Int. J. Hum. Comput. Interact.* **2020**, *36*, 67–81. [CrossRef]
17. Nguyen, M.T.; Nguyen, H.K.; Vo-Lam, K.D.; Nguyen, X.G.; Tran, M.T. Applying virtual reality in city planning. In Proceedings of the Virtual, Augmented and Mixed Reality: 8th International Conference, VAMR 2016, Held as Part of HCI International 2016, Toronto, ON, Canada, 17–22 July 2016; pp. 724–735.
18. Frank, J.A.; Kapila, V. Mixed-reality learning environments: Integrating mobile interfaces with laboratory test-beds. *Comput. Educ.* **2017**, *110*, 88–104. [CrossRef]
19. Pribeanu, C.; Balog, A.; Iordache, D.D. Measuring the perceived quality of an AR-based learning application: A multidimensional model. *Interact. Learn. Environ.* **2017**, *25*, 482–495. [CrossRef]
20. Arino, J.J.; Juan, M.C.; Gil-Gómez, J.A.; Mollá, R. A comparative study using an autostereoscopic display with augmented and virtual reality. *Behav. Inf. Technol.* **2014**, *33*, 646–655. [CrossRef]
21. Zhao, M.Y.; Ong, S.K.; Nee, A.Y.C. An Augmented Reality-Assisted Therapeutic Healthcare Exercise System Based on Bare-Hand Interaction. *Int. J. Hum. Comput. Interact.* **2016**, *32*, 708–721. [CrossRef]
22. Huang, T.L.; Liao, S.L. Creating e-shopping multisensory flow experience through augmented-reality interactive technology. *Internet Res.* **2017**, *27*, 449–475. [CrossRef]
23. Fonseca, D.; Martí, N.; Redondo, E.; Navarro, I.; Sánchez, A. Relationship between student profile, tool use, participation, and academic performance with the use of Augmented Reality technology for visualized architecture models. *Comput. Hum. Behav.* **2014**, *31*, 434–445. [CrossRef]
24. Huang, H.M.; Rauch, U.; Liaw, S.S. Investigating learners' attitudes toward virtual reality learning environments: Based on a constructivist approach. *Comput. Educ.* **2010**, *55*, 1171–1182. [CrossRef]
25. Wu, H.; He, Z.; Gong, J. A virtual globe-based 3D visualization and interactive framework for public participation in urban planning processes. *Comput. Environ. Urban Syst.* **2010**, *34*, 291–298. [CrossRef]
26. Celikcan, U.; Askin, M.B.; Albayrak, D.; Capin, T.K. Deep into visual saliency for immersive VR environments rendered in real-time. *Comput. Graph.* **2020**, *88*, 70–82. [CrossRef]
27. Jamei, E.; Mortimer, M.; Seyedmahmoudian, M.; Horan, B.; Stojcevski, A. Investigating the role of virtual reality in planning for sustainable smart cities. *Sustainability* **2017**, *9*, 2006. [CrossRef]
28. Roumpani, F. *Developing Classical and Contemporary Models in ESRI's City Engine*; Working Papers Series; UCL: London, UK, 2013; p. 191.
29. Zhang, C.; Zeng, W.; Liu, L. UrbanVR: An immersive analytics system for context-aware urban design. *Comput. Graph.* **2021**, *99*, 128–138. [CrossRef]
30. Belaroussi, R.; Dai, H.; González, E.D.; Gutiérrez, J.M. Designing a Large-Scale Immersive Visit in Architecture, Engineering, and Construction. *Appl. Sci.* **2023**, *13*, 3044. [CrossRef]
31. Jo, H.I.; Jeon, J.Y. Perception of urban soundscape and landscape using different visual environment reproduction methods in virtual reality. *Appl. Acoust.* **2022**, *186*, 108498. [CrossRef]

32. Gómez-Tone, H.C.; Martin-Gutierrez, J.; Bustamante-Escapa, J.; Bustamante-Escapa, P. Spatial skills and perceptions of space: Representing 2D drawings as 3D drawings inside immersive virtual reality. *Appl. Sci.* **2021**, *11*, 1475. [CrossRef]
33. Zaini, A.I.; Embi, M.R. Virtual Reality for Architectural or Territorial Representations: Usability Perceptions. *Int. J. Built Environ. Sustain.* **2017**, *4* . [CrossRef]
34. Suh, A.; Prophet, J. The state of immersive technology research: A literature analysis. *Comput. Hum. Behav.* **2018**, *86*, 77–90. [CrossRef]
35. Lee, H.G.; Chung, S.; Lee, W.H. Presence in virtual golf simulators: The effects of presence on perceived enjoyment, perceived value, and behavioral intention. *New Media Soc.* **2013**, *15*, 930–946. [CrossRef]
36. Jensen, L.; Konradsen, F. A review of the use of virtual reality head-mounted displays in education and training. *Educ. Inf. Technol.* **2018**, *23*, 1515–1529. [CrossRef]
37. Yoon, H.J.; Kim, J.; Park, S.W.; Heo, H. Influence of virtual reality on visual parameters: Immersive versus non-immersive mode. *BMC Ophthalmol.* **2020**, *20*, 1–8. [CrossRef] [PubMed]
38. Gomez-Tone, H.C.; Chávez, M.A.; Samalvides, L.V.; Martin-Gutierrez, J. Introducing Immersive Virtual Reality in the Initial Phases of the Design Process—Case Study: Freshmen Designing Ephemeral Architecture. *Buildings* **2022**, *12*, 518. [CrossRef]
39. Plechatá, A.; Sahula, V.; Fayette, D.; Fajnerová, I. Age-related differences with immersive and non-immersive virtual reality in memory assessment. *Front. Psychol.* **2019**, *10*, 1330. [CrossRef] [PubMed]
40. Portman, M.E.; Natapov, A.; Fisher-Gewirtzman, D. To go where no man has gone before: Virtual reality in architecture, landscape architecture and environmental planning. *Comput. Environ. Urban Syst.* **2015**, *54*, 376–384. [CrossRef]
41. Lehtola, V.V.; Koeva, M.; Elberink, S.O.; Raposo, P.; Virtanen, J.P.; Vahdatikhaki, F.; Borsci, S. Digital twin of a city: Review of technology serving city needs. *Int. J. Appl. Earth Obs. Geoinf.* **2022**, *114*, 102915. [CrossRef]
42. Dodge, M. Towards the virtual city: VR & internet GIS for urban planning. *GIS Eur.* **1998**, *6*, 26–29.
43. Rehan, R.M. Sustainable streetscape as an effective tool in sustainable urban design. *HBRC J.* **2013**, *9*, 173–186. [CrossRef]
44. Verma, D.; Mumm, O.; Carlow, V.M. Identifying Streetscape Features Using VHR Imagery and Deep Learning Applications. *Remote. Sens.* **2021**, *13*, 3363. [CrossRef]
45. Wang, R.; Rasouli, S. Contribution of streetscape features to the hedonic pricing model using Geographically Weighted Regression: Evidence from Amsterdam. *Tour. Manag.* **2022**, *91*, 104523. [CrossRef]
46. Delhoum, Y.; Belaroussi, R.; Dupin, F.; Zargayouna, M. Analysis of MATSim Modeling of Road Infrastructure in Cyclists' Choices in the Case of a Hilly Relief. *Infrastructures* **2022**, *7*, 108. [CrossRef]
47. Brownson, R.C.; Ramirez, L.K.B.; Hoehner, C.M.; Cook, R.A. Analytic Audit Tool and Checklist Audit Tool. 2003. Available online: https://bel.uqtr.ca/id/eprint/3348/1/Audittoolanalyticversion.pdf (accessed on 20 January 2023).
48. Moudon, A.V.; Lee, C. Walking and Bicycling: An Evaluation of Environmental Audit Instruments. *Am. J. Health Promot.* **2003**, *18*, 21–37. [CrossRef]
49. Gardner, K.; Johnson, T.; Buchan, K.; Pharaoh, T. Developing a Pedestrian Strategy for London. Transport Policy and its Implementation. In Proceedings of Seminar B Held at the 24th European Transport Forum, Brunel University, Londen, UK, 2–6 September 1996.
50. Ewing, R.; Handy, S.; Brownson, R.C.; Clemente, O.; Winston, E. Identifying and Measuring Urban Design Qualities Related to Walkability. *J. Phys. Act. Health* **2006**, *3*, S223–S240. [CrossRef] [PubMed]
51. Levitte, A. Piéton et perception visuelle: Le design visité par les sciences cognitives. *Actes INRETS (Arcueil)* **2008**, *115*, 227–235.
52. Brown, B.B.; Werner, C.M.; Amburgey, J.W.; Szalay, C. Walkable route perceptions and physical features: Converging evidence for en route walking experiences. *Environ. Behav.* **2007**, *39*, 34–61. [CrossRef]
53. Thibaud, J.P. *En Quête D'ambiances: éprouver la Bille en Passant*; MétisPresses: Genève, Switzerland, 2015.
54. Tyler, N. Next-generation infrastructure for next-generation people. *Proc. Inst. Civ. Eng. Smart Infrastruct. Constr.* **2020**, *173*, 24–28. [CrossRef]

MDPI
St. Alban-Anlage 66
4052 Basel
Switzerland
www.mdpi.com

Sustainability Editorial Office
E-mail: sustainability@mdpi.com
www.mdpi.com/journal/sustainability

9 783725 804108